Wettability

Wettability

Erle C. Donaldson
and
Waqi Alam

Gulf Publishing Company
Houston, Texas

Wettability

Gulf Publishing Company
2 Greenway Plaza, Suite 1020
Houston, TX 77046

ISBN-10: 1-933762-29-2
ISBN-13: 978-1-933762-29-6

Library of Congress Cataloging-in-Publication Data

Donaldson, Erle C.
Wettability / Erle C. Donaldson and Waqi Alam.
p. cm.
Includes bibliographical references and index.
ISBN 1-933762-29-2 (978-1-933762-29-6 : alk. paper)
1. Wetting. I. Alam, Waqi. II. Title.
QD506.D66 2008
547'.83—dc22

2008016213

Text design and composition by TIPS Technical Publishing, Inc.

We dedicate this book to Professor Teh Fu Yen of the University of Southern California for his support, encouragement, and contributions to wettability, microbial enhanced oil recovery, and petroleum engineering.

—Erle C. Donaldson and Waqi Alam

Table of Contents

Preface

The subject of wettability has not received enough attention in the petroleum industry despite its significant role in defining the ultimate recovery of a reservoir. The reason has primarily been the lack of knowledge of this rather complex subject. Initial description of wettability was covered by Amyx, Bass and Whiting's classic book *Petroleum Reservoir Engineering Physical Properties* in 1960, and the lead author of the present work developed what is known as the USBM method of measuring wettability in the 1970's. The first serious study of wettability was undertaken by the world's largest E&P company, the former Arabian-American Oil Company (now called Saudi Aramco, post-nationalization in 1988) when they invited the lead author, in the late 1980's, to serve as a consultant in a seven-year-long R&D project to help understand the wettability of the world's largest oilfield.

In using the wealth of research and experiences garnered since then and in several other projects world-wide, the book provides a simple approach to understanding a complex phenomenon. Based on fundamental physics and chemistry of wettability, the concept has been extended to petroleum engineering to understand the mechanics of petroleum reservoirs with respect to interfacial tension and capillary pressure. The book also analyzes effects of wettability on primary and secondary recovery mechanisms. It compiles the diverse aspects of wettability, serving as a one-stop shopping experience for students and researchers alike.

By interspersing the theory of wettability with detailed physics while including interesting engineering concepts and applications, the authors have provided a well-planned presentation sure to keep engineering students and instructors thoroughly engaged.

The book opens the subject of wettability by briefly detailing the concept using contact angle and interfacial tension and including well-established experimental methods of measurements. By discussing the effects of wettability on relative permeability and dispersion, the authors demonstrate the paramount importance of wettability on oil recovery. The following applications elaborate on these broader concepts.

Chapter 2 visits at length the fundamental sciences of wettability, a monumental compilation of findings in physics and physical chemistry that is markedly absent in the current knowledge base on wettability. This information will immensely benefit serious researchers, and the examples given at the end will amply help students understand the complex subject.

In **Chapter 3**, the focus is again shifted from seemingly 'boring science' to a more interesting engineering subject: oil production. The chapter covers the entire spectrum of petroleum engineering, including how wettability relates to the petroleum geology of hydrocarbon accumulation and migration through the reservoir engineering theories of fractional flow and frontal advance. The discussion immediately addresses the practical issues of waterflooding—the most widespread enhanced oil recovery technique—and the great impact of wettability on the success or failure of waterflood projects. Again, example problems should prove useful to complete the understanding of the concepts.

The next chapter (**Chapter 4**) returns to the intricacies of physics and chemistry working at pore-scale level, exploring the different ways wettability works in the two principal rock types, sandstone and carbonates. As a bonus the authors also explain wettability vis-à-vis asphaltene solvency of crude oils and adsorption on reservoir rocks.

Chapter 5 captures rare field examples brought in from various oilfields of the world, a feat made possible by the authors' unique experiences gained as expert consultants. The case histories cap the meticulously presented theories of wettability in the preceding chapters, illustrating the need to more earnestly consider wettability as a key to unlocking the huge reserves of oil that will otherwise be left unexploited. Finally, the FORTRAN programs in **Chapter 6** are real prizes for researches in both the E&P industry and academia.

To facilitate classroom instructions, a full semester graduate course can be developed entirely with this book, drawing heavily from Chapters 1, 4, 3 and 5—in that order. Chapters 1 and 4 can be easily integrated in an undergraduate course on petrophysics.

—M. Rafiqul Awal, PhD
Petroleum Engineering Department
Texas Tech University, Lubbock, TX

Nomenclature
(Some symbols are defined in the text as representing different terms)

A	(1) area [m^2]; (2) ampere [Cs^{-1}]; or (3) Hamaker constant [J]
Å	10^{-8} cm (10^{-1} nm)
a	(1) local acceleration of gravity [m/s^2, cm/s^2, ft/s^2] or (2) van der Waals pressure correction constant [atm*liters^2/mol^2]
atm	atmosphere [0.1013 Mpa; 14.696 psi]
B	constant, as defined in the text
b	van der Waals volume correction constant [liters/mol]
C	(1) electrical charge [Coulomb; ampere second] or (2) concentration [mg/L]
C'	capacitance [C/V]
cal	calorie [4.284 J]
cP	viscosity [centipoises]
D	dipole moment (Debye) [$3.336*10^{-30}$ C·m]
D_L	coefficient of dispersion
D_r	pore size distribution function [m^2]
d	distance [nm, m]
dm	decimeter
E	(1) energy [J]; (2) electric field [$Vm^{-1} = JC^{-1}m^1$]; or (3) free energy of interaction between liquid and solid

E_d	displacement efficiency $[1 - S_{or}/(1 - S_{wi})]$
e	electron charge [$1.602*10^{-19}$ C]
eV	electron volt [$1.602*10^{-19}$ J]
Ex	Example problem at end of chapter
erg	10^{-7} J
F	force [$N = Jm^{-1}$]
F_r	formation resistivity factor
F(r)	interactive force between atoms or molecules [N]
f_w	fraction of water flowing
G	free energy [J]
G_H	hydrostatic gradient [Pa/m, psi/ft]
g_c	gravitational constant (9.8 m/s^2)
g^f	gram-force = g^m * 980 cm/s^2 = 980 dynes
g^m	gram-mass
gg	geothermal temperature gradient
H	height [of liquid rise in a capillary, for example]
h	(1) height [nm, mm or m]; (2) film thickness [nm]; or (3) Plank's constant [$6.626*10^{-34}$ Js]
I	electric current [amperes]
I_A	Amott wettability index
I_U	USBM wettability index
iep	iso-electric point of an electrolytic solution
K	absolute (single phase) permeability [Darcy, m^2, μm^2]
KE	kinetic energy [J]
K_B	Boltsmann constant (R/N_o)[$1.381*10^{-23}$ $J \cdot mol^{-1}K^{-1}$; $1.381*10^{-16}$ $erg \cdot mol^{-1}K^{-1}$]
kT	$4.144*10^{-21}$ J (at 298 K ≈ 25°C); $4.045*10^{-21}$J (at 293 K ≈ 20°C)
k_e	effective permeability
k_r	relative permeability [k_e/K]

k_{rn}	relative permeability of the nonwetting phase
k_{rw}	relative permeability of the wetting phase
k_s	Carman-Kozeny shape factor
L, l	length or liter
L_e	effective length of the fluid flow paths through a porous medium
L_{iw}	effective length of the wetting phase flow path
lb^F	pounds-force
lb^m	pounds-mass
M	(1) mobility ratio or (2) solution molarity [mol]
N	(1) force [Newton] or (2) revolutions per minute (centrifuge)
N_c	capillary number
N_o	Avogadro's number [$6.022*10^{-23}$/mol]
n	(1) refractive index; (2) exponent in the Lenard-Jones eq.; or (3) molar
n_i	number of pores
nm	nanometer [10^{-9} m]
P	pressure [Pa, psi, atm]
Pa	pressure [Pascals]
P_B	pressure exerted by buoyant force [Pa]
P_c	capillary pressure [Pa]
$P_{c\text{-}area}$	area under capillary curves
Q	rate (ml/s)
Q_i	pore volumes injected (W_i/V_p)
q	(1) electrical charge [Coulombs] or (2) volumetric flow rate of water, oil or gas [m^3/s, cm^3/s, ft^3/s, bbl/day]
gm	gram
gms	grams
q_{rel}	relative flow rate

q_T	total flow rate ($q_w + q_o$)
R	(1) gas constant, PV/nT [0.082056 L·atm/deg·mol; 8.3143 J/deg·mol; 1.9872 cal/deg·mol]; (2) electrical resistance [ohm]; or (3) resistivity [ohm-m, Vm/amp]
R_e	oil recovery [percent of original oil in place]
r	(1) radius; (2)distance of separation between interacting atoms; or (3) molecules [nm, m]
r_B	first Bohr radius of an electron [0.053 nm (where energy = 2hv)]
r_F	point of incipient attractive force between atoms
r_m	point of minimum well depth of attractive force between atoms
S	fluid saturation
S_{area}	surface area [m_3/g]
S_c	connate, or initial water saturation
S_{hy}	empirical constant of hydration forces (3 – 30 mJ/m^2)
S_i	(1) initial fluid saturation or (2) saturation of an individual capillary (as used in text)
S_{or}	residual oil saturation
S_{rn}	effective saturation of the non-wetting phase
S_{rw}	effective saturation of the wetting phase
S_w	water (brine) saturation
S_{wbt}	water saturation at water break-through
$S_{w\alpha}$	water saturation in the stabilized flow zone behind the advancing front of a waterflood
$S_{w\beta}$	water saturation in the non-stabilized zone in front of the advancing front of a waterflood
S_{wi}	irreducible water saturation
S_{wor}	water saturation at the residual oil saturation
S_z	water saturation at zero capillary pressure
s	second
T	temperature [deg or K] (K = °C + 273.15)

T_A	adhesion tension [mN/m]
T_F	temperature of the subsurface formation
T_S	temperature at the surface
t	time [s]
U	(1) intermolecular pair potential energy or (2) work of separation [N·m; J]
u	linear fluid velocity [m/s]
u_e	effective fluid velocity
V	(1) volume [m^3] or (2) volts [JC^{-1}; J·amp·s; NmC^{-1}]
V_b	bulk volume
V_i	initial interstitial pore volume
V_p	pore volume
W	work [N-m; Joules]
W_i	water injected
$\underline{W}$	wettability
x	distance
x_i	ratio of equivalent flow path to the core length [L_e/L]
y_i	group of specific pore size ranges
z	(1) valence of an ion or (2) vertical distance [nm, m]
zpc	zero point charge of the zeta potential

Greek symbols

α	(1) angle of inclination of a core or reservoir; (2) polarizability [C^2m^2/J]; or (3)average production deline constant (Aranosky Eq.)
γ	specific weight [lb/ft^3]
δ	distance (as defined in text)
ε	dielectric constant (relative permittivity, dimensionless)
ε_o	permittivity of free space [$8.854*10^{-12}$ $C^2/J{\cdot}m = C^2/N{\cdot}m^2$]
ζ	zeta potential [volts]
θ	contact angle

κ^{-1}	Debye length [m]
λ	separation distance of atoms, molecules or particles [nm]
μ	(1) potential due to molecular interaction [C·m] or (2) viscosity [Pa-s, N-s/m^2]
μ_{in}	induced electrical potential in a molecule [C·m]
ν	orbiting frequency of an electron [$3.3*10^{15}$ s^{-1}]
ξ	electric field strength [N/C]
ξ_s	liquid spreading coefficient [mN/m]
Π	pressure drop of an inclined porous medium ($\Delta P = \rho gh$)
ρ	density [kg/m^3, g/cm^3, lb/ft^3]
ρ_n	number density [ions/m^3]
σ	interfacial tension, mN/m
σ_c	electric surface charge [C/m^3]
φ	porosity
τ	tortuosity of pores
Ψ	electrostatic potential [V]
Ω	canonical potential [dimensionless]

Superscripts

f	force
m	mass

Subscripts

c	(1) capillary or (2) electric charge
ex	exterior
g	gas
in	interior
o	oil
os	oil-solid
ow	oil-water
r	radius (meter)

sf	surface free energy
T	total
w	water
ws	water-solid

Units

atm	14.696 psi = $0.101325(10^6)\mu m^2$
cubic foot	28.316 liters, 7.481 gallons
Darcy	$0.986923(10^{-12})m^2 = 0.986923\mu m^2$
dyne	$1\ g{\cdot}cm/s^2$
g_c	gravitational constant ($981{\cdot}g^m{\cdot}cm/g^F{\cdot}s^2$, $g^m{\cdot}cm/dyne{\cdot}s^2$, $32.174{\cdot}lb^m{\cdot}ft/lb^F{\cdot}s^2$)
gallon	3.7853 liters, 231 cubic inches
mile	1.60935 km
pound	453.5924 g
temp.	°F = 1.8°C + 32

Acronyms

DLVO	Derjaquin, Landau, Verwey, Overbeek theory (Chap. 2)
DTAB	dodecyl-trimethyl-ammonium bromide
EOR	enhanced oil recovery
Ex	example problem
IFT	interfacial tension (mN/m)
IW	intermediate-wet
MEOR	microbial enhanced oil recovery
NSO	nitrogen, sulfur and oxygen (referring to compounds in crude oils)
OOIP	original oil in place
OW	oil-wet
SEM	scanning electron microscope
USBM	U.S. Bureau of Mines
WOC	water-oil contact
WW	water-wet

About the Authors

Erle C. Donaldson

Erle C. Donaldson began his career as a pilot plant project manager for Signal Oil and Gas Research in Houston, Texas. Later he joined the U.S. Bureau of Mines Petroleum Research Center in Bartlesville, Oklahoma (the laboratory was later absorbed by the U.S. Department of Energy as the Bartlesville Energy Technology Center). Dr. Donaldson was project manager of subsurface disposal of industrial wastes and petroleum reservoir characterization.

When the laboratory was detailed to private industry for operation in 1983, Dr. Donaldson joined the faculty of the School of Petroleum and Geological Engineering at the University of Oklahoma as associate professor. Since retiring from the university in 1990, he has consulted for various universities, companies and U.S. agencies including universities in Bolivia and Thailand, King Fahd Research Institute in Saudi Arabia, and Tetrahedron, Inc.

Dr. Donaldson has earned four degrees: a PhD in chemical engineering from the University of Tulsa, an MS in organic chemistry from the University of South Carolina, a BS in chemical engineering from the University of Houston and a BS in chemistry from The Citadel. He has served as editor for the *Journal of Petroleum Science and Engineering* and as chairman of committees and sessions for the Society of Petroleum Engineers and the American Chemical Society, as well as other national and international conferences.

Waqi Alam

Dr. Waqi Alam is the president of Tetrahedron, Inc., a petroleum and environmental engineering consulting firm based in Maryland. He has conducted research and published several papers in the field of fluid flow through porous media, enhanced oil recovery, and environmental management systems. He has provided consulting services to several public and private sector clients including the US and foreign governments in the areas of reserve estimates, reliability analysis, environmental mitigations, and energy management. He has worked in various geographic locations of the world including the Americas, Asia, Africa, and Europe. Dr. Alam earned his doctoral degree in petroleum engineering from the University of Oklahoma. He also has a master's degree in engineering management and a bachelor's degree in chemical engineering.

CHAPTER 1

Wettability

1.1 The Four States of Wettability

Wettability is the relative adhesion of two fluids to a solid surface. With respect to two immiscible fluids in a porous media, wettability is the measure of the preferential tendency of one of the fluids to wet (spread or adhere to) the interstitial surfaces of the porous medium in the presence of the other fluid. The surfaces of the pores in rocks contain a wide variety of exposed minerals that have preferential affinities for water, hydrocarbons, or constituents suspended and dissolved in the fluids. Thus for a water/oil/rock system, the term wettability refers to the average, overall, relative wetting preference of the interstitial surfaces of the rock.

Four general states of wettability have been recognized: (1) water-wet, (2) fractional-wettability, (3) mixed-wettability, and (4) oil-wet.

1.1.1 Water-wet System

A water/oil/rock system is considered to be *water-wet* when more than 50% of its surface is wet by water. Water occupies the smaller pores and dead-end cul-de-sac pores, and exists as a film covering the surfaces of the preferentially water-wet larger pores of the rock. Oil is lodged in the larger pores as droplets resting on a film of water; oil globules may extend through two or more of the larger pores and coat random areas of the pore surfaces containing minerals that are preferentially oil-wet. Therefore, water exists as a continuous phase throughout the porous system and the nonwetting phase (oil) is a discontinuous phase consisting of globules in the larger spores surrounded by water. If the water saturation is reduced to its irreducible saturation (S_{wi}), water remains as a continuous phase in the small pores and crevices through the porous medium. At S_{wi} the oil saturation is high enough for it to also exist as a continuous phase through the larger pores of the rock. As the water saturation increases, the nonwetting phase quickly becomes discontinuous, with globules and fingers of oil in larger pores completely surrounded by water. The wetting phase saturation exists as a continuous phase at all saturations equal to or greater than S_{wi}. If a preferentially water-wet core is saturated with oil then contacted with water, water will spontaneously imbibe into the rock displacing the oil until a state of static equilibrium is attained between the capillary and surface energy forces of the fluids and rock surfaces. If another sample of the same rock is saturated first with water and then contacted with oil, oil will not imbibe into the rock.

1.1.2 Fractionally-wet

The term *fractional wettability* was proposed by Brown and Fatt (1956) to characterize heterogeneous wetting of the pore surfaces where the preferential wetting is randomly distributed throughout the rock. In some cases, the random distribution of minerals (with a variety of chemical properties) exposed to the surfaces in the pores is such that areas which are either preferentially water-wet or oil-wet are scattered through the rock and there are no continuous oil networks through the rock.

1.1.3 Mixed-wet

The term *mixed wettability*, as defined by Salathiel (1973), is a condition where the small pores in the rock are water-wet and saturated with water, but the larger pores are oil-wet and filled with oil in contact with the pore walls that form a continuous path through the length of the rock. Salathiel reasoned that this condition could occur during the original accumulation of oil in a reservoir if oil-containing surface active compounds displaced connate water from the larger pores; the surface active compounds would gradually displace the remaining films of water on the pore surfaces. Oil would not enter the smaller pores where the threshold capillary pressure for displacement of water is too large.

The terms fractional-wet and mixed-wet are included in the frequently used general term *neutral-wettability*. This term only implies that half of the rock surface is water-wet and the other half is oil-wet; it does not distinguish the type of wettability condition.

1.1.4 Oil-wet

When the system is preferentially *oil-wet*, the positions of water and oil in the rock are reversed. Oil occupies the smaller pores to the exclusion of water, and oil is in contact with the rock surfaces of the larger pores. Where water is present in the larger pores, it is generally in the center of the pores resting on a film of oil. Water (the nonwetting phase in an oil-wet system) also exists as a continuous phase distributed through the larger pores when it is present as a high saturation (near the residual oil saturation, S_{or}). If the water saturation is decreased (by injection of oil), it rapidly loses continuity and becomes isolated in the larger pores as pockets and fingers of water that are surrounded by oil. Thus in an oil-wet system, oil is a continuous phase for all saturations equal to and greater than S_{or}.

If the preferentially oil-wet rock is saturated with water and contacted with oil, the oil will imbibe into the rock displacing water until a state of equilibrium is attained. Water will not spontaneously imbibe into an oil-wet rock.

Most of the world's oil reserves are found in either sandstone or carbonate rocks. The wettability of sandstones generally ranges from neutral to strongly water-wet; however, carbonates exhibit oil-wet tendencies ranging from neutral to strongly oil-wet. Some sandstone oil reservoirs are oil-wet because the oils contain surface-active polar compounds that are expelled from the bulk oil to the interfacial surface where they are physically and chemically attached to the surfaces of the sandstone: the well-known oil-wet sandstone reservoirs in the U.S. are the Bradford, Pa., sand (Marsden and Khan, 1962); Wilcox, Okla., sand (Katz, 1942); and Tensleep, Wyo., sand (Nutting, 1934).

Wettability is the most important parameter affecting: (1) the microscopic distribution of water and oil in rock pores, (2) the characteristics of capillary pressure curves, (3) the fluid flow properties of oil displacement by water, and (4) the amount and distribution of residual oil saturation. Donaldson and Crocker (1977) present color photographs of residual oil saturation that are visual verification of the four states of wettability.

1.2 Interfacial Tension

When two immiscible fluids (gas-liquid or liquid-liquid) are in contact, the two fluids are separated by a definite interface that is only a few molecular diameters thick. Within the liquid (away from the interface) the molecules exert an attractive force on each other in all directions, Fig. 1–1a. At the interface, however, there is an inward directed force attempting to minimize the surface by pulling it into the shape of a sphere, Fig. 1–1b. The molecular activity at the surface creates a film-like layer (or membrane) of molecules that are in tension; the tension is the surface free energy per unit area that develop at the interfaces: gas-liquid, gas-solid, liquid-liquid, and liquid-solid.

Liquids that wet the walls of a capillary tube inserted into the bulk liquid (by adsorption of a thin film of molecules on the capillary wall) will rise in the capillary tube in order to reduce the total surface area, Fig. 1–2a. At equilibrium between the attractive forces and gravity, the free energy of the surface is reduced to a minimum value. The decrease of an element of surface area is $2\pi r \cdot dh$ which corresponds to a decrease in surface energy equal to $\sigma \cdot dA$, thus

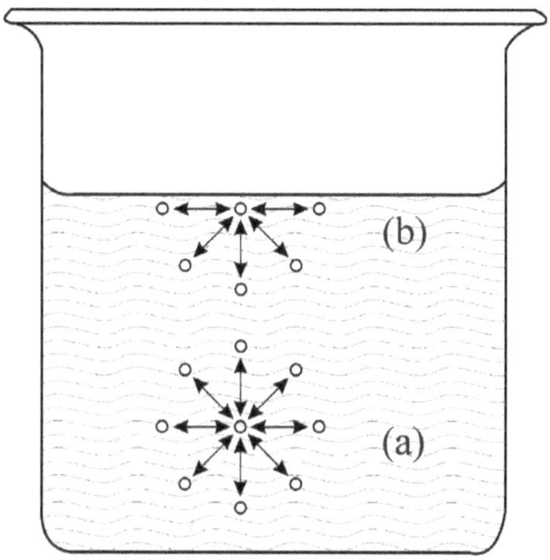

Figure 1–1 *Interaction of attractive molecular forces in the bulk liquid (a) and at the surface (b).*

$$\Delta G_{sf} = \sigma dA = \sigma 2\pi r dh \tag{1.1}$$

The free energy required to raise the elemental volume of liquid (with density ρ) to the height *h* against the downward force of gravity is

$$\Delta G = \pi r^2 \rho g_c dh \tag{1.2}$$

At equilibrium the two energies are equal (Ex. 1–1), therefore

$$\sigma = \tfrac{1}{2} r \rho g_c dh \tag{1.3}$$

If the liquid does not wet (adsorb to) the wall of the capillary, the liquid will be depressed in the capillary because the mutual attraction of the liquid's molecules is greater than the glass-liquid molecular attraction, Fig. 1–2b. The pressure of the nonwetting fluid in the capillary, P_2, is greater than the pressure of the fluid just above the interface, P_1; and

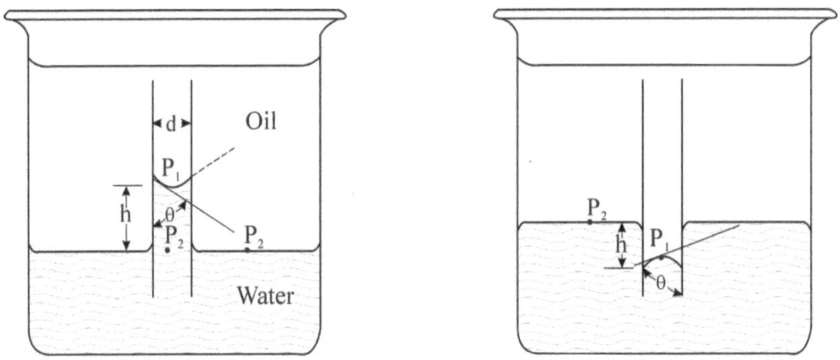

Figure 1–2 *(a) Capillary rise of a liquid wets the walls of a tube. (b) Depression of a liquid that does not wet the walls: d = diameter, h = height, σ = interfacial tension, and θ = contact angle.*

the nonwetting liquid is depressed a distance, *h*, below the level of the liquid outside the capillary.

The du Noüy (1919) method for measuring interfacial tension (IFT) uses an analytic balance beam to measure the force required to pull a platinum ring through the interface. The surface tension exerts a force on the inner and outer circumferences of the ring. The difference between the inner and outer diameters of the ring is negligible; therefore, the length of the ring in contact with the fluid is $2{*}{\cdot}2\pi r$ (where r is the average radius of the ring) and the IFT is $\sigma = g^f/4\pi r$ (Ex. 1–2).

The Wilhelmy plate method for measuring interfacial tension uses a plane rectangular plate that is pulled upward or pushed downward through an interface. The force acting on the perimeter of the plate and the advancing or receding contact angle of the fluid against the plate are used to calculate the interfacial tension (Ex. 1–3). As the plate is moved into or out of the liquid (Fig. 1–3), the change of the force, *F*, due to the adhesion of the fluid to the surface of the plate is equal to the length of the interface in contact with the plate (2× width + thickness) times the interfacial tension and cosine of the contact angle (θ).

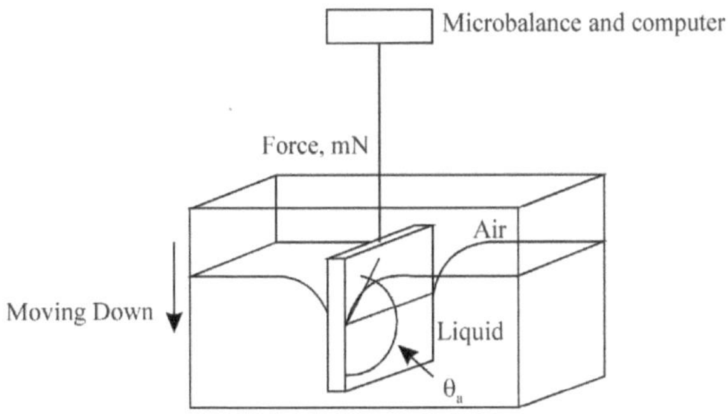

(a)

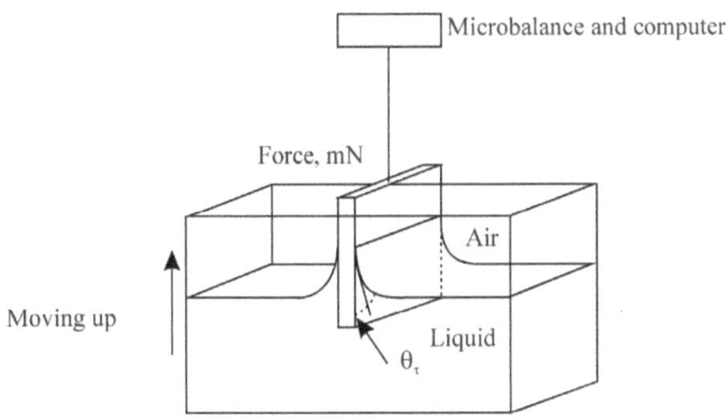

(b)

Figure 1–3 *(a) Wilhelmy plate method for measurement of advancing contact angle. (b) Receding constant angle (after Tiab and Donaldson, 2004).*

$$F = l \cdot \sigma \cos\theta \tag{1.4}$$

Other methods for measuring IFT that take the effect of gravity into account are discussed by Adamson and Gast (1997): drop weight, sessile drop or bubble, maximum bubble pressure, and pendant drop (for a detailed description of the pendant drop apparatus refer to Donaldson and Pavelka [1967]).

The air/oil interfacial tensions and other properties at 25°C of several crude oils are listed in Table 1–1. The values are one third, or less, than the IFT of air/water at 25°C (72.8 mN/m). Chu (1982) found that the air-water IFT decreases almost linearly with temperature from 72.5 mN/m at 21.1°C to 60.1 mN/m at 93.3°. Organic solutes, surface active compounds, and caustics decrease the air-water IFT while sodium chloride and calcium chloride produce a small increase of IFT at any specific temperature. The air-liquid interfacial tension of various liquids at 20°C is listed in Table 1–2.

1.3 Contact Angle

Consider a drop of water resting on a horizontal surface immersed in oil, Fig. 1–4. The drop of water will adopt a position between completely spreading on the surface (infinitely water-wet with a contact angle equal to 0°) or a round drop resting lightly on the surface because its density is greater than the oil's density (infinitely oil-wet surface, contact angle = 180°). Between these two extremes, the general shape of the drop will exhibit a measurable contract angle (measured through the denser phase). The force exerted by the oil-water interfacial tension acts tangent to the surface of the drop with its horizontal component, $\sigma_{ow} Cos\theta$, pulling the circumference of the drop toward its center, Fig. 1–4. The water-solid IFT pulls the circumference of the water drop area of contact with the solid toward the center. At equilibrium these two forces are balanced by the oil-solid IFT of an adsorbed layer of oil on the solid surface acting to pull the circumference of the drop's area of contact away from its center (Lowell and Shields, 1981).

Fig. 1–5a–c show the configuration of sessile drops and the water/oil interface for different conditions of wettability. The surface is water-wet (hydrophilic) when the contact angle is less than 90° and the pressure difference across the interface (capillary pressure, P_c) is positive; neutral-wet at contact angles close to 90° ($P_c = 0$); and oil-wet (hydrophobic) for angles greater than 90° (P_c is negative).

The contact angle and interfacial tensions of a sessile drop are related according to Young's equation (1855)

$$\sigma_{os} - \sigma_{ws} = \sigma_{ow} Cos\theta \qquad (1.5a)$$

Table 1–1 Air-oil interfacial tensions, viscosity and density of crude oils measured at 25°C (McGhee et al., 1979).

Oil	State	IFT (25°C)	Density (25°C)	Viscosity, cP (25°C)
Bradford	Pennsylvania	9.9	0.807	3
Clearfork	Texas	12.5	0.835	44
Rio Bravo	California	17.8	0.831	4
Toborg	Texas	18.0	0.936	154
Dakota	Colorado	19.9	0.819	39
Nugget	Utah	21.9	0.790	35
Squirrel	Oklahoma	22.3	0.879	33
San Andreas	Texas	22.4	0.866	99
Woodbine	Texas	23.9	0.544	48
Berea	W. Virginia	27.4	0.393	47

Table 1–2 Air-liquid interfacial tensions and densities of various liquids at 20°C (Weast, 1970: pages F30–F33).

Liquid	Density	Air/Liquid IFT
Benzene	0.8787	28.85
Dodecane	0.7487	25.44
Ethyl alcohol	0.7895	24.05
n-Hexane	0.6603	18.43
Mercury	13.5939	486.5
n-Octane	0.7025	21.62
Toluene	0.8669	28.5
Water	0.9982	72.75

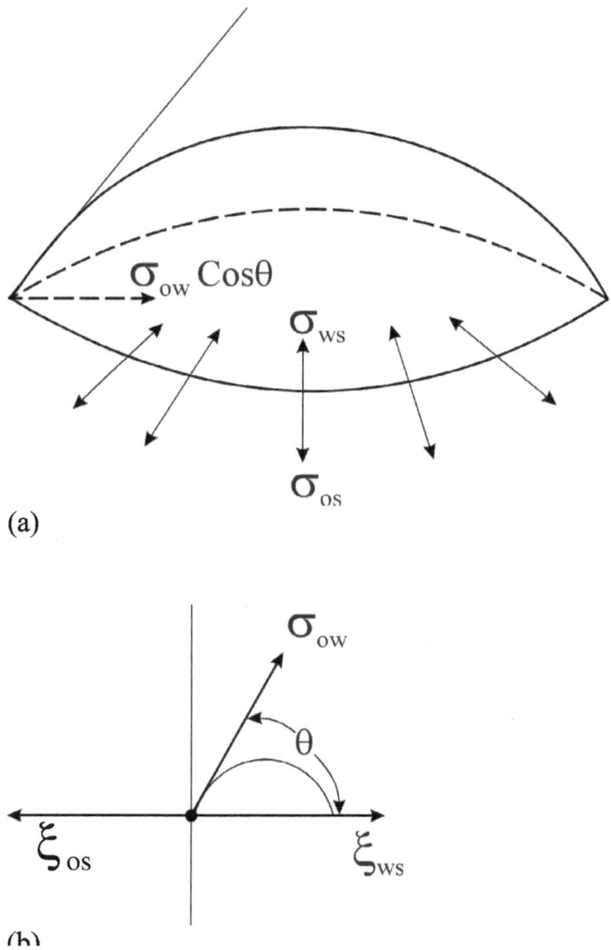

Figure 1–4 *Drop of liquid resting on a solid surface. (a) The tensile interfacial forces pull the drop into a spherical shape. (b) Two-dimensional view of the forces acting on the drop and the contact angle.*

Gibbs (1928) generalized Young's equation by considering the spreading tendency of a liquid onto a solid surface as follows (Fig. 1–4b):

$$Cos\theta = \frac{E_{f,ws} - E_{f,os}}{\sigma_{ow}} \tag{1.5b}$$

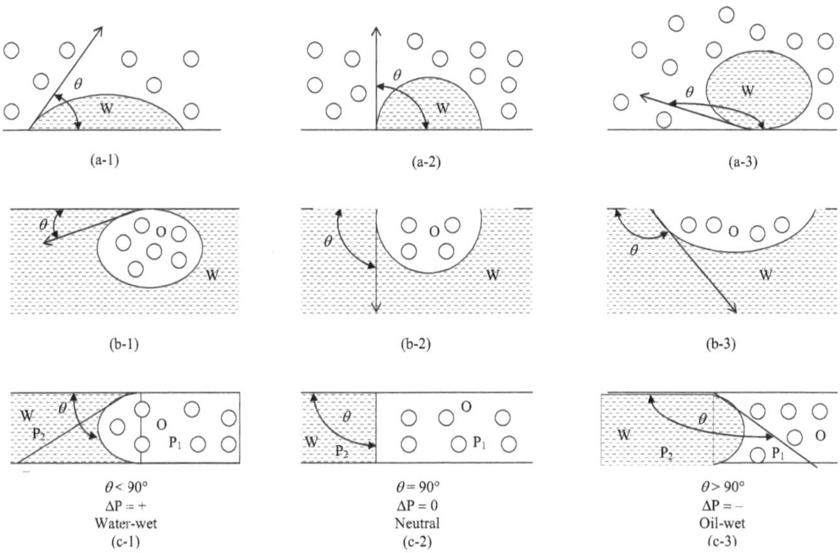

Figure 1–5 *Contact angles for various wetting conditions of water and oil: (a) drops of water in oil on a plane surface, (b) drops of oil on a plane surface, and (c) water and oil in a capillary tube.*

The terms $E_{f,os}$ and $E_{f,ws}$ are the derived free energies (in Newtons/meter) of interactions between the liquids and the solid

$$\begin{aligned} E_{f,os} &= \sigma_{s,vac} - \sigma_{so} \\ E_{f,ws} &= \sigma_{s,vac} - \sigma_{ow} \end{aligned} \tag{1.5c}$$

where $\sigma_{s,vac}$ is the excess free energy of the solid surface in a vacuum. Substitution of Eq. 1.5c into Eq. 1.5a yields the Gibbs-Young equation (Eq. 1.5b), showing that the solid interacts with the liquid drop to either enhance or oppose spreading of the drop of water onto the solid surface. The contact angle of a very low energy surface (such as a fluorocarbon) is a function only of the liquid surface tension. Furthermore, a liquid will not adsorb on a solid that has a lower surface tension than the liquid.

Consider a nonwetting spherical drop resting on a surface. The IFT is equivalent to the surface free energy per unit area (erg/cm^2), which is the work necessary to form one square centimeter of the drop's surface.

As the tensile forces pull the drop into a spherical shape, the radius decreases which in turn decreases the surface free energy.

$$dG_{sf} = 8\pi r \sigma_{ow} dr \tag{1.6}$$

The decrease of the drop's radius is opposed by an increase of the fluid's internal pressure. Thus the work performed by the IFT in decreasing free energy is equal to the work of compression produced by the difference between the internal and external pressures:

$$W = 4\pi r^2 (P_{in} - P_{ex}) dr \tag{1.7}$$

The resulting difference of surface free energy is equal to the work of compression; therefore, Eq. 1.6 and 1.7 are equivalent at equilibrium and yield the special case of the capillary pressure equation where $Cos\theta$ is equal to 1.0.

$$\Delta P = 2\sigma / r \tag{1.8}$$

A more general form of Eq. 1.8 was developed for capillaries and applied to porous media by Leverett (1941) (also Tiab and Donaldson, 2004: 314–318).

$$P_c = 2\sigma Cos\theta / r_c \tag{1.9}$$

Trieber, et al. (1972) examined the wettability of crude oils from 55 reservoirs using contact angles measured on silicate and carbonate surfaces. They adopted the following ranges of contact angles to represent three states of wettability: water-wet 0°–75°, intermediate-wet 76°–105°, and oil-wet 106°–180°. They found that for the oils from sandstone reservoirs, 43% were water-wet, 7% were intermediate-wet, and 50% were oil-wet. Oils from the carbonate reservoirs produced very different results: 8% were water-wet, 8% were oil-wet, and 84%

Table 1–3 Measurement of contact angles of crude oils on carbonate and silicate surfaces.

Wettability	Contact Angle	Silicate Rocks	Carbonate Rocks
Trieber et al. (1972)			
Water-wet	0°–75°	43%	8%
Intermediate-wet	75°–105°	6	4
Oil-wet	105°–180°	51	88
Morrow et al. (1976)			
Water-wet		40	8
Intermediate-wet		33	64
Oil-wet		27	28
Chilingar and Yen (1983)			
Water-wet	0°–80°		8%
Intermediate-wet	80°–100°		12
Oil-wet	100°–180°		80

were oil-wet. Later, Chilingar and Yen (1983), working with crude oils and carbonate surfaces, selected 90°±10° for the range of intermediate-wetting, Table 1–3. The table shows that the surfaces of carbonate rocks are preferentially oil-wet and that silicate surfaces are almost equally divided between water-wet and oil-wet.

1.4 Advancing and Receding Contact Angles

The characteristics of advancing and receding contact angles were studied by Leach et al. (1962) who placed a drop of oil between two plates; one of the plates was moved parallel to the stationary plate allowing water to advance against the oil drop in the direction of movement and to recede on the other side of the drop, Fig. 1–6. The angles gradually changed after the plate was moved, reaching equilibrium after 24–48 hours. The angles were measured and the plate advanced again. As the system aged, it eventually attained constant

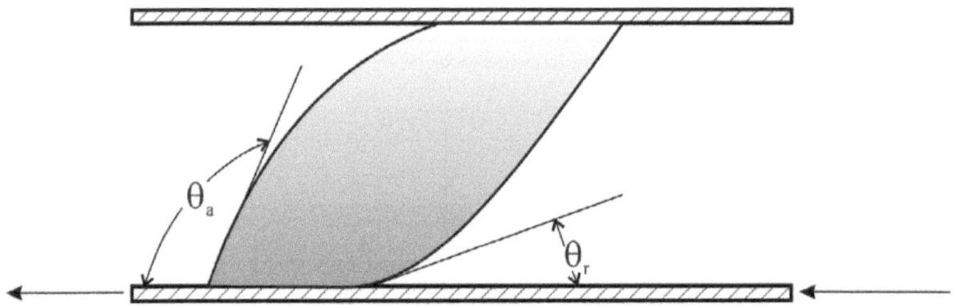

Figure 1–6 *Advancing (θ_a) and receding (θ_r) angles of a drop of liquid between a stationary top plate and displacing bottom plate.*

contact angle values with respect to time, requiring thousands of hours in some cases.

Trieber et al. (1972) examined the characteristics of 55 crude oils and pure samples. Crude oils had variable equilibrium contact angle values that apparently were functions of natural surface active compounds dissolved in the oils. Crude oils containing surfactants and asphaltenes gradually deposit them on the solid surface rendering the surface progressively more oil-wet with respect to time as the system aged.

Several oils, solid surfaces, and water compositions were used by Trieber et al. (1972) to examine the behavior of advancing contact angles clearly indicating various wettability conditions. Fig. 1–7 shows four of their results that clearly illustrate the usefulness of the contact angles measured as functions of aging time:

1. Curve A is the flat zero contact angle of a completely water-wet system: pure decane/quartz/demineralized water,
2. Curve B, equilibrium contact angle = 15°, is strongly water-wet: Alaska crude oil/quartz/formation water,
3. Curve C, equilibrium contact angle = 88°, is intermediate-wet: San Andres crude oil/calcite/formation water,
4. Curve D, equilibrium contact angle = 160°, is oil-wet: Kareem crude oil/quartz/formation water.

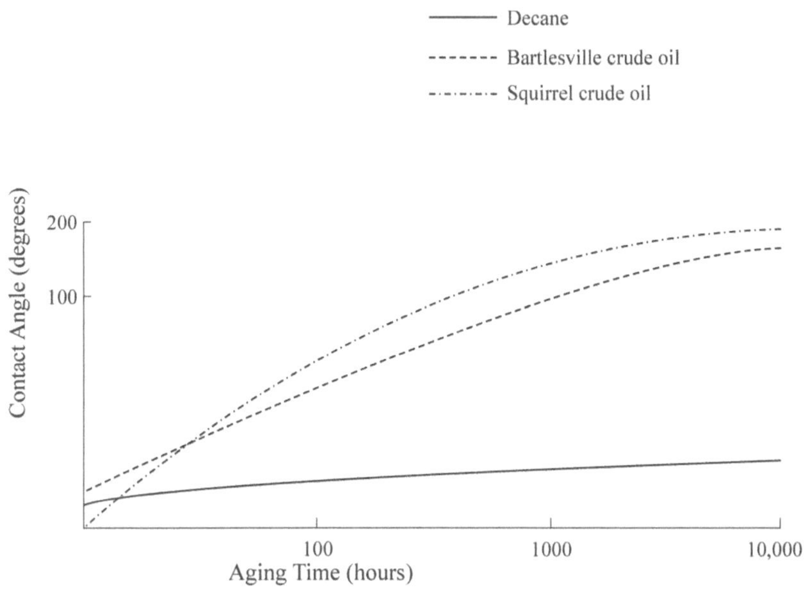

Figure 1–7 *Change of contact angle with respect to time for three crude oils and decane (after Trieber et. al, 1972).*

The equilibrium advancing contact angle clearly delineates the variation of wetting conditions. The disadvantages are that: (1) the aging time required for equilibrium may take 30–60 days (the solid surface used has a distinct affect because of its preferential wetting properties), (2) quartz is preferentially water-wet while calcite is a preferentially oil-wet surface, and (3) compounds dissolved in the water chosen for the test also contribute to differences of wettability. Therefore the test surface and fluids should be tailored to match the reservoir of interest as close as possible for proper interpretation of the results.

The advancing contact angle is greater than the receding angle by as much as 75°; this difference is representative of the hysteresis (delay of the equilibrium values when the forces acting on the oil drop are changed).

Recently Yan et al. (2002) used sessile drops of oil on mica to examine advancing contact angles and hysteresis. A drop of oil was placed on a plane surface from a fine capillary tube and allowed to rest on the surface for aging. The capillary tube was used to expand and contract the size of the drop to obtain advancing and receding contact angles. Mica surfaces were equilibrated with crude oil diluted with heptane to decrease the solvency of the asphaltenes. The diluted crude oil exhibited larger contact angles (~140°) than undiluted crude oils (~75°). The decrease of asphaltene solvency resulted in deposition of asphaltenes on the surface, producing an oil-wet surface that was evident from the higher values of the contact angles.

Measurements of contact angles are made on polished crystals of silica or mica to avoid complication introduced by roughness and heterogeneity of the surface. Morrow (1970) and Erick et al. (1975) found that the sharp edges of surface roughness and heterogeneity of surface, caused by mineral inclusions, are responsible for a wide range of contact angles measured under identical conditions.

Contact angle measurements are instructive for examination of the tendencies of crude oils to wet a surface because they contain surface active compounds that migrate to the surface (asphaltenes and aggregates of other polynuclear compounds) that precipitate from the bulk oil. The contact angles measured on smooth or polished surfaces are not representative of the conditions imposed by the complex geometery and surface heterogeneity of rocks. The constituents of crude oils affect the overall equilibrium wettability of the water/oil/rock system that can be determined by several methods: spontaneous imbibition, the centrifuge USBM, and the relative imbibition Amott methods.

1.5 Core Preservation

In many circumstances it is important to work with *restored state* cores, but complete removal of oil-wetting compounds adsorbed from crude oil during the long periods of geologic time present difficult procedural problems. Extraction with solvents (usually toluene followed by alcohol and water) still leaves a rock with minerals coated with strongly adsorbed coatings of organic compounds. Chilingarian and Yen (1992) identify four items that can affect the wettability of rocks used in laboratory procedures: (1) cores thoroughly cleaned, followed by experiments using refined oils that do not have polar petroleum compounds; all such cases using sandstones and carbonates result in water-wet systems that do not reflect the native state condition of reservoir rocks; (2) trace metals dissolved in the water used for the experiments produce preferentially water-wet systems; (3) the use of produced-formation water containing trace metals dissolved from metal tubes, casing, etc., also can change the wettability toward more a water-wet condition; (4) exposure of the rocks to air can render them more water-wet.

Donaldson et al. (1991) recommended preserving cores at the wellhead by placing them in glass bottles filled with the produced oil and keeping them in a cool room. A cleaning method for complete removal of the surface coatings is treatment of the core with steam after extraction by toluene. This procedure was accidentally discovered when humic acids were extracted from outcrop cores after the standard cleaning procedure using toluene and alcohol was implemented. The clay and other minerals are not affected by the steam, and the cores can be repeatedly restored to their original condition of wettability (Donaldson et al., 1991).

Cores extracted from a reservoir should be preserved immediately using one of the following methods that have been proven by many laboratories to be effective in preserving the native state wettability of the cores: (1) wrap the cores in plastic and cover them with a coating of wax to seal the fluids in place, or (2) seal the cores in nonmetal containers filled with crude oil. Exposure of the cores to air during handling and processing will allow diffusion of oxygen into the fluids to take place as volatile compounds evaporate from the cores. The evaporation of volatile compounds and oxidation of active chemical compounds in the crude oil will produce changes in wettability. If the crude oil contains polar compounds, or asphaltenes, the decrease of temperature and pressure that results as the cores are brought to the surface can result in precipitation of the high molecular weight compounds which will change the wettability toward a more oil-wet condition in spite of attempts to preserve the native state wettability, Richardson et al. (1955) and Mungan (1966).

Deposition and adsorption of organic compounds from crude oils onto the reservoir rock around a producing well can produce an oil-wet zone. An oil-wet condition of the reservoir rock around the wellbore will reduce the capillary end effects that produce high water saturations (water coning) at many producing wells. This effect would decrease the water/oil production ratios of such wells.

1.6 Capillary Pressure

When the interface between immiscible fluids is curved (Fig. 1–5c), the pressure difference across the interface will be greater on the convex side to balance the forces creating the interfacial tension. The pressure difference is defined as the capillary pressure where, by convention, it is:

$$P_c = P_{nw} - P_w = \sigma\left(\frac{1}{r_1} - \frac{1}{r_2}\right) = 2\sigma Cos\theta / r_t \tag{1.10}$$

where

P_{nw} = pressure of the nonwetting fluid

P_w = pressure of the wetting fluid

r_1, r_2 = radii of curvature of the interface

r_t = radius of the capillary tube

Displacement processes in capillary systems are described as *drainage* and *imbibition*. Drainage means that the wetting fluid is displaced by the nonwetting fluid, and imbibition is the opposite (the displacement of the nonwetting fluid by the wetting fluid). *Spontaneous imbibition* of the wetting phase occurs as the capillary pressure declines to zero, Fig. 1–8 (dashed curve 2). The terms drainage and imbibition can be confusing because they apply to the specific wettability condition: for a water-wet rock the terms refer to water, but for an oil-wet rock they apply to oil. For an intermediate-wet system the terms apply to both phases because they are both equally wetting the rock surfaces. Therefore, instead of using drainage and imbibition, the fluid leaving the core will be referred to as the *displaced fluid* and the fluid entering the core will be referred to as the *injected fluid* regardless of the wettability condition of the porous medium.

As the saturation varies in the core, there is a certain amount of hysteresis in the capillary pressure behavior: the displaced and injected fluid curves are different (Fig. 1–8, curves 2 and 4,5). A reason for this capillary hysteresis is the difference between the advancing and receding contact angles. The advancing contact angle dominates during injection as the wetting phase moves into the porous system, but the receding contact angle becomes dominant during displacement as the wetting phase is moved away from its contacts with the solid surfaces.

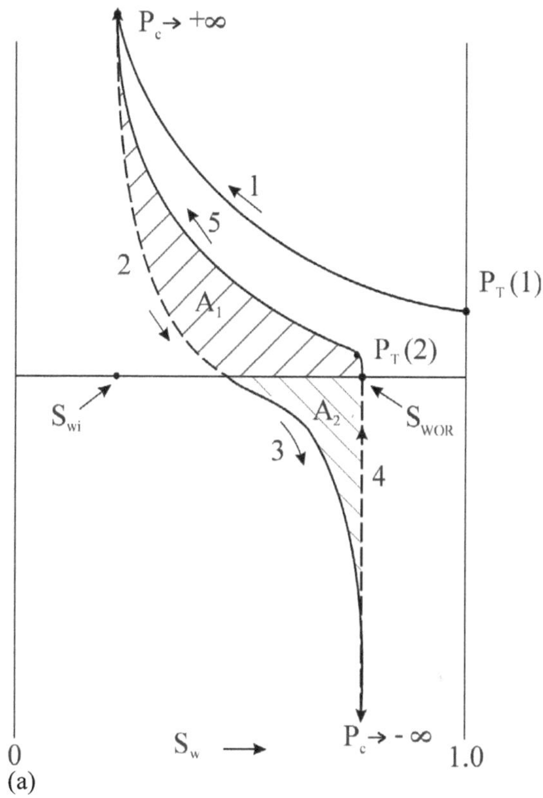

Figure 1–8 *Water-wet capillary pressure curves: (1) oil displacing water from S_w=1.0, (2) spontaneous imbibition of water, (3) forced displacement of oil to S_{wor}, (4) threshold pressure $P_{T(2)}$ must be exceeded before oil can enter, (5) forced displacement of water by oil.*

Fig. 1–8 illustrates the capillary pressure behavior of a water-wet system in which the core was initially saturated with water. Curve 1 is the initial displacement curve that results when the water-saturated core is contacted by oil; an initial elevated pressure ($P_{to(1)}$) known as the *threshold pressure* of oil is required before oil will enter the water-wet core. As soon as $P_{to(1)}$ is exceeded, oil enters the core, displacing water. For capillary pressure versus saturation measurement this displacement is conducted with step increases of the oil injection: quasi-static, pressure-dominated, oil injection, yields curve 1 that reaches a practical irreducible water saturation (S_{wi}) as the capillary pressure approaches an infinite value. Continued injection of oil will result in continued displacement of a small quantity of water, at a diminishing rate, as the relative permeability to water approaches zero. When oil injection is stopped, the capillary pressure relaxes to zero and molecular surface forces acting between the rock and the two fluids gradually distribute the oil and water saturations evenly throughout the core.

When the core is contacted by water at S_{wi} (at zero capillary pressure), water will imbibe spontaneously into the core displacing oil. If one end of the core is exposed to water and the other is closed, the movement of the fluids will be countercurrent to each other and the saturation distribution in the core will vary from high water saturation at the contacted face to low water saturation at the closed face. Given sufficient time the saturation distribution will adjust to an evenly distributed saturation within the core. If both ends of the core are in contact with water, both countercurrent and co-current movement of the fluids can occur because of the action of molecular forces.

When the spontaneous imbibition reaches a limiting value, pressure must be applied to force water into the core and displace oil to a practical residual oil saturation (S_{or}) as the capillary pressure approaches a negative infinite value (Fig. 1–8, curve 3). This displacement is conducted in quasi-static steps of increasing water pressure. Continued injection of water after a practical value for the residual oil has been attained will continue to displace a diminishing amount of oil by frictional drag (film drainage) as the relative permeability of the oil approaches zero; this process can be examined in greater detail using a semi-log plot of relative permeability versus saturation.

If the water-wet core at residual oil saturation is contacted with oil, oil will not imbibe into the core. A threshold pressure ($P_{to(2)}$) must be exceeded before oil will enter the core displacing water (curve 5), and another period of quasi-static, pressure-dominated oil injection is imposed to measure the capillary pressure.

The area under the capillary pressure curves represents the work required for fluid displacement:

$$W(Area) = \phi V_b \int Pc \cdot dS_w \ \ [m^3 * N / m^2 = N - m(Joules)] \qquad (1.11)$$

In a water-wet system, the work required for displacement of water from S_{or} to S_{wi} (A_1) is greater than the work required for displacement of oil from S_{wi} to S_{or} ($A_1 > A_2$). For an intermediate-wet (I-W) core, an equal amount of work is required for displacement of the two fluids ($A_1 = A_2$); for an oil-wet system, a greater amount of work is required for displacement of oil than for displacement of water ($A_1 < A_2$). Donaldson et al. (1969), used the logarithm of the ratio of the areas for a wettability index. Values increasing from zero are progressively more water-wet, and when the areas are equal [$\log(A_1/A_2) = 0$] the system is at intermediate wettability; for oil-wet systems, $\log(A_1/A_2)$ assumes increasing negative values as the system becomes more oil-wet.

Fig. 1–9 shows the capillary pressure versus saturation relationships of an intermediate-wet core initially saturated with water. Some oil may imbibe into the core when it is contacted by oil at zero capillary pressure, after which pressure is required for injection of oil to displace water to S_{wi} (curve 1).

With the saturation at S_{wi}, if the core is contacted with water, a small amount of water will spontaneously imbibe into the core (Fig. 1–9, position 2). As the water pressure is increased, oil is displaced to its residual saturation (S_{or}). The capillary pressure is negative because the water injection pressure is greater than the oil pressure [$P_c = P_o - P_w < 0$].

If an intermediate-wet core is contacted with oil when the saturation is at rest (at S_{or}), a very small amount of oil will imbibe into the system (Fig. 1–9, position 4). After the initial period of spontaneous imbibition, an oil injection pressure is required for displacement of water back to S_{wi} (Fig. 1–9, curve 5). Spontaneous imbibition of both oil and water was reported for fractional or mixed wettability cores by Mohanty and Salter (1983).

The areas under the capillary pressure curves (A_1 and A_2) for an I-W system are equal, thus the logarithm of the area ratios is zero. The I-W system can be the result of several types of wettability conditions: neutral, fractional, or mixed.

In general, I-W is distinguished from the W-W and O-W systems by: (1) a threshold pressure is not required for injection of water or oil, (2) the system will spontaneously imbibe a very small amount of water when the saturation is a S_{wi}, or oil at S_{or}, and (3) the work required for displacement of water and oil (A_1, A_2) are equal.

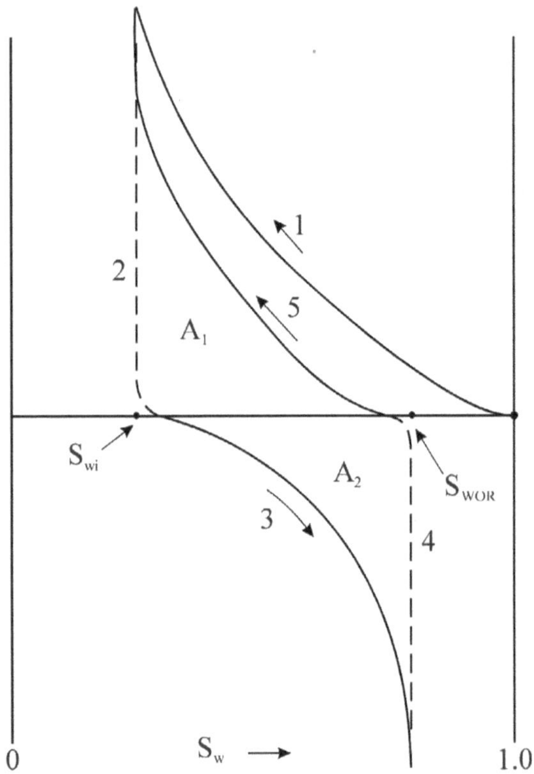

(b)

Figure 1–9 *P_c curves for a neutral wettability system: (1) a small amount of oil will imbibe at the beginning of the water drainage curve, (2) a small amount of water will imbibe when P_c reduces to zero, (3) forced displacement of oil to S_{wor}.*

Fig. 1–10 depicts the capillary pressure behavior of an oil-wet system initially saturated with water. When the water-saturated core is contacted by oil, it will spontaneously imbibe some oil and displace water by either co-current or countercurrent movement of the fluids in the rock (position 1).

Injection of oil will displace the water to S_{wi}. At this point (when the core is contacted by water) water will not spontaneously imbibe into the core. A threshold pressure (P_{tw}) is required to overcome the repulsive surface forces to water. After P_{tw} is exceeded, oil will be displaced to S_{or} (Fig. 1–10, curve 3). At S_{or}, if the core is contacted with oil, oil

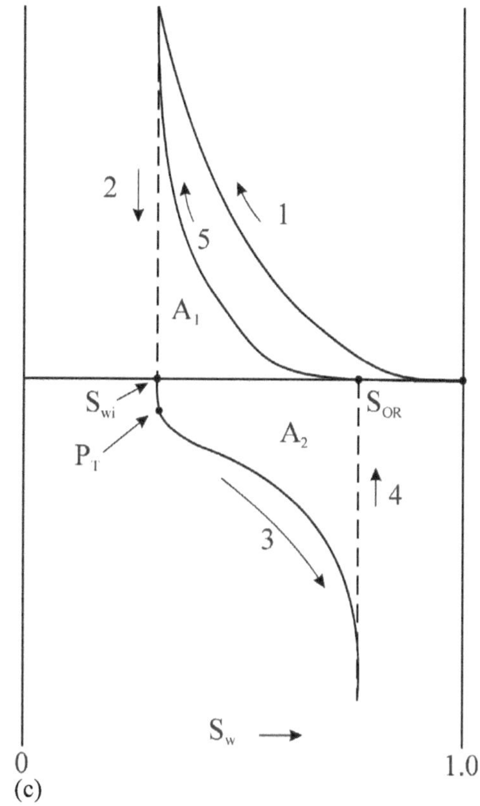

Figure 1–10 *P_c curves for an oil-wet porous medium: (1) oil spontaneously imbibes, (2) water will not imbibe, (3) a threshold pressure (P_T) must be exceeded before water will displace oil, (4) oil will imbibe, (5) oil displacement of water.*

will spontaneously imbibe into the core. When imbibition is complete, an injection pressure is required for displacement of water back to S_{wi}. The work required for displacement of water from an oil-wet core (A_1) is less than the work required for displacement of oil; thus the logarithm of the area ratio is negative.

Capillary pressure curves at various states of wettability were found to fit a least square solution of a hyperbola (see *Program Listing 6.24 WET.FOR*). The derived hyperbolic equation is then used to accurately measure the areas under capillary pressure curves, determine the USBM wettability index, and obtain the pore-size distribution

(Anderson, 1986a, b; Donaldson et al., 1991; Donaldson and Tiab, 2004; and Singh, 1990):

$$P_c = \frac{1 + AS_w}{B + CS_w} \tag{1.12}$$

$$Area = \phi V_b \left[\frac{BS_w}{C} + \frac{AB - C}{C^2} Log(1 = CS_w) \right] \tag{1.13}$$

The details of equipment design, procedures, and calculation of capillary pressure obtained by centrifuge are presented by Tiab and Donaldson (2004) in the text and in the appendix under Measurement of Rock and Fluid Properties. Using a centrifuge, capillary pressure at the *inlet* of the core is calculated from:

$$(P_c)_i = (1.096 * 10^{-6})(\Delta\rho)N^2 \left(r_e - \frac{L}{2}\right)L \tag{1.14}$$

L= core length, cm

N= revolutions per minute of the centrifuge rotor

$(P_c)_i$= capillary pressure at the inlet face of the core, kPa

r_e= distance from the center of the centrifuge arm to the end of the core situated in the centrifuge, cm

$\Delta\rho$= density difference between the fluids, g/cm^3

The saturation calculated from the amount of fluid displaced at each step-increase of centrifuge revolutions per minute is the average saturation of the core (S_{av}). The capillary pressure calculated using Eq. 1.14, however, is the capillary pressure at the *inlet face* of the core. In many cases $(P_c)_i$ is plotted versus S_{av} for determination of core properties such as wettability; when reported in literature, it should always be noted as $(P_c)_i$ versus $S_{av.}$

Hassler and Brunner (1945) derived a relationship for calculation of the inlet saturation as a function of the measured inlet capillary pressure:

$$S_i = S_{av} + (P_c)_i * \frac{dS_{av}}{d(P_c)_i} \tag{1.15}$$

The derivative of the average saturation with respect to the inlet capillary pressure $[dS_{av}/d(P_c)_i]$ is determined from the experimental data and used to calculate the saturation at the inlet of the core corresponding to the inlet capillary pressure.

A least squares solution of the hyperbolic function (Eq. 1.12) can be fit to the $(P_c)_i$ versus S_{av} to determine the constants *A, B* and *C* and then differentiated to yield:

$$S_i = S_{av} + (P_c)_i * \left[\frac{(1 + CS_{av})^2}{B - AC}\right] \tag{1.16}$$

Details of the equation derivations and sample calculations using them are presented by Tiab and Donaldson (2004). The computer program *WET.FOR* (*Program Listing 6.24*) may be used to calculate the wettability index, capillary pressure curves, areas under the curves, and threshold pressures.

1.7 Amott Wettability Index

The Amott (1959) wettability index, I_w, is based on spontaneous imbibition and forced displacement of water and oil. Beginning at S_{or}, the core is immersed in oil and the amount of water displaced by spontaneous imbibition of oil is recorded as V_{wsp}; oil is then injected until S_{wi} is attained and the total amount of water displaced (by imbibition and forced displacement of oil) is recorded as S_{wt}. The ratio of the amount of water displaced by oil imbibition (V_{wsp}) to the total amount of water displaced (S_{wt}) is the displacement-by-oil ratio, δ_o.

The core (at saturation S_{wi}) is immersed in water and the amount of oil displaced by spontaneous imbibition of water is recorded as V_{osp}. Water is injected into the core until the residual oil saturation is attained; the total amount of oil displaced (by imbibition and forced displacement) is recorded as V_{wt}. The ratio of the amount of water displaced by imbibition (V_{osp}) and the total amount of oil displaced (by water imbibition and forced displacement) is the displacement-by-water ratio, δ_w.

The Amott wettability index is the displacement-by-water ratio minus the displacement-by-oil ratio:

$$I_A = \frac{V_{osp}}{V_{ot}} - \frac{V_{wsp}}{V_{wt}} = \delta_w - \delta_o \tag{1.17}$$

1.8 Combined Amott-USBM Wettability Test

According to several authors (Anderson, 1986b; Sharma and Wunderlich, 1989), the USBM wettability index is improved if it is modified to account for changes in saturation that occur by imbibition. At each point where the capillary pressure is equal to zero, the sample is immersed in the displacing fluid for 20 hours and the amount of fluid imbibed is recorded and used to determine the Amott wettability ratios. The capillary pressure data are plotted versus the average saturation (not the saturations at the inlet face of the core). The six steps in the combined Amott-USBM test are:

1. complete saturation of the core with water,
2. initial displacement of water to S_{wi} (oil drive),
3. spontaneous imbibition of brine,
4. displacement of oil by brine to S_{or} (water drive),
5. spontaneous imbibition of oil, and
6. displacement of water by oil to S_{wi} (final oil drive).

The Amott wettability index is then calculated from the spontaneous imbibitions of water and oil and the total water and oil displacements. Then the USBM index is calculated from the areas under the curves.

Sharma and Wunderlich reported that the USBM index gave values greater than 1.0 for very strongly water-wet cores, and that imbibition of oil at S_{wi} = 1.0, or when the water saturation is at S_{wi}, clearly defined intermediate-wet cores.

1.9 Spontaneous Imbibition

In very strongly water-wet systems, spontaneous imbibition of water from S_{wi} may increase the water saturation all the way to S_{or} (the imbibition capillary pressure curve, Fig. 1–8 curve 2, is greater than zero until it reaches S_{or}). When this occurs, the USBM wettability index, I_u, cannot be determined; and only the displacement-by-water ratio of the Amott wettability index can be determined. Thus both indices only indicate that the system is very strongly water-wet; however, the rate of imbibition and the capillary pressure curve versus water saturation (measured at incremental increases of capillary pressure as the saturation changes from S_{wi} to S_{or}) can be used to examine the wettability of very strongly water-wet systems (Ma et al.,1997, 1999).

When water is displaced to S_{wi} the capillary pressure begins a rapid increase toward an infinite value. If the fluid pressure is relaxed at this point in a water-wet or intermediate-wet system, the core will imbibe water as the capillary pressure declines to zero because of a decrease of the surface free energy. As imbibition takes place, the wetting phase saturation increases from S_{wi} to S_z (the point at which the imbibition pressure is equal to zero). If the wettability condition is W-W or I-W, a characteristic amount of water will imbibe at a rate that depends to a large extent on the wettability.

1.10 Oil Recovery by Imbibition

Mattax and Kyte (1962) proposed a relationship to express oil recovery by imbibition from fractured reservoir rocks as a single curve based on dimensionless time. The dimensionless time incorporated the fluid and rock properties proposed earlier by Rapoport (1955). The correlation gave approximate results for groups of rocks with similar properties. Modifications of the equation over a period of time by Lefevre Durey (1978), Cuiec (1994), Zhang et al. (1996), and Ma et al. (1999) lead to development of an equation that has been shown to correlate all of the previously published data on various types of fluids and cores into a single curve that can be approximately fit by the Aranosky equation using dimensionless time as the single mass transfer function.

The relationship developed by Ma et al. (1997) for correlation of oil recovery by spontaneous imbibition as a function of dimensionless time is:

$$\frac{R_e}{R_{e(\max)}} = t_D = Ct\sqrt{\frac{k}{\phi}}\frac{\sigma}{\sqrt{\mu_w \mu_o}}\frac{1}{L_c^2}$$

$R_e/R_{e(max)}$ = ratio of recovery to maximum recovery by imbibition

C = constant of unit conversion – 0.018849

K = permeability, md

L_c = characteristic length of the core, cm

t = time, minutes

t_D = dimensionless time

μ = viscosity, cP

σ = interfacial tension, dyne/cm

ϕ = fractional porosity

The characteristic length (L_c) compensates for the differences of sample size, shape and boundary conditions (Fig. 1–11).

$$L_c = \sqrt{V_b \Big/ \left(\sum_{i=1}^{n} A_i / x_i\right)} \tag{1.18}$$

V_b = bulk volume of the core

A_i = the open area of the ith face

x_i = distance from the open face to the no-flow boundary (center of the core for configurations *a*, *b* and *c* and length of the core for *d*, Fig. 1–11)

Fig. 1–12 shows the variation of oil recovery obtained by spontaneous imbibition for four boundary conditions depicted in Fig. 1–11. The data from 20 cores were correlated using dimensionless time to obtain a general fit, Fig. 1–13 and the derived Eq. 1.17.

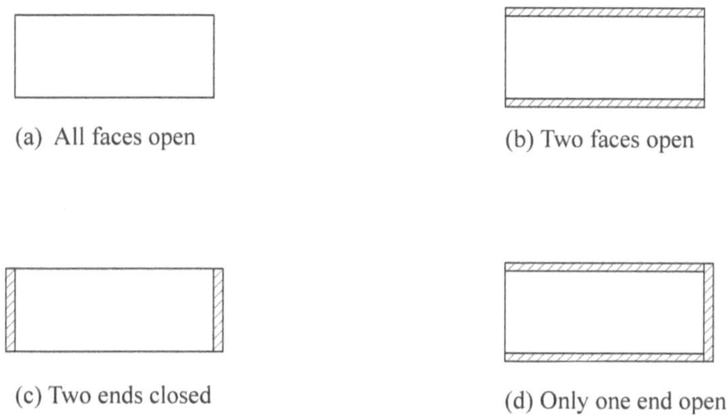

Figure 1–11 *Different arrangements of cores used for spontaneous imbibition experiments.*

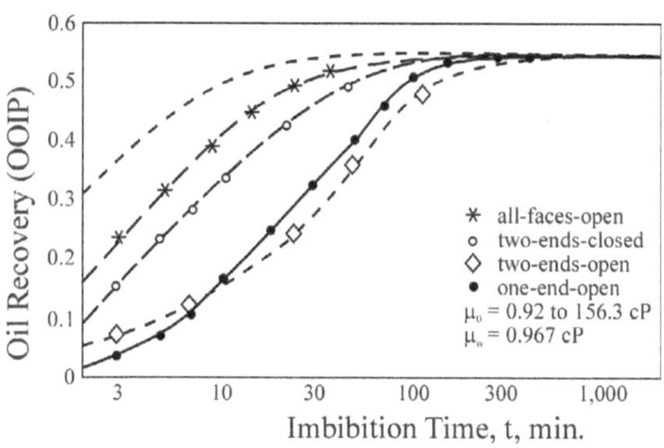

Figure 1–12 *Recovery of original oil in place by imbibition (after Ma et al., 1997).*

A general correlation to published imbibition data was then made as shown in Fig. 1–14. The correlation is close considering the wide variations of porous media: sandstone and alundum, Mattax and Kyte (1962); aluminum silicate, Harmon and Vidal (1986); and Berea sandstone, Zhang et al. (1996). Furthermore, variations of fluid and rock properties, core size, and maximum recovery by imbibition, did not have discernable affects on the fractional recovery when plotted as a function of dimensionless time.

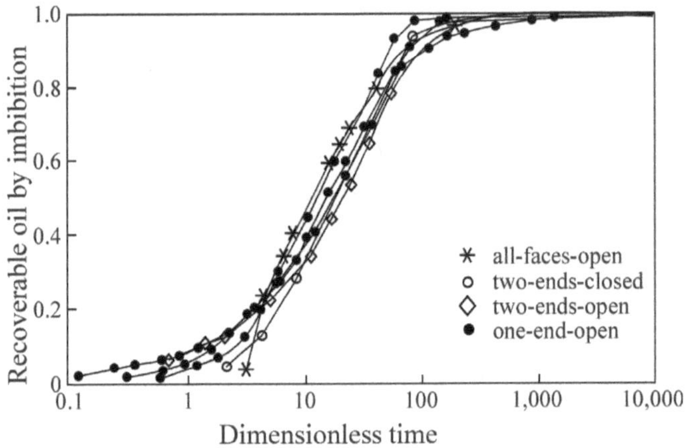

Figure 1–13 *Oil recovery with respect to dimensionless time (after Ma et al., 1997).*

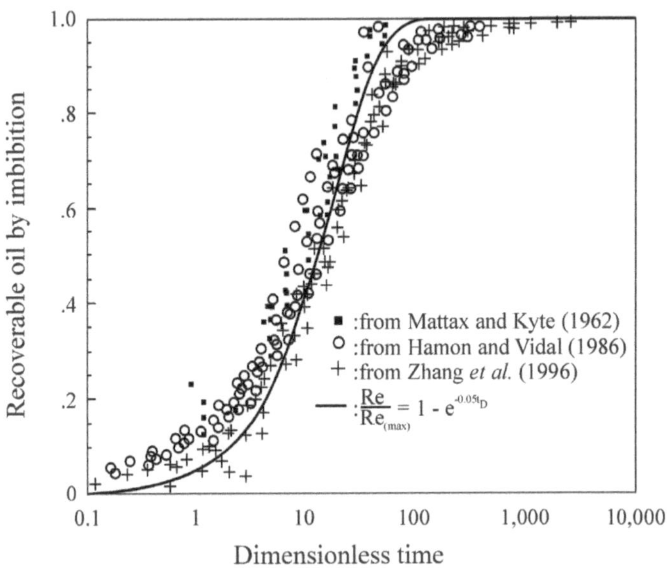

Figure 1–14 *Correlation of data from Mattax and Kyle (1962) and Zhang et al. (1996) and a single parameter model fit of the data (after Ma et al., 1997).*

The recovery equation proposed by Aranosky et al. (1958) using dimensionless time as the single mass transfer parameter and an average production decline constant (α) equal to 0.05, gave an approximate fit to the curve (Fig. 1–14).

$$\frac{R_e}{R_{e(\max)}} = 1 - e^{-\alpha t_D} \tag{1.19}$$

Later, Ma et al. (1999) developed a new wettability index to describe the wetting conditions of systems that are strongly water-wet and thus cannot be adequately characterized by the USBM or Amott wettability indices.

1.11 Spreading

It has not been possible to measure the IFT of a liquid-solid interface; but, its existence is used in the discussions of contact angles between liquids and solids. The adhesion tension (T_a) is defined as the difference between the water-solid and oil-solid IFT, Fig. 1–4.

$$T_a = \sigma_{os} - \sigma_{ws} = \sigma_{wo} Cos\theta_{wo} \tag{1.20}$$

In cases where the water-solid IFT is equal to zero the contact angle also is zero, and thus the adhesion tension is equal to the water-oil interfacial tension $\sigma_{ws} = 0;\ \cos\theta = 1.0\ \therefore T_a = \sigma_{wo}$. This condition occurs when the solid surface exhibits a complete water-wet condition and water spreads over the surface as a film.

The imbalance of forces that occur in a completely water-wet condition (where $\theta = 0$ and water spreads on the surface as a film) defines the spreading coefficient:

$$\xi_s = \sigma_{os} - \sigma_{ws} - \sigma_{wo};\ \ \theta = 0;\ \ Cos\theta = 1 \tag{1.21}$$

Thus far discussion of contact angles has been related to water, oil, and a solid because these are the general three-phase systems that describe petroleum reservoirs. Another three phase relationship involves gas, water, and oil associated with wetting films covering the solid matrix. Zhou and Blunt (1998) used the contact angle relationships and spreading coefficient to explain the behavior of gravity drainage of oil from sand packs treated to make them range in wettability from strongly water-wet to strongly oil-wet.

Consider a completely water-wet system where the contact angle is equal to zero and water is spread over the solid as a film; if a drop of oil is placed on the film of water in the presence of gas, Fig. 1–15a:

$$Cos\theta_{go} = \frac{\sigma_{gw} - \sigma_{wo}}{\sigma_{go}} = 1 + \frac{\xi_{s,o}}{\sigma_{go}} \tag{1.22}$$

When $\sigma_{gw} \gg \sigma_{wo}$; $\theta \to 0$; $Cos\theta_{go} \leq 0$ and $\xi_{s,o} = \sigma_{gw} - \sigma_{wo} - \sigma_{go}$

Gas is the most nonwetting phase and therefore oil will spread on the film of water; thus the film of water prevents oil from contacting the solid. In the pores, oil can move more easily within a continuous filament resting on water until snap-off occurs, thus isolating filaments and globular clusters of oil. The isolated oil globules resist further displacement and become the residual oil saturation. This is in agreement with the energy required for displacement of water and oil expressed by the capillary pressure relationships of a water-wet system. The oil displacement curve (the negative curve, Fig. 1–8) shows that spontaneous imbibition of water occurs as the capillary pressure falls from a high value at S_{wi} to zero.

At the end of imbibition, forced displacement of the oil to S_{or} requires only a small amount of energy. In cases of very strongly water-wet systems, spontaneous imbibition can drive the saturation all the way to S_{or}. The curve for water displacement from S_{or} to S_{wi} shows that imbibition of oil will not take place and that a large amount of energy is required for water displacement.

When the system is strongly oil-wet, oil spreads over the solid as a film, Fig. 1–15b. If a drop of water is placed on the oil film, water will not spread on the oil surface and this implies that water is a nonwetting phase under these conditions. The spreading coefficient is:

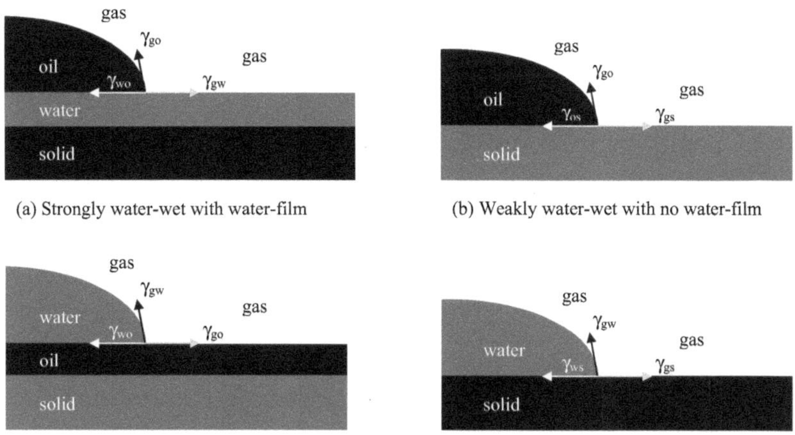

(a) Strongly water-wet with water-film

(b) Weakly water-wet with no water-film

(c) Strongly oil-wet with oil-film

(d) Weakly oil-wet with no oil-film

Figure 1–15 *(a) Water-wet surface covered by a film of water, (b) very slightly water-wet surface without a film of water, (c) oil-wet surface with a film of oil, (d) very slightly oil-wet surface without a film (after Zhou and Blunt,1998).*

$$Cos\theta_{gw} = \frac{\sigma_{go} - \sigma_{wo}}{\sigma_{gw}} = \frac{\sigma_{gw} + \sigma_{go} - \sigma_{wo} - \sigma_{gw}}{\sigma_{gw}} = 1 + \frac{\xi_{s,w}}{\sigma_{gw}} \qquad (1.23)$$

For most systems, $\sigma_{gw} >> \sigma_{go}$, and thus $\theta_{gw} > \pi/2$ and $\xi_{s,w} < 0$, and gas wets the oil surface preferentially to water which affects the distribution of the fluids in a capillary.

The capillary pressure of each phase for an oil-wet system (Fig. 1–10) shows that more energy is expended for forced displacement of oil by water. Oil is in intimate contact and adhering to the solid while water is resting on an oil film as a nonwetting phase and can be displaced with a much smaller amount of energy. In fact, the spontaneous imbibition of oil may displace most of the water before forced displacement is required to drive the saturation to S_{wi}.

If the solid surface and liquid interfacial molecular forces are not strong enough to cause complete wetting of either water or oil, both water and oil can contact the solid forming mixed layers of films. This resulting system is an intermediate-wet (mixed or fractional) system and the relative spreading of the two liquids results in a complex system that can no longer be described as a single function of the spreading coefficient.

The corners of pores, where grains are in contact in the three dimensional structure of the matrix, are very important in the stable configuration of fluids with respect to the overall relative wetting characteristics. Zhou and Blunt (1998) used equilateral triangles to represent two dimensional cross sections of pores to illustrate and explain how wetting can control the stable, static, relative positions of gas, water, and oil as well as their dynamic behavior when a global pressure drop, such as a gravity field, is imposed across a porous medium composed of an aggregate of pores with the same conditions of wettability. In a water-wet system (Fig. 1–16a) water occupies the smaller crevices, the corners of the triangle, while oil spreads as globules and films on the water, and gas (the least wetting phase) occupies the center of the pores. The different phase pressures of the three fluids in a water-wet condition ($P_g > P_o > P_w$) contribute to the stability of the distributions of the three phases within the pores.

Donaldson and Crocker (1977) used a microscope to study the water and oil distributions of residual oil saturations. Color photomicrographs of water-wet, intermediate-wet, and oil-wet systems clearly show the differences of the positions of water and oil in the corners of the pores. In the water-wet system, water occupies the small pores and corners and wets the sand grain surfaces. Small spherical drops of oil were evident, indicating the nonwetting behavior of oil and small pools, or globules, of oil seen resting on films of water as illustrated by Zhou and Blunt (1998), Fig. 1–16a. When water was injected to displace oil from its saturation as S_{wi} to S_{or}, each fluid moved through individual networks of pores until the oil network was depleted and broke into discontinuous trapped pockets of oil that could only be displaced by increasing the pressure, and thus the flow rate, of water to very high values. Water invaded the system as a uniform front resulting in a high displacement efficiency, but when the front reached the end of the sand pack and water broke through, oil production immediately declined to almost zero. The same result was observed for numerous water-wet core flood experiments.

An oil-wet condition is clearly shown in Zhou and Blunt's triangles (Fig. 1–16b). Oil instead of water occupies the corners, gas is in contact with oil, and water (which is the least wetting fluid for this condition)

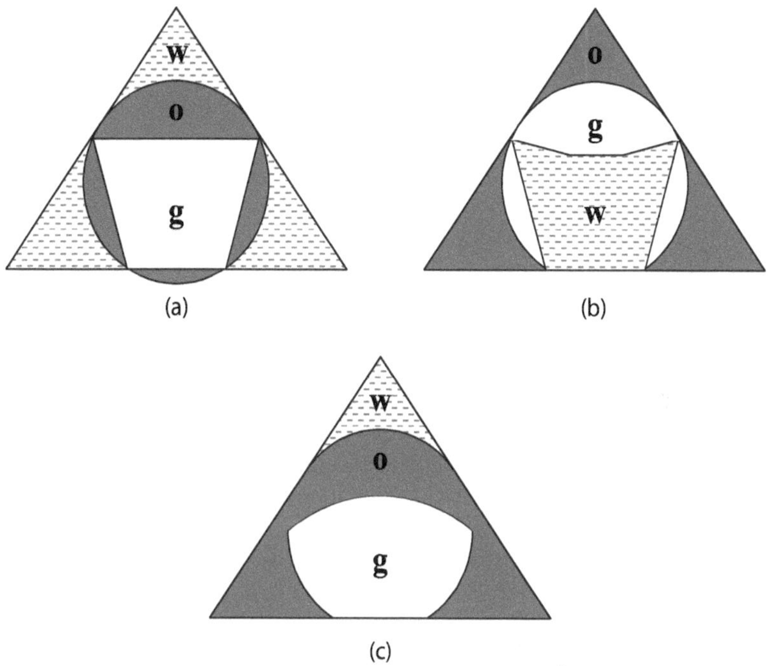

Figure 1–16 *Possible static postions of gas, water, and oil in three overall conditions of wettability. (a) Water-wet pores with water in the corners and oil spread over the water, (b) oil-wet with oil occupying the corners and gas (more wetting to oil than to water) in contact with the oil, (c) intermediate-wet where corners contain either water or oil depending on the preferential wetting conditions of exposed forces of individual grains (after Zhou and Blunt, 1998).*

occupies the center, isolated from the oil surface by a buffer of more wetting gas when the gas-oil interfacial tension is low. With this distribution of fluids, the phase pressures are in an entirely different relationship than those of the water-wet condition: $P_w > P_g > P_o$. The system will remain stable with this distribution of phase pressures.

The photomicrographs of residual oil saturation of oil-wet systems show that oil occupies the crevices (corners) and small pores and wets the grain surfaces as a film (Donaldson and Crocker, 1977). Water was present only in the larger pores and resting on films of oil. The residual oil was present as a continuous filamentary phase through the small pores, analogous to the irreducible water saturation of a water-wet system. The waterflood displacement of oil from S_{wi} occurred with a fragmented advancing front of water containing numerous water fingers branching out to the end of the sand pack. This behavior caused early breakthrough of water followed by a long period of oil production at a high water/oil production ratio as oil slowly drained through the continuous filaments of oil in the corners and crevices. This behavior was confirmed by waterfloods of oil-wet systems.

The intermediate-wet system depicted in Fig. 1–16c is more complex and can assume multiple configurations of the three-phase system. Gas generally occupies the center of the pores and water and oil occupy the crevices in a random manner. In the three-phase system, gas within the porous matrix can have very different distributions of the phase pressures that largely stabilize the relative locations of water and oil in the pores: (1) $P_g > P_o > P_w$, (2) $P_g > P_w > P_o$, and (3) $P_g > P_w = P_o$ where there is an approximately even distribution of water and oil in the crevices of the grains.

The color photomicrographs of the residual oil saturations of intermediate-wet systems show that water and oil are almost evenly distributed in the small pores and crevices between the grains. Small areas of oil films (smaller than the sand grains) coat random sand grains. The systems imbibe small amounts of water and/or oil depending on whether they are slightly more water-wet or oil-wet. Pockets of oil in random areas of the sand drain through continuous filaments of oil that are connected through the corners and crevices of the pores. The oil drains until the filaments are broken leaving small pockets and strings of oil filaments surrounded by water. In addition, the residual oil saturations of intermediate-wet systems were lower (greater ultimate oil production) than those of either water-wet or oil-wet systems.

The configuration of the three phases in the porous medium is thus governed by (1) wettability of the system (the relative liquid-solid adhesion characteristics), (2) the phase pressure relationships, and (3) the fluid-fluid surface energies (interfacial tensions).

The spreading coefficients (the differences between the work of adhesion and cohesion) of the liquids (ξ_w and ξ_o) express the imbalance of the phase IFTs acting at the lines of contact of the phases. An $\xi_o > 0$ indicates that oil will form a film on the gas/water interface; and if $\xi_o < 0$, spreading of oil will not occur. Similarly if $\xi_w > 0$, water will spread out on an oil film or solid surface; but if $\xi_W < 0$, water will not spread, Grattoni et al. (1977); Grattoni and Dawe (2003); Zhou and Blunt (1995).

1.12 Effect of Wettability on Relative Permeability

Absolute permeability of a porous medium (K) is the ease with which a fluid will flow through it when saturated with that fluid; it also may be defined as the reciprocal of the resistance to fluid flow of a single fluid. Darcy's law describes the laminar flow of a fluid through a porous medium subjected to a pressure difference. The velocity of the fluid is a vector quantity with both magnitude and direction, and it is impeded by the tortuosity of the path, the wetting properties of the solid, and the surface exposed to the fluid. When more than one fluid is present (gas, water, oil) the permeability of each phase is its *effective permeability*, or a measure of the ease with which each fluid will flow through the porous system in the presence of other immiscible fluids (k_g, k_w, k_o). The *relative permeability* of a specific fluid is the ratio of its effective permeability to the absolute permeability of the porous medium ($k_{rg} = k_g/K$; $k_{rw} = k_w/K$; $k_{ro} = k_o/K$). A large number of factors have simultaneous effects on the absolute and relative permeabilities: (1) the composition of the porous medium (shapes and sizes of sand grains, inclusions of clay and minerals, degree of cementation, tortuosity of the paths for flow through the matrix, etc.), (2) fractures in the matrix, (3) laminations of sediments, (4) distribution of the fluids within the pores caused by preferential wetting throughout the matrix, (5) the overall saturations of the fluids, (6) saturation history (the order in which the fluids entered the porous medium), and (7) viscosities of the fluids. For a given set of fluids and rock properties, however, wettability largely governs the relative permeabilities. Detailed discussions of permeability, its general magnitudes, and methods for its measurement are presented by Tiab and Donaldson (2004).

Wettability affects the relative permeabilities by controlling the distributions of the immiscible fluids. The oil relative permeability increases while the water relative permeability decreases as wettability is varied from water-wet to oil-wet, Fig. 1–17. In addition, the crossover point of the curves (the point where the relative permeabilities are equal) moves from left to right (lower to higher water saturations).

When three phases are present in a core, the two nonwetting phases compete for the larger pores and thus they mutually interfere with the movement of each other as they are displaced. Thus, in a water-wet core, gas will reduce the relative permeability to oil; and in an oil-wet core the water relative permeability will be lower than it would be in the absence of gas. In both cases the relative permeability to the nonwetting phase is unaffected. It was generally observed that the

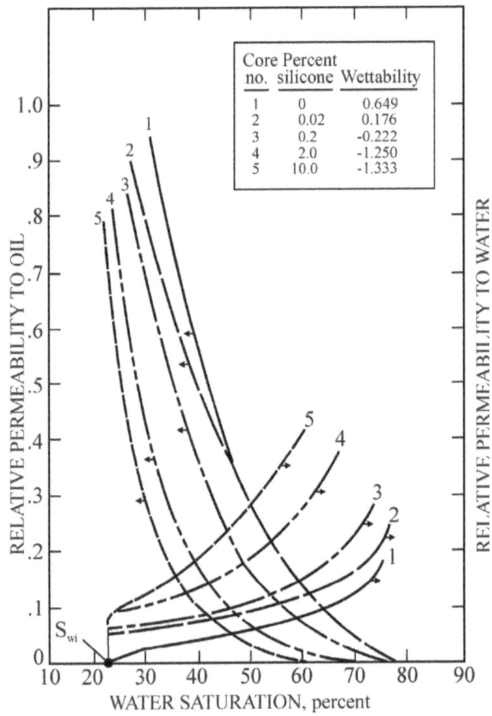

Figure 1–17 *Oil and water relative permeabilities of sandstone cores for water-wet to oil-wet conditions.*

wetting phase relative permeability is a function of the wetting phase saturation, while the permeabilities of the two nonwetting phases are functions of the fluid saturation distributions of the two phases (Leverett and Lewis, 1941; Donaldson and Dean, 1964; Schneider and Owens, 1970; Emmett et al.,1971).

Visual studies using microscopes have shown conclusively that in a water-wet system, water will occupy the crevices and small pores of the matrix and will spread on the grain surfaces as a film. Oil occupies the large pores resting on the water film. During displacement by water, a fairly even distribution of oil develops in front of the water, and each phase moves through a separate set of pores. Small globules of oil left behind the front remain as residual oil. Thus when water breaks through at the producing end, the production of oil declines rapidly toward almost zero production, Fig. 1–18(1). In an oil-wet

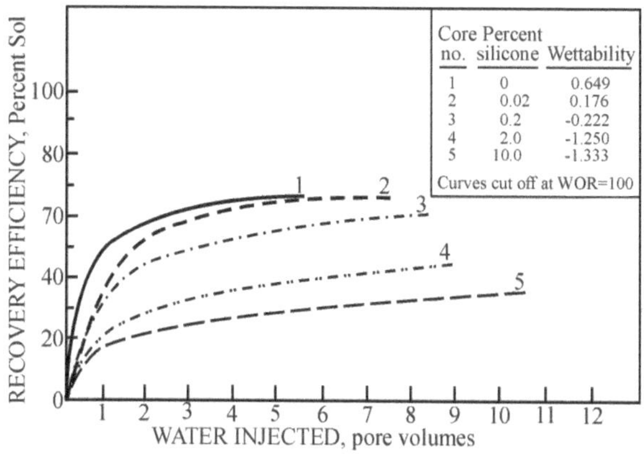

Figure 1–18 *Recovery efficiency for water- to oil-wet conditions using sandstone cores.*

system the roles of water and oil are reversed, and injected water forms continuous fingers throughout the larger pores displacing some oil, Fig. 1–18(5). More oil is displaced in front of the water as the fingers of water grow through the matrix until water flows freely through the established channels. At this point the oil production declines rapidly while water production increases. Production of oil continues at a slowly declining rate (Donaldson and Thomas, 1971; Donaldson and Crocker, 1977; Hornof and Morrow, 1986; Yadav et al., 1987).

1.12.1 Effect of Wettability on Waterfloods

To determine the effects of wettability on waterfloods, Donaldson and Crocker (1977) used cores, 25 cm long and 4 cm in diameter, from a sandstone with a fairly uniform permeability of 300 ± 50 md (tortuosity of the interconnected pores has an influence on relative permeability; therefore, the cores were selected on the basis of the least variation of absolute permeability). Long cores (25 cm) were used in order to diminish capillary end effects. The cores were first saturated with water; then the water saturation was reduced to S_{wi} by displacement with the oil selected for the tests. The oil recovery test (from S_{wi} to S_{or}) begins when the inlet face of the core is contacted by water. Water is the wetting phase; therefore, water imbibition begins as soon as the inlet face is contacted by water. If countercurrent flow of oil (the nonwetting phase) is restricted at the inlet because of the rapidly increasing injection water pressure, oil is forced out of the exit by co-current flow of oil because of imbibition and the pressure drop imposed by the injected water. This produces a short period of unstable flow which is responsible for the greater than 100% relative permeability to oil that has been observed (Donaldson et al., 1966). The water saturation at the inlet increases rapidly from S_{wi} to the saturation where the imbibition capillary pressure reduces to zero (S_z) at the outlet. Water breakthrough occurs when $S_w = S_z$ and the water/oil producing ratio increases as the water saturation approaches S_{or} $(1 - S_{wi})$. If this is the case, production of oil will diminish to almost zero as oil becomes a discontinuous phase in the core where oil is trapped in large pores and as immobile globules occupying several connected pores but completely surrounded by water. Wettability affects relative permeabilities and waterflood behavior because it is a major factor in controlling the location, flow, and special distribution of fluids in the rock. In water-wet systems with a moderate viscosity ratio water moves through the cores in a fairly uniform front and the injected water tends to imbibe into the small pores moving oil into the larger pores. Where both water and oil are moving, a portion of the oil exists as continuous channels (with some dead-end branches) while the remainder is trapped as discontinuous globules. After the passage of the water front, displacing oil in front of it, the remaining oil is immobile. Therefore, for water-wet cases there is little or no oil production after water breakthrough. The oil remains as spherical globules in the centers of large pores and large patches of oil extending over many pores surrounded by water. Oil-wet rocks exhibit different water and oil flow characteristics because oil in is contact with the rock surfaces and is located in the small pores and as a film on the surface of the rock walls, while water is in the large pores. In this case, the waterflood is less efficient for the water-wet system and water injected to displace the oil forms continuous fingers through the centers of the larger pores. Oil production before water breakthrough is small and in most cases most of the oil is produced after water breakthrough and total recovery is strongly dependent on the amount of water used for the waterflood. For waterfloods of systems near neutral wettability, water breakthrough occurs at an earlier time than for the water-wet cases and the water/oil producing ratio rises gradually. The economic residual oil saturation, however, is lower (production is greater) than for either water-wet or oil-wet systems.

From Darcy's law:

$$q_w = -\frac{k_w A \delta P}{\mu_w \delta x} = -\frac{k_w A}{\mu_w}\left(\frac{\delta P}{\delta S_w} \bullet \frac{\delta S_w}{\delta x}\right) \tag{1.24}$$

$$\frac{\delta S_w}{\delta x} = -\frac{q_w \mu_w}{k_w A\left(\frac{\delta P}{\delta S_w}\right)} \tag{1.25}$$

$$\frac{\delta P}{\delta S_w} \rightarrow 0; \quad S_w \rightarrow S_z \text{ and } \lim_{x \rightarrow L}\left(\frac{\delta S_w}{\delta x}\right) \rightarrow 0 \tag{1.26}$$

As long as the wetting phase saturation at the effluent face is less than S_z, only the nonwetting phase will be produced. The wetting phase begins to flow when the saturation at the effluent face reaches S_z; thereafter, the wetting phase saturation increases until the practical, or ultimate (for waterfloods of many pore volumes), is attained. When waterflood tests are conducted at high flow rates in order to minimize the effluent end effects, the inlet end effects become more severe (Collins, 1961; Dullien, 1979; Marle, 1981).

1.12.2 Procedures

In order to change the wettability in steps from water-wet to oil-wet, the cores were treated with toluene solutions of a silicone compound (General Electric Dri-Film 144, organochlorosilane) ranging in concentration from 0.001 to 10%. The USBM wettability index was measured for each core by removing a piece of the core large enough to use in the centrifuge. The cores were then saturated with water and a crude oil to S_{wi} and flooded with water to examine the production of the percentage of original oil in place as a function of the amount of water injection and time (Fig. 1–18) and the relative permeabilities (calculated using the Johnson et al., 1959, method, *Program Listing 6.20 RELPERM.FOR*), Fig. 1–17.

Morgan and Gordon (1970) reported large differences in the behavior of relative permeabilities between rocks with small and large pores. The rocks with smaller pores had higher values of S_{wi} and the crossover point of the relative permeability curves (where $k_{rw} = k_{ro}$) occurred at higher water saturations. The crossover point and irreducible water saturation also moved from lower to higher water saturations when the wettability was constant, but the tortuosity and pore-size distributions of the series of cores changed from smaller to larger pore size distributions.

1.12.3 Production Curves

The production curves clearly show how oil production is affected by differences of wettability. The water-wet system, curve 1 ($I_u = 0.649$), shows the rapid decline of oil production that occurs just after water breakthrough (WBT) when about one pore volume of water had been injected. Such is the result of the almost piston-like displacement that occurs in strongly water-wet systems.

In intermediate-wet (fractional wettabilty) cores the relative permeability to oil decreases while the water relative permeability increases as the number of fractionally oil-wet surfaces increases (change toward a more oil-wet system). In a mixed wettability system the small pores are water-wet and the large pores are oil-wet with a continuous network of oil. Thus the relative permeability to oil is shifted up and to the right, allowing oil displacement to low residual oil saturations (lower than the S_{or} of water-wet systems).

A waterflood of a water-wet core is more efficient than the waterflood of an oil-wet core. Fig. 1–17 shows that 50% of the original oil in place was produced at WBT, whereas only 18% was produced from an oil-wet core at WBT. This occurs because k_{ro} is at its highest values for a water-wet core while k_{rw} exhibits very low values for any given saturation.

Fig. 1–17 (curve 2) ($I_u = 0.176$) represents production from a core that may be considered close to intermediate-wet ($I_u = 0$) as indicated by the behavior of the production and relative permeability curves. The production curve indicates that production of oil would surpass that of the water-wet core if the runs had been continued for a longer period of time. Other work by Amott (1959) and Lorenz et al. (1974) has shown that production from intermediate wet cores leaves behind a much lower residual oil saturation, Fig. 1–19. Water breakthrough, however, occurs earlier in the intermediate-wet system and production of oil is slower until about 5 pore volumes of water were

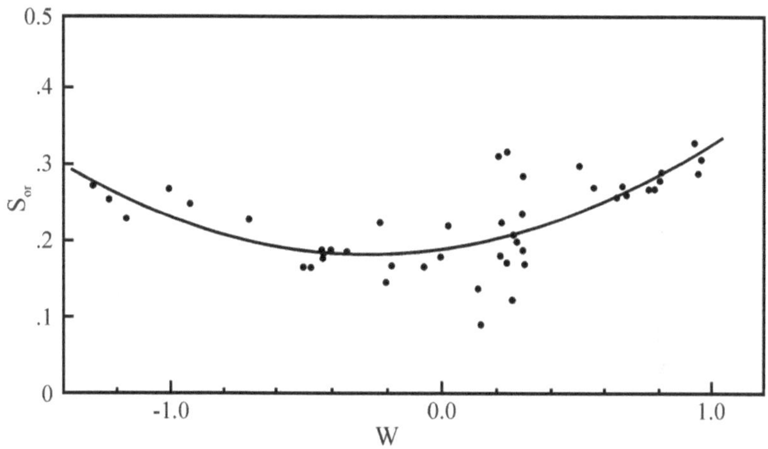

Figure 1–19 *Ultimate recovery as a function of wettability. Maximum oil recovery apparently occurs in neutral or slightly oil-wet rocks.*

injected. At this point, production is the same or greater than production from the water-wet core. The relative permeability curve indicates that production from the I-W core becomes the same after a water saturation of about 46% is attained and ends at the same point as the W-W core flood.

Fig. 1–18 (curves 3–5) shows that the amount of oil production decreases as the wettability changes from water-wet to more oil-wet. Water breakthrough occurs earlier for the oil-wet cores. This is in accord with the visual observations using a microscope; the early water breakthrough of the oil-wet systems was caused by the fingering of water ahead of the major producing front. The relative permeability to water is low for the water-wet cores and increases as the wettability becomes more oil-wet while the corresponding oil relative permeability decreases as the system changes from W-W to O-W. The changes of relative permeability to water and oil corresponding to the change of wettability from W-W to O-W are accompanied by increases in the producing water/oil ratio. The same results were obtained by Kyte et al. (1961) and Emery et al. (1970). The relative permeability curves for water-wet and oil-wet systems are plotted as functions of the specific single phase saturation (generally water), but the relative permeability of each phase is a complex function of the saturation distribution of the two phases in the rock which is governed by wettability, the saturation history of the water/oil/rock system, and pore geometry including the pore-size distribution of the rock.

Most core studies reported in the literature are conducted at ambient conditions of pressure and temperature; however, the fluid flow properties of the cores and fluids are influenced by other important parameters. An increase of the difference between the overburden pressure and the pore fluid pressure (applied to cores in the laboratory to match reservoir conditions) decreases the overall pore-size distribution and porosity, reducing the size of pore throats and the larger pores; these changes increase the irreducible water and residual oil saturations and thus decrease the range of the mobile oil saturation (between S_{wi} and S_{or}). An increase in temperature changes the wettability toward a more water-wet condition. Thus, if core floods are run to examine the characteristics of a reservoir, the tests should be conducted at simulated reservoir conditions of pressure and temperature using reservoir fluids (Mungan, 1972; Longeron et al., 1989).

1.12.4 Effects of Viscosity

The water/oil viscosity ratio (μ_w/μ_o) has an influence on relative peremeability curves which has the same appearance of the effects of wettability. A high μ_w/μ_o is accompanied by a high relative permeability curve and WBT at a relatively high water saturation, similar to the behavior of a water-wet system. If the μ_w/μ_o is low, WBT will occur early and production will continue with a high water/oil producing ratio and water fingering will occur ahead of the water/oil displacement front which is the behavior of an oil-wet system. This behavior is evident from the fractional flow equation (see *Program Listing 6.6 FRAFLOW.FOR*):

$$f_{w(S_w)} = \frac{1}{1+\left(k_{ro}/k_{rw}\right)\left(\mu_w/\mu_o\right)} \tag{1.27}$$

$$\left(k_w/k_o\right)_{(S_w)} = \left(\mu_o/\mu_w\right)\left(f_w/(1-f_w)\right) \tag{1.28}$$

Production before water breakthrough is controlled by both viscosity and the relative permeability ratio. An increase or decrease of μ_w/μ_o results in an equal response of the relative permeability ratio. Thus the fluid viscosity ratio mimics the behavior of wettability. In addition, oil viscosity is very sensitive to temperature; therefore, waterflood tests conducted to evaluate the behavior of a specific oilfield must be conducted at reservoir pressure and temperature with reservoir rocks and fluids. Tests conducted at ambient conditions must match the subsurface pore geometry and fluid properties for correlation to oilfield behavior (Buckley and Leverett, 1942; Loomis and Crowell, 1962; Jennings, 1966; and Anderson, 1987c).

1.13 Effects of Wettability on Electrical Properties

Wettability and the saturation history of a core establish the fluid distributions within the pore spaces and therefore control the resistivity of the electrical pathways through the porous system. Resistivity is proportional to the impressed voltage and cross-sectional area and inversely proportional to the current flowing and the length of the electrical conductive path:

$$R = \left(\frac{E}{I}\right)\left(\frac{\phi A}{L}\right) \tag{1.29}$$

Rocks are generally classified as insulators, but inclusions of minerals such as some clays and metals can lower their resistivity. Hydrocarbons also are treated as insulators. The resistivity of a solution containing dissolved salts (brine) placed in a container with the same volumetric dimensions of a rock sample has a lower resistivity than the rock saturated with brine. The insulating rock grains diminish the cross-sectional area of the conducting brine, and the tortuosity of the porous pathways through the rock increases the distance that the current must traverse.

The formation resistivity factor (F_r) is defined as the ratio of the resistivity of the rock saturated with brine (R_o) to the resistivity of the brine (R_w). Archie (1942) discovered that the formation factor is inversely proportional to porosity and Wyllie and Gardner (1958) related the formation factor to the tortuosity:

$$F_r = \frac{R_o}{R_w} = \frac{a}{\phi^m} = \left(\frac{L_a}{L}\right)\left(\frac{1}{\phi}\right) = \frac{\sqrt{\tau}}{\phi} \tag{1.30}$$

Table 1–4 is a list of the empirical, statistically derived, values of parameters in Eq. 1.30.

When a hydrocarbon is present in the core, the nonconductive hydrocarbon reduces the cross-sectional area and blocks flow-paths (fluid and electrical current) in the rock; this effect increases the resistivity of the rock. Wettability controls the distribution of fluids in the pores and therefore it has a significant influence on the total resistance to an electric current. According to Archie's equation, relating fluid saturation to resistivity:

$$S_w^n = \frac{R_o}{R_t} = \frac{F_r R_w}{R_t} = \frac{aR_o}{\phi^m R_t} \tag{1.31}$$

In general, brine will have a continuous path through the porous system and resistivity will decrease as the brine saturation is increased while oil is displaced in a waterflood. In an oil-wet system resistivity will decrease at a greater rate (with respect to brine saturation) than it does in a water-wet system because brine will occupy areas that were completely blocked to the electrical current as oil is removed. The saturation exponent (n) in Archie's equation depends on the distribution of the conductive fluid throughout the porous medium and therefore it is not a single-valued function of saturation. The overall statistical average value of the saturation exponent from laboratory and field experience has established its value at 2.0 for quartz-type, water-wet sands, Pirson (1963). Experimental work, however, has shown that the value of n is influenced by wettability. Different values of resistivity are obtained for the same brine saturation if wettability is changed. Keller (1953) reported values of n from 1.5 to 11.7 as wettability of the same rock is changed from strongly water-wet to strongly oil-wet.

Moore (1958) examined 150 cores from two wells of the Bradford sandstone (Pennsylvania) and found that n varied from 1.6 to 2.7, but when the cores were cleaned with toluene he attained a value of 2.0. Sweeny and Jennings (1960) found that $n = 1.6$ for definitively water-

Table 1–4 Parameters used for calculation of the formation resistivity factor (F) for different types of matrices.

Criteria for Equation	Equation
Carbonate (limestones, dolomites)	$F = 1.0/\varphi^{2}$
Calcareous sands	$F = 1.45/\varphi^{1.70}$
Consolidated sandstones	$F = 0.81/\varphi^{2}$
Humble (consolidated sandstones)	$F = 0.62/\varphi^{2.15}$
Unconsolidated sands	$F = 1.0/\varphi^{1.5}$
Unconsolidated sands (US Gulf Coast)	$F = 1.97/\varphi^{1.29}$
Shaly sands	$F = 1.65/\varphi^{1.33}$

wet cores of carbonate rocks that had been treated with acid. Cores that were cleaned with toluene exhibited intermediate wettability and the saturation exponent values averaged 1.9. Cores that were oil-wet had values of n that ranged from 1.5–2.5 at brine saturation greater than 35% where continuous pathways for brine were established through the core; however, at water saturations less than 35%, the values of n changed rapidly ranging from 8–12 because of oil blocking the pathways of brine. Morgan and Pirson (1964) reported the largest variation of n: from 2.5–25.2 for packs of glass beads whose wettability was systematically changed from water-wet to strongly oil-wet by treatment with organochlorosilicones.

Donaldson and Siddiqui (1989) examined the effect of wettability on the saturation exponent and temperature. For both sandstones, the wettability apparently increased linearly as wettability was changed from water-wet to strongly oil-wet (as indicated by the decrease of the USBM wettability index) at temperatures of 25°C and 78°C (Fig. 1–20). For both cases, the systems became more water-wet at the higher temperatures. Differences of the slopes of the lines may be indicative of differences in pore geometry between the two types of rocks. Note that an increase in temperature also produces an increase in the electrical resistance of conductors.

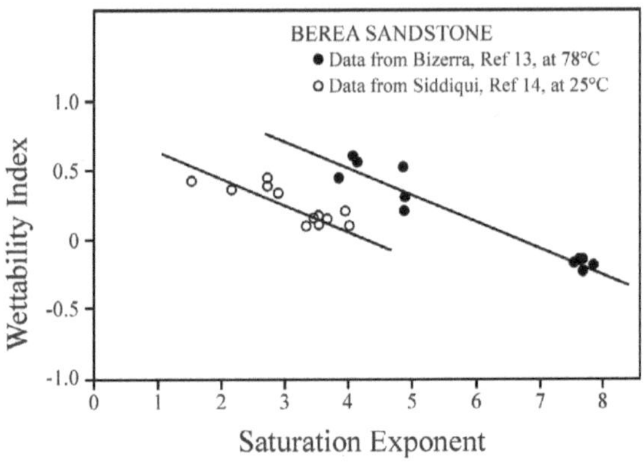

(a)

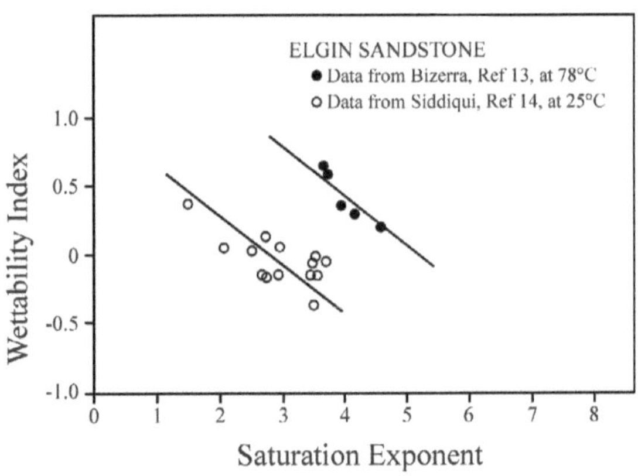

(b)

Figure 1–20 *(a) Relationship between wettability and the Archie saturation exponent at two temperatures for Berea cores (K ~ 200 md). (b) Relationship between wettability and the Archie saturation exponent at two temperatures for Elgin sandstone cores (K > 1,000, large grain sizes, relatively large pores) (Tiab and Donaldson, 2004).*

Lorenz et al. (1974) used different outcrop sandstone cores and crude oils to show the effect of temperature on the wettability of the same systems. In each case, the wettability exhibited an unambiguous change toward a more water-wet system when the temperature was increased, as indicated by the change in the USBM index, Fig. 1–20.

Without considering the complications introduced by changes of wettability and temperature, Archie's equation implies that the saturation exponent (n) is a simple valued constant for any given brine/hydrocarbon system. This means that only one value of resistivity will be obtained for any specific saturation; however, wettability controls the distribution and tortuosity of the fluid pathways within the porous medium. A change of wettability in the system at constant saturation will change the core's cross-sectional area and length of the brine pathways through the rock. In addition, as the system becomes more oil-wet continuous filaments of brine are broken and isolated by encroachment of oil. A water-wet rock will have more conductive paths, and thus a lower resistance, than an oil-wet rock at the same saturation. This phenomenon, illustrated in Fig. 1–20, where a change in wettability (at constant temperature) from water-wet to oil-wet produces a significant increase in the value of n. The change of wettability, produced by the change of temperature, results in a wettability change toward a more water-wet condition with an accompanying increase of electrical conductivity because the continuous brine flow paths are increased and there is less blocking of the pathways by oil.

Therefore, for laboratory measurements of electrical properties to be correlated to reservoir data, formation cores and fluids must be restored to reservoir wettability and temperature before the electrical measurements are made.

1.14 Dispersion and Wettability

1.14.1 Introduction

The dispersion (or mixing) of a solvent injected within an oil reservoir is of considerable importance to the various methods used for secondary and enhanced oil recovery. The dispersion of solvents has been analyzed for various types of environmental and chemical engineering problems. Environmental effects are principally concerned with the dispersion of pollutants—such as smoke from burning fuels, exhausts of vehicles within cities, water-soluble waste products entering surface streams and shallow subsurface aquifers, and many other localized conditions—and general dilution and distribution within the atmosphere. It enters almost all aspects of chemical engineering from the distribution of heat in solids, liquids, and gas, the passage and mixing of fluids in reactors, ion exchange and filtration columns, chromatographic transport through adsorption columns, etc. Thus the topic is broad, but it has largely been neglected with respect to petroleum reservoir, especially in regard to the effects of wettability on dispersion within a two-phase system.

Early studies of dispersion in flowing streams and porous media were made by Danckwerts (1953) and Taylor (1953). Brigham (1961, 1974) began the application to programs related to petroleum and the solution of the equations of dispersion. Perkins and Johnston (1963) published a review of the literature on the subject and concluded that dispersion is the principal parameter causing the dilution, or dissipation, of miscible slug injection used for enhanced oil recovery (EOR). Donaldson et. al (1976 and 1980) determined the coefficients of dispersion for several sandstone cores and compared the calculation results using an analytic solution and two different finite difference solutions. Poulin (1985) conducted a comprehensive literature review and compared the calculation of dispersion coefficients by the analytic, finite difference and graphical methods for experimental data obtained on 25-cm long cores of the following sandstones: Bandera (Kansas); Berea (Ohio); Cottage Grove (Noxie); and Torpedo (Oklahoma). The petrophysical properties, 12 sets of experimental data, and the computer programs are included in Poulin's MS thesis. The finite difference solution, using Crank-Nicolson discretisation, was developed as computer program *DIS2.FOR* (see *Program Listing 6.1*) for this work.

1.14.2 Applications

For the interpretation of well logs after or while drilling, the dispersion of mud fluid into the porous zones encountered in the subsurface is of utmost importance. If logging is delayed after drilling has been completed, the pressure-driven dispersion of mud fluid reduces to the slower process of molecular diffusion, and eventually the concentration of mud fluid around the wellbore is reduced to almost zero. Drillers adjust the density of the drilling mud column so that the pressure at the bottom of the column is about 1 MPa or greater than the pressure of the reservoir fluid. Therefore, the liquid part of the mud invades the porous rocks while the mud solids deposit as a filter cake on the wall of the bore-hole. The correct interpretation of the well logs depends to a large extent on the resistivity of the mud fluid and its penetration into the zone around the wellbore.

Donaldson and Chernoglazov (1986) present a simulation of mud fluid invasion using a finite difference solution of the radial form of the diffusivity equation:

$$Dl\frac{\delta^2 C}{\delta r^2} - \frac{Q}{r}\frac{\delta C}{\delta r} = \frac{\delta C}{\delta t} \tag{1.32}$$

Using an experimentally evaluated coefficient of dispersion and a mud fluid filtration rate (Q), the mud fluid invaded zone can be simulated. The results show that when drilling stops, static filtration of mud fluid continues at a diminishing rate until it finally stops. Molecular diffusion then continues the slow dispersion (dissipation) of the fluid into the formation brine while the oil saturation around the wellbore gradually returns. When Q becomes zero, Eq. 1.32 becomes Fick's second law of diffusion because the coefficient of dispersion includes the coefficient of molecular diffusion: $Dl = D_e + D_m$.

The flow pattern of injected fluids toward production wells is strongly influenced by dispersion. Hence, fluid-flow simulations with inclusion of dispersion is important to the application of miscible-phase displacement technology to recovery of petroleum, as well as the injection of mobility control agents (polymers) and wettability alteration reagents. Anomalies in the simulated predictions of the time and concentration of a water soluble tracer can be used for modifications of the reservoir parameter until the simulated results match the observed data. This leads to analyses of variations caused by fractures, faults, pore size distributions, and thus better reservoir characterization.

Enhanced oil recovery processes involve the injection of wettability-altering chemicals, heat, caustics, surfactants, polymers, gases, microbes or microbial nutrients, etc. The patterns of the flow of these extraneous constituents in reservoirs are influenced by the geological and petrophysical properties of the specific reservoir, in addition to a sometimes overriding influence of the dispersion (or mixing) between the multiple injected fluids and the miscible reservoir fluids.

The dispersion and mixing of industrial wastes injected into deep wells is important to the planning and monitoring of well performance. The feasibility of injection of low-level radioactive wastes into deep wells (greater than 1 km in depth) also has been considered; in this case the pattern of waste dilution and long-term dispersion into the subsurface brine must be a part of the feasibility study.

1.14.3 Theory

In the absence of adsorption, an injected miscible-phase solution containing a solute which is different from the reservoir fluid composition gradually spreads into the reservoir brine by the simultaneous interactions of convection dispersion and molecular diffusion. Several properties influence the rate of dispersion: (1) fluid flow velocity, greater viscosity of the injected fluid, and oil-wet reservoir conditions result in the formation of advancing fingers of the injected fluid resulting in inefficient displacement and rapid advance of the injected solutes; (2) injection of a fluid with greater density than the resident brine will result in stratification of the brine and injected fluid with considerable advance along the lower portion of the reservoir; (3) fractures in the reservoir rock will conduct the injected fluids rapidly into the reservoir without mixing with the formation brines; and (4) the pore size distribution and tortuosity of the matrix pores also can cause unstable displacement at the advancing front.

Molecular diffusion resulting from the motions of the molecules and concentration differences is present at all stages but is more evident when injection of fluid is stopped and the pressure differences within the reservoir equalize. Molecular diffusion will cause the eventual complete mixing of the injected solutes with the reservoir brine, and rate of dissipation also can be simulated, especially when the reservoir rock is assumed to be homogeneous.

When linear, laminar flow displacements are made in laboratory cores, eddy diffusion (caused by turbulence) and transverse diffusion, taking place at right angles to the direction of flow, are both neglected because their contributions to the overall convective dispersion are very small. The experimental determination of the coefficients of dispersion for several very different sandstones, reported by Donaldson et al. (1976, 1980) and Poulin (1985), were conducted using an aqueous solution having approximately the same density and viscosity as the brine originally saturating the cores. This was done so that the measured coefficient of dispersion would only be a function of the petrophysical properties of the rock and the flow rate. The flow rate used was approximately 0.0015 cm/s (~1 m/D) to represent a nominal flow rate of a reservoir. At this rate, however, molecular diffusion becomes an important component to the overall dispersion (Ebach and White, 1958; Blackwell et al., 1959; and Burghardt and Zaleski, 1968).

The residence time distribution of the concentration of a tagged solvent may be measured if the solvent is introduced into a flowing system as a step increase in concentration (for example, the sudden increase of the concentration of a salt). If the coefficient of linear dispersion (*Dl*) is equal to zero, and the velocity is high enough to make molecular diffusion insignificant, no convective dispersion will take place. A condition referred to as "plug flow" occurs and the effluent concentration would, theoretically, register a straight vertical line as the step-increased concentration arrived. Increasing values of *Dl* produce tracer concentration response curves, with respect to time (or pore volumes injected), that are S-shaped. At higher values of *Dl* the curves are extended in length because of the increase of mixing between two miscible fluids.

The diffusivity equation applied to one-dimensional miscible displacement (neglecting adsorption) is:

$$\frac{\lambda}{L}\frac{\delta^2 C}{\delta\xi^2} - \frac{\delta C}{\delta\xi} = \frac{\delta C}{\delta\tau} \tag{1.33}$$

where

$$\lambda = \frac{Dl \cdot A \cdot \phi}{Q} \qquad \xi = \frac{\phi \cdot A \cdot X}{V_p} = \frac{X}{L}$$

$$\tau = \frac{Q \cdot t}{V_p} \qquad C = \frac{C_i}{C_o}$$

The initial and boundary conditions for finite difference solutions to Eq. 1.33 are:

$$C_{(\xi,0)} = 0 \qquad 0 \le \xi \le 1 \qquad X = L; \quad \xi = 0 \tag{1.33a}$$

$$C_{(0,\tau)} = C_0 + \frac{\lambda}{L}\frac{\delta C}{\delta\xi} \qquad 0 \le \tau \le \tau_1 \tag{1.33b}$$

$$C_{(o,\tau)} = \frac{\lambda}{L}\frac{\delta C}{\delta \xi} \qquad \tau > \tau_1 \tag{1.33c}$$

$$C_{(0,\tau)} = 0 \tag{1.33d}$$

$$\left(\frac{\delta C}{\delta L}\right)_{\xi=1} = 0 \qquad \tau > 0 \tag{1.33e}$$

The inlet boundary condition for constant injection (C_i/C_o = 1) is Eq. 1.33b. However, when the input is decreased to zero after a short period of injection (slug injection), the entrance boundary condition becomes Eq. 1.33d and Eq. 1.33 must be solved with the new boundary conditions after the initial slug input.

The equipment and experimental procedures are presented, in detail, by Donaldson et al. (1980) and experiments and data are presented by Donaldson et al. (1976) and Poulin (1985).

1.15 Influence of Wettability

A quantitative relationship between wettability and dispersion has not been developed. Qualitative observations (Donaldson and Thomas, 1971; Donaldson et al., 1977; Donaldson and Crocker, 1977) have shown that dispersion increases as the wettability of the system varies from water-wet to oil-wet.

1.16 Example Problems

Example 1–1

Calculate the interfacial tension (IFT) of water at 20° (density = 0.9982) if the water rises 2 cm in a capillary with a radius of 0.0743 cm.

$$\sigma = \tfrac{1}{2}(0.0743\,cm)\left(0.9982\frac{g^m}{cm^3}\right)\left(981\frac{cm}{s^2}\right)(2\,cm)$$

$$\sigma = 72.76\left(\frac{g^m \cdot cm}{s^2}\right)\left(\frac{1}{cm}\right) = 72.76\ dyne\,/\,cm = 72.76\,mN/m$$

Example 1–2

Calculate the interfacial tension of water at 20°C using a ring with an average radius of 2.0 cm if the force required to pull the ring through the interface is 1.865 g^f.

$$\sigma = 72.75\,dyne/cm = 72.75\,mN/m$$

Example 1–3

Calculate the IFT of water at 20° for a Wilhemy plate 3 cm long and 0.4 cm thick if the force acting on the plate is 0.2521 g^f and the contact angle is 60°.

$$\sigma = F\big/ l \cdot Cos60^o = (0.2521)\left(981\frac{g^m \cdot cm}{s^2}\right)\Big/ 2(3.0+0.4)cm(0.5)$$

$$= 72.74\,mN\,/\,m$$

CHAPTER 2

Surface Forces

2.1 Introduction

Surface forces are the basis for the various states of wettability and are the intermolecular forces that control the attraction, repulsion and adhesion of condensed states of matter. The subject is very broad when it is viewed as the global interactions of all molecules and is beyond the scope of this single chapter. Therefore, the topic that encompasses the interactions of molecules is confined to the preferential adhesion of water and hydrocarbons on surfaces of geological materials: specifically, the preferential wetting of water and oil on rock surfaces.

When the pressure, molar volume, and temperature of a gas or mixture of gases exceed a critical value, the gaseous molecules have been compressed to distances where attractive forces dominate, causing the gas to condense to a liquid. The gas is easily compressed; however, the compressibility of the liquid is relatively minute because further compression is opposed by strong repulsive forces as the molecules are brought closer together. It is thus apparent that the molecular interaction is a function of the distance of separation of the molecules.

When two atoms (with symmetric electron clouds) are separated by a large distance, they do not interact. The total energy of the pair is the sum of the kinetic energies of each atom. If the atoms are brought close enough to experience an attractive force between them, the interacting energy between them adds to the total energy of the atomic pair system, thus:

$$E_T(r) = E_1 + E_2 + U(r) \quad or \quad U(r) = E_T(r) - E_1 - E_2 \tag{2.1}$$

The energy of interaction, $U(r)$, is labeled the *intermolecular pair potential energy*. The pair potential energy is the difference between: (1) the total energy of the system at infinite separation and (2) the total energy of the system when the atoms are brought close enough for the forces of interaction to develop. This energy difference can be expressed as the work required to bring the atoms from infinite separation to the short distance where interactive forces are present:

$$U(r) = \int_r^{\infty} F(r)dr \qquad [Newton - meters]$$
$$F(r) = -\frac{dU(r)}{dr} \tag{2.2}$$

The molecular interactive force, $F(r)$, is negative when it is attractive and positive when repulsive. Fig. 2–1 illustrates the typical change of work energy that occurs as two spherically symmetrical atoms or molecules approach each other from a separation distance where there is zero interactive force to a point where the negative attractive force begins to exert its influence (r_F). As the atoms continue to approach each other, the force increases in magnitude to a maximum value of attractive force between them (or minimum negative interactive potential [$-U_m$] at separation r_m). When the atoms are brought closer together, the attractive force decreases to zero once more at a separation distance that is labeled λ_o. Any attempt to decrease the distance of separation beyond λ_o is opposed by a rapidly increasing repulsive force and the work required increases to an infinitely large value. Thus the interactive energy is attractive at relatively long range, and it is a repulsive force at short distances, which is the reason for the very low compressibility of liquids and solids.

The Lennard-Jones equation describes the behavior of the potential energy:

$$U(r) = U_m \left[\frac{6}{n-6} \left(\frac{r_m}{r} \right)^n - \frac{n}{n-6} \left(\frac{r_m}{r} \right)^6 \right] \tag{2.3}$$

The parameters U_m and λ_o are determined experimentally and then Eq. 2.3 is fit to the experimental data. The value of n is generally taken as equal to 12 and $r_m = 2^{1/6}\lambda$ to yield:

$$U(r) = U_m \left[\left(\frac{r_m}{r} \right)^{12} - 2 \left(\frac{r_m}{r} \right)^6 \right]$$
$$= 4U_m \left[\left(\frac{\lambda}{r} \right)^{12} - \left(\frac{\lambda}{r} \right)^6 \right] \tag{2.4}$$

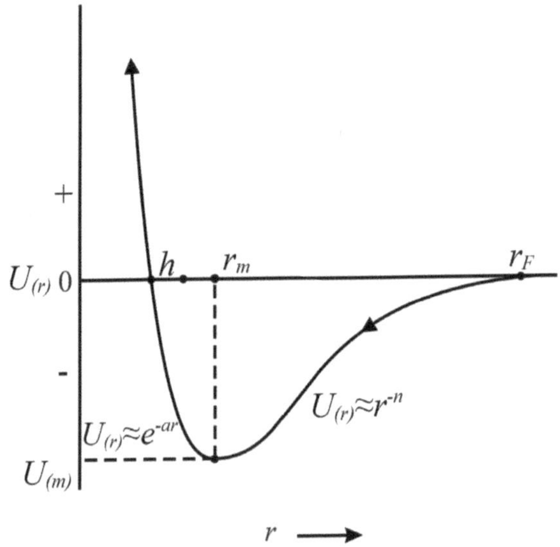

Figure 2–1 *Characteristic change of the intermolecular pair potential energy as two hard spherically symmetric atoms or molecules approach from infinite separation.*

The equations are used with experimental data to estimate the performance of potential energy (Castellan, 1971). The negative attractive term is based on the behavior of dispersion energy (London interactive forces) and is theoretically justified, but the positive repulsive term is entirely empirical. For unlike atoms, a geometric mean of the potential well, $-U_{(m)}$, and an average of λ_o are used for estimates of the best values for these parameters:

$$U_{m,ab} \approx \sqrt{U_{m,aa} U_{m,bb}} \quad \text{and} \quad \lambda_{o,ab} = \frac{1}{2}\lambda_{o,aa} = \frac{1}{2}\lambda_{o,bb} \tag{2.5}$$

A practical example of the behavior of intermolecular forces is to consider the intermolecular forces of the liquid state which is related to the heat of vaporization: the heat required to transform a liquid to a vapor at constant temperature and pressure. For water at 100°C, this quantity is 40.7 kJ/mol; the heat of vaporization of a liquid is a function of temperature since the heat of vaporization approaches zero as the liquid approaches complete transformation to its equilibrium vapor.

The analogy to the pair potential is the interaction of a single molecule in the liquid with its neighbors at a separation distance where the attractive forces are dominant. The potential energy between the molecules bound to their neighbors in the liquid state and the completely free molecules in the vapor has been defined as the heat of vaporization, where:

$$\Delta H_{vapor} = H_{vapor} - H_{liquid} \quad and$$
$$\left(\frac{\delta H}{\delta T}\right)_p = C_{p(vapor)} - C_{p(liquid)} = \Delta C_p \tag{2.6}$$

The minimum interactive potential is then approximately equal to the heat of vaporization ($U_m \approx \Delta H_v$).

2.2 Van der Waals Equation of State

To explain the deviation of real gas from ideal behavior predicted by the kinetic theory ($PV = nRT$), the Dutch chemist Johannes Diderik van der Waals introduced corrections to the volume and pressure. He reasoned that the volume occupied by a gas in a container also included a volume, b, which is the incompressible volume of the molecules themselves, leaving the volume available for free movement of the gas equal to $V - b$. The value of the constant b is a characteristic of each specific gas and is determined experimentally (Table 2–1).

The attractive forces between the molecules reduce the pressure exerted by the gas and lead to condensation of the gas molecules to a liquid when the temperature is reduced. As the temperature is reduced, the kinetic energy of the gas is reduced to a point where the long-range attractive forces can overcome the lowered kinetic energy and hold the molecules in a liquid state, Ex. 2.1. Thus the attractive forces act in consort with the confining pressure to hold the molecules together. The attractive forces of the molecules acting on each other is proportional to the square of the moles per unit volume, $a(n/V)^2$, which is added to the pressure. Thus van der Waals' equation is expressed as:

$$\left(P + \frac{an^2}{V^2}\right)(V - nb) = nRT \tag{2.7}$$

Table 2–1 Van der Waals constants.

Compound	a atm(L/mol)	b L/mol
H_2	0.16	0.0217
N_2	0.81	0.030
O_2	0.82	0.0248
CO	0.85	0.30
H_2O	1.32	0.015
CH_4	1.33	0.329
CO_2	2.01	0.0319
C_6H_6	9.44	0.0855
n-C_5H_{12}	9.52	0.01034

2.3 Electrostatic Interactive Energies

Coulomb's law describes the interactive force between the charges, q_1 and q_2, separated by a distance r in a vacuum:

$$F = -\frac{dG}{dr} = \frac{q_1 q_2}{4\pi\varepsilon\varepsilon_o r^2} = \frac{z_1 z_2 e^2}{4\pi\varepsilon\varepsilon_o r^2} \tag{2.8}$$

where ε_o is the permittivity of free space (vacuum). An electric field surrounds all charges, and is defined as the force exerted by a point charge:

$$F(q) = \xi q \tag{2.9}$$

where: $F(q)$ = force, (N); ξ = field strength (N/C) and q = electric charge (C).

A unit positive charge in a vacuum at some distance, *d*, from a second unit positive charge is repelled by a force equal to:

$$F_{(q^+,q^+)} = \frac{q}{4\pi\varepsilon_o r^2} \tag{2.10}$$

If the two positive charges are separated by a dielectric material, the dielectric material diminishes the force between the charges by an amount equal to the dielectric constant of the material, thus from Eq. 2.9:

$$F_{(q^+,q^+)} = \frac{q_1 q_2}{\varepsilon 4\pi\varepsilon_o r^2} \tag{2.11}$$

where ε is the dielectric constant, or relative permittivity.

The electric field developed by a single charge as a function of distance is:

$$E_{(d)} = \frac{q}{4\pi\varepsilon\varepsilon_o r^2} \quad [Vm^{-1}] \tag{2.12}$$

For ionic interactions, the electric charge and its magnitude is expressed as the valency, *z*, multiplied by the elementary electron charge, *-e*. For monovalent cations and anions the valency charges are $z_1 = +1$ and $z_2 = -1$; divalent ions have corresponding charges of +2 and –2. The term ε represents the relative permittivity, or dielectric constant, of the surrounding medium, and *r* is the distance of separation.

The free energy of interaction for two charges separated by a distance equal to the Bohr radius yields the Bohr energy which is the electrostatic free energy of an ion in a medium with a dielectric constant equal to ε:

$$G_{(r)} = \frac{q_1 q_2}{4\pi\varepsilon\varepsilon_o r_b} = \frac{z_1 z_2 e^2}{4\pi\varepsilon\varepsilon_o r_b} \tag{2.13}$$

If the ionic species have unlike charges, the interactive force, $F_{(r)}$, and free energy, $G_{(r)}$, are negative, yielding an attractive force; if like charges are considered, they will be positive and the force is repulsive, Ex. 2.2 (Castellan, 1971; Stone, 1996; Israelachvili, 1997).

2.4 Dielectric Media

If two plates are separated in a vacuum by a short distance, x, and connected to a one-volt battery, electrons will be attracted to the positive charge leaving the plate with a positive charge. An equal number of electrons will be repelled from the negative pole of the battery to accumulate as a negative charge on the facing plate. The movement of electrons will take place until the potential across the plates is equal to the electric potential across the positive and negative poles of the battery, Fig. 2–2a. This is a capacitor which is used for storing an electric charge.

The electric field in a vacuum, because of a positive charge per unit area, is:

$$E_o = \frac{q}{A\varepsilon_o} = \frac{\sigma_c}{\varepsilon_o} \quad \left[\left(\frac{C}{m^2}\right)\left(\frac{Jm}{C^2}\right) = \frac{V}{m}\right] \tag{2.14}$$

Therefore, the potential across the plates of the capacitor (in a vacuum) is:

$$V = qx/A\varepsilon_o \tag{2.15}$$

The charge that accumulates on the plates of a condenser is proportional to the applied voltage; and the proportionality constant, C'_o, is the capacitance in a vacuum:

$$q = C'_o V \quad \left[\frac{C}{V} V\right] \tag{2.16}$$

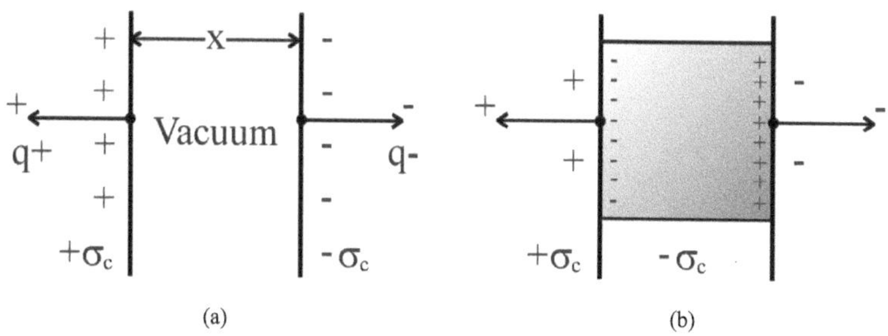

Figure 2–2 *(a) A capacitor (or condenser) is represented by two conductive plates separated by a distance, x, supporting opposite electric charges, q^+ and q^-, introduced from a direct current source; where σ_χ is charge density on the plates [C/m^2]. (b) A dielectric material between the plates of a capacitor. Charges on the plates produce dipoles in the dielectric resulting in oppositely charged surfaces on the dielectric adjacent to the capacitor plates.*

The capacitance can be related to the physical properties (geometry) of the condenser by substituting relationships expressed in Eq. 2.15 into Eq. 2.16:

$$C'_o = \frac{\varepsilon_o A}{x} \tag{2.17}$$

If a dielectric material (an insulator: air, oil, mica) is placed in the space between the plates of the condenser with the same applied voltage used in a vacuum, Fig. 2–2b, the charges on the plate produce dipole moments within the molecules of the dielectric material resulting in an opposite surface-charge density ($-\sigma_c$) on the dielectric medium. The dielectric medium's surface charge partially balances the charge on the condenser and the electric field is diminished from the value in a vacuum (Eq. 2.14) by a factor equal to the dielectric constant of the material:

$$E = \frac{q}{A\varepsilon\varepsilon_o} \tag{2.18}$$

The dielectric constant of any material can be measured by comparing the capacitance of a condenser in a vacuum to the capacitance with a dielectric medium:

$$C' = C'_o\left(\frac{\varepsilon}{\varepsilon_o}\right) = \frac{C'_o \bullet x}{\varepsilon_o A} \tag{2.19}$$

The dielectric constant, ε, is greater than 1.0 for any material; therefore, Eq. 2.19 shows that the capacitance is increased by the presence of a dielectric material between the plates.

Fluids with low dielectric constants (hydrocarbons, Table 2–2) exert a repulsive force on polar and ionic particles, expelling them from solution. Fluids with high dielectric constants (for example, water) exert an attractive force on polar and ionic constituents and thus tend to retain them in solution.

Table 2–2 Dielectric constants of various materials at 25°C, unless otherwise specified (Weast, 1970; various other sources).

Material	Dielectic Constant	Material	Dielectic Constant
Acetic acid	6.2	Vacuum	1.0
Acetone	20.7	Air	1.0006
Benzene	4.8	Methane	1.7
Cyclohexane	2.0	Aluminum oxide	8.4
Ethanol	24.3	Calcium carbonate	6.1
Methanol	32.6	Micas	5–7
Phenol	9.8	Polystyrene	2.5
Water	78.5	Quartz	4.5
Water (0°C)	88.0	Calcium carbonate	6.1

The force and energy of attractive interaction of a hypothetical mono-ionic molecule with a separation equal to 0.4 nm was calculated in Ex. 2.2. The energy required for separation of ions in a vacuum ($\varepsilon = 1.0$) versus water ($\varepsilon = 78.5$) was shown to be 348 kJ/mol and 4.47 kJ/mol, respectively.

2.5 Principle Interactive Forces

The forces that affect wettability are all electrostatic interactions between molecules that are arbitrarily separated into long-range and short-range forces acting on atoms, molecules, and particles. This separation is one of convenience for discussion of the intermolecular forces of wettability.

The important forces with respect to wettability are: (1) short-range forces where electron clouds interact, (2) London dispersion forces that develop from asymmetric charge distributions of molecules, (3) Keesom forces originating from Coulombic forces between static charges and permanent dipoles, and (4) Deybe interactions resulting from induced dipole moments.

2.5.1 Short-Range Forces

Short-range forces arise from interactions occurring at distances that are near atomic or molecular contacts (less than one nanometer). When the molecular separation becomes less than r_m (Fig. 2–1), the forces change exponentially from the minimum negative interactive potential ($-U$) becoming repulsive at λ_o; thus the particles (atoms or molecules) exhibit exponential pair-potential behavior with respect to the decreasing distance of separation:

$$U_r \approx e^{-ar} \tag{2.20}$$

A repulsive short-range force develops when the particles approach close enough that their electron clouds are distorted. If the electron wave functions have the same spin symmetry, the interactive force is one of electrostatic repulsion. The Pauli exclusion principle forbids electrons that have the same spin from occupying the same region of space. Conversely, if the electrons have opposite spin they can become coordinated when the electrons overlap both atoms, producing an internal increase of the electron density between the two nuclei resulting in attraction of the positive protons toward the negative region. This develops a short-range attractive force similar to covalent bonding (Castellan, 1971; Mailtand, 1981; Stone, 1996).

2.5.2 London Dispersion Forces

London forces are long-range interactions that refer to those where the energy of separation (U_r) exhibits a characteristic pair potential behavior which is some inverse power of the radius of separation:

$$U_r \approx r^{-n} \tag{2.21}$$

The important long-range forces with respect to wettability are: (1) electrostatic, originating from Coulombic forces between static charges and permanent dipoles (they can be attractive or repulsive); (2) dipole moments, induced in molecules by electric fields of adjacent molecules (these are always attractive); and (3) dispersion forces that develop from charge distributions of molecules. The long-range forces are constantly fluctuating in response to the movements of the electrons as the molecules approach each other where the motions of the electron clouds become coordinated, favoring lower energy configurations that become stronger as the molecular separations decrease.

Dispersion forces act between all interactions of particles (atoms and molecules). The London dispersion forces contribute up to a third of the total interactive forces that are known as the van der Waals force. The London forces are effective at relatively long-range (0.2 to 10 nm).

The attractive interaction of nonpolar compounds is unusual because they do not have permanent dipoles or electrostatic properties, thus the time average dipole moment of nonpolar molecules is zero; yet they exert an attractive force toward each other. The origin of this attractive force is quantum mechanical. The simplest model of a nonpolar molecule can serve to explain the source of the dispersion forces. Consider an electron distribution around the nucleus of a spherical molecule. The time average distribution of the electron's positions is a spherical electron cloud. The molecule, however, is composed of positive and negative charges with the negative charges oscillating about the positive with an angular frequency. Therefore, at any given instant the molecule experiences a separation of charges that corresponds to an instantaneous dipole moment.

When two molecules approach, the electron motions are coupled by the instantaneous dipoles and the electron clouds synchronize momentarily with the dipoles oriented in an attractive configuration that lowers the total energy of the engaged molecules.

The dispersion forces are an important contribution to all interactions of atoms and molecules, affecting the overall properties and important phenomena such as wetting, adsorption, etc. In some cases they are the only forces extant; large molecules such as hexane are liquids that are held together only by dispersion forces.

For an atom with a single electron, the minimum distance between the electron and proton is the first Bohr radius, r_B (0.053 nm, Ex. 2.3). At this radius, the Coulomb energy is equal to 2 $h\nu$, and the instantaneous dipole moment of the simple Bohr atom is $\mu = r_B e$. The instantaneous dipole moment is thus accompanied with an instantaneous electrical field which is capable of polarizing a neighboring atom and gives rise to an attractive dipole interaction energy (London, 1937; Mailtland, 1981; Israelachvili, 1997):

$$E^L_{(r)} = -\frac{\alpha^2 h\nu}{(4\pi\varepsilon\varepsilon_o)^2 r^6} \tag{2.22}$$

2.6 Dipole Forces

2.6.1 Keesom Dipole-Dipole Forces

The dipole moment of two equal but opposite charges (q^+, q^-) is defined as the product of charges and the distance separating them; thus it is a vector quantity where the magnitude of the dipole moment vector is q_r and the direction is from the negative charge to the positive. Fig. 2–3 illustrates the effect of two charges on a nearby point where the charge separations are $r = r_1$ and $r_2 = 2r_1$.

$$\mu = -\sum_{i=1}^{n} q_i r = -(+q)r_1 - (-q)r_2 = qr \tag{2.23a}$$

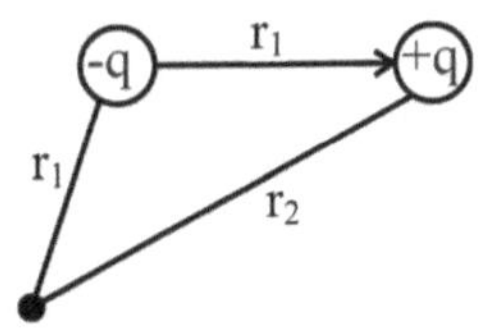

Figure 2–3 *Dipole moment about a point for two equal, but opposite charges where the charge separate r is equal to r_1 and $r_2 = 2r_1$, : $\mu = qr$.*

If the charges are equal to unit electronic charges and the separation distance is equal to 0.1 nm, the dipole moment is:

$$\mu = (1.602 * 10^{-19}\ C)(0.1 * 10^{-9}\ m) = 1.602 * 10^{-29}\ Cm \tag{2.23b}$$

The standard SI unit for the dipole moment is the Debye ($3.336*10^{-30}$) Cm; hence, the dipole unit (for a unit charge) is 4.802 D.

Permanent dipole moments occur in molecules where the electron distributions about the covalent bonds are asymmetrical. The skewed electron distribution establishes a slightly greater negative charge at one of the nuclei, resulting in a permanent dipole moment for the molecule.

Approximate covalent bond dipole moments for a few molecular groups that can be used to estimate the dipole moments of molecules by using vector summation when the molecular dipole moment is treated in isolation from the interaction with other molecules are listed in Table 2–3. Thus, taking water as a vapor where the angle of separation between the hydrogen and oxygen is known to be 104.5°, the bond moment of the molecule can be estimated (Fig. 2–4):

$$\mu_{H_2O} = 2\mu_{OH} Cos(52.25^\circ) = 2(1.51)(0.612) = 1.85\ D \tag{2.24}$$

Measurements of dipole moments provide a sense of the separation of positive and negative charges in molecules with permanent dipoles interacting with each other exerting a net attractive force because the

Table 2–3 Dipole moments of various molecules (Weast, 1970); and covalent bonds (the plus sign indicates the polarity) (Weast (1970); Israelachville (1992)).

Compound	Dipole Moment Debyes	Covalent Bond Groups	Dipole Moment Debyes
Benzene	0.0	C-C	0
Carbon tetrachloride	1.94	C^{+}-Cl	1.5
Ethane	0.0	C-H^{+}	0.4
Ethanol	1.32	C^{+}-N	0.2
Hydrochloric acid	1.08	C^{+}-O	0.7
Methane	0.0	C^{+}=O	2.3
Methanol	1.71	H^{+}-Br	0.8
Phenol	1.45	H^{+}-Cl	1.1
Nitric acid	2.16	N-H^{+}	1.3
Water	1.87	O-H^{+}	1.5

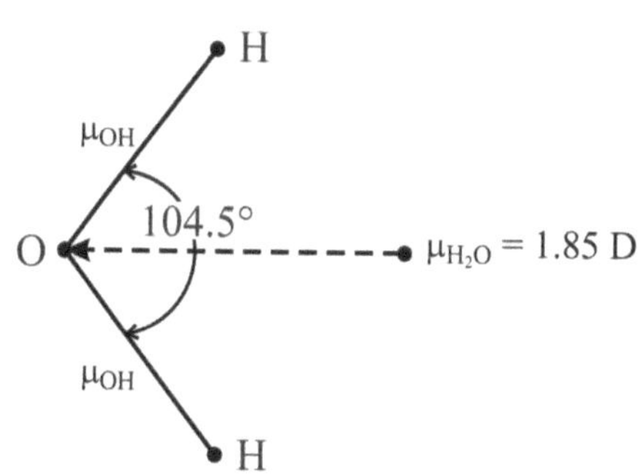

Figure 2–4 *The angle of separation between hydrogen and oxygen (in a vapor phase water molecule) is 104.5°.*

low, negative potential is favored over the high-energy repulsion configurations. As the temperature is increased, thermal motion decreases the dipole-dipole physical attraction.

Consider two particles with opposite charges separated by a distance (d) equal to two times the radius of ions that come together to form a dipolar molecule. The free energy of the dipole that is formed is equal to the sum of the individual energies due to the charges, q_1 and q_2, minus the energy required to move the ions together. Referring to Eq. 2.11:

$$E^K_{(r)} = \left(\frac{1}{4\pi\varepsilon\varepsilon_o}\right)\left(\frac{q_1}{2r} + \frac{q_2}{2r} - \frac{q^2}{d}\right) = \frac{\mu^2}{4\pi\varepsilon\varepsilon_o d^3} \tag{2.25}$$

If the dipoles have moments of one Debye each, the interaction energy for a separation of 0.35 nm is only 39 J/mole in a vacuum and considerably less in the condensed state. In liquids where hydrogen bonding is extant (water, ammonia, acids), however, the bond moments are significant, causing attractive associations between the molecules.

When two dipoles are aligned, the dipole-dipole interaction energy is proportional to r^{-3} and will vary from an attractive to a repulsive force depending on the orientation of the molecules. For an attractive, inline orientation:

$$E^K_{(r)} = \frac{2\mu^2}{4\pi\varepsilon\varepsilon_o r^3} \tag{2.26}$$

Analysis of the interactions in three dimensions, including the rotation of the dipoles, yields an angle-averaged dipole-dipole *Keesom interaction energy* (Barrow,1973; Mailtand, 1981; Israelachvili, 1997):

$$E^K_{(r)} = -\frac{2\mu^2}{3(4\pi\varepsilon\varepsilon_o)^2 k_B T r^6} \tag{2.27}$$

2.6.2 Debye-Induced Dipole Interactions

Molecules with permanent electrical dipoles may interact with the electron distributions of adjacent nonpolar or weakly polar molecules in such a way that polarization is induced by distortion of the adjacent molecule's electron distribution. This interaction also is one that results in an attractive, low pair potential electrical field between the permanent dipole and the resulting induced dipole.

The electron distribution of all molecules can be distorted to some degree by the presence of an electric field. The ease by which this distortion of the normal electron distribution of a molecule can take place is defined as the polarizability, α, where:

$$\mu_{induced} = \alpha E \qquad \left[Cm = \left(\frac{C^2 m^2}{J} \right) \left(\frac{J}{Cm} \right) \right] \tag{2.28}$$

The polarizability is the dipole moment induced in a molecule by a unit strength electric field (1 volt/m). Symmetrical molecules such as carbon tetrachloride and carbon dioxide have zero permanent dipole moments, but they have induced polarizabilities of 2.6 and 10.5 C^2m^2/J. Table 2–4 lists the polarizabilities of several compounds that are of interest to wettability.

When a polar molecule induces polarization of a nonpolar molecule, the electric field results from the charge of the permanent dipole. The interaction energy for a nonpolar molecule oriented at an angle (θ) with a permanent dipole is:

$$E_{(r,\theta)} = -\frac{1}{2}\alpha E^2 = -\frac{1}{2}\frac{\alpha\mu^2(3Cos^2\theta + 1)}{(4\pi\varepsilon\varepsilon_0)^2 r^6} \tag{2.29}$$

When the interaction energy is averaged over all orientations, using Boltzmann weighting ($e^{-E_{ind}/k_B T}$), the induced energy is:

$$E_{ind} = -\frac{\alpha\mu^2}{(4\pi\varepsilon\varepsilon_0)^2 r^6} \tag{2.30}$$

Table 2-4 Polarizability of various compounds (Barrow, 1973; Stone, 1996; Israelachvili, 1997).

Compound	α $C^2m^2J^{-1}$
Benzene	10.39
Carbon tetrachloride	10.5
Ethane	4.5
Hydrogen	0.8
Hydrogen sulfide	3.64
Hydrochloric acid	2.6
Methane	2.6
Methanol	3.2
Nitrogen	1.73
Oxygen	1.60
Water	1.44

The net dipole-induced dipole interaction for two or more molecules with permanent dipoles (known as the Debye-induced dipole moment) is:

$$E_{ind} = -\frac{2\alpha\mu^2}{(4\pi\varepsilon\varepsilon_o)^2 r^6} \quad (2.31)$$

2.6.3 Van der Waals Forces

The three long-range interactive forces (London, Keesom, Debye) are known collectively as the *van der Waals molecular interactive forces.* Each force has an energy of interaction that varies with respect to the inverse sixth power of the distance of separation, thus:

$$E_{vdW} = E_{disp} + E_{elec} + E_{ind} \tag{2.32}$$

And for two molecules with polarizability, α, and dipole moment, μ:

$$E_{vdW} = \frac{1}{(4\pi\varepsilon\varepsilon_o)r^6}\left(\frac{3}{4}\alpha^2 h\nu + \frac{2\mu^4}{3k_B T} + 2\mu^2\alpha\right) \tag{2.33}$$

2.7 Zeta Potential and the Electric Double Layer Interaction

There is a region at the surface of solids that has a difference in electrical potential across just a few molecular diameters. When two dissimilar metals are contacted, electrons at the surface of each metal redistribute in such a manner that one of the metals acquires a positive charge while the other becomes negatively charged. If a liquid and solid are brought together, an electrical potential develops across a distance of a few molecular diameters at the interface. The changes that are established are characteristic of the specific phases and are the underlying cause of many natural phenomena recognized as electrophoresis, electroosmosis, colloid stability, fluid flow behavior, adsorption, catalysis, corrosion, crystal growth, etc.

The separation of charges is known as the interfacial *electrical double layer.* It is a complex association of charges illustrated schematically in Fig. 2–5a. There is a potential charge (negative or positive) at one or two molecular distances from the surface. This charge may originate from several sources such as: (1) inclusions of extraneous atoms in the lattice structure, (2) dissolution of slightly soluble atoms at the surface of water, (3) chemical reaction (chemisorption) of ions in water with surface atoms forming complex polar molecules on the surface, or (4) exposure of metallic oxides at the surface which react with water to form surface ions. These are some of the major causes of surface charges; others are recognized in suspensions of particles and flocculants in water (Hunter, 1981).

Counterions from the water solution balance the charges at the solid surface and form the immobile Stern layer, Fig. 2–5a. The thickness of the Stern layer is only one or two molecular diameters consisting of ions that are adsorbed strongly enough to form an immobile layer. The

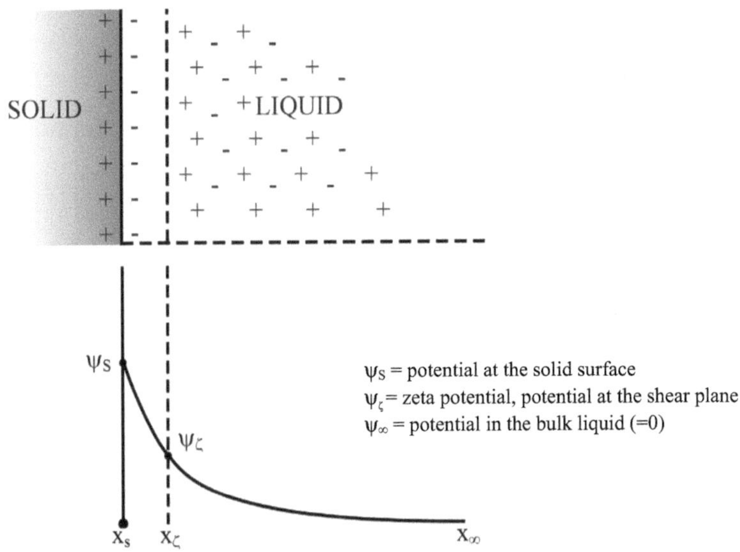

Figure 2–5a *Electric double layer at the interface between a solid and liquid: x_s = surface of the solid, x_ξ = shear plane, x_∞ = bulk liquid, $x_\xi - x_s$ = Stern layer, $x_\infty - x_\xi$ = electrical diffuse (Gouy) layer. (Debye length, l/k)*

outer edge of the Stern layer where the ions are mobile is known as the *shear plane.* There is a linear potential drop across the width of the Stern layer ($\psi_s - \psi_\varsigma$), followed by an exponential potential difference across the diffuse layer between the shear plane and the bulk solution ($\psi_\varsigma - \psi_\infty$); the bulk solution is designated as the reference zero potential. This potential difference between the shear plane and the bulk fluid is known as the *zeta potential.*

Cations, anions, and molecules with electrical dipoles can be adsorbed by nonelectrical forces. Grahame has observed that anions are adsorbed by nonelectrical forces with the centers of the negative charges lying on an inner plane (within the Stern layer) from the surface known as the Helmholtz inner plane (IHP at x_i distance from the solid surface, Fig. 2–5b). The IHP is followed by the outer Helmholtz plane (OHP) drawn through the charges of the hydrated counterions. The thickness of the Helmholtz layers thus reflects the size of the adsorbed anions and counterions within the Stern layer and is observed by the differences of the measured linear potential differences within the Stern layer.

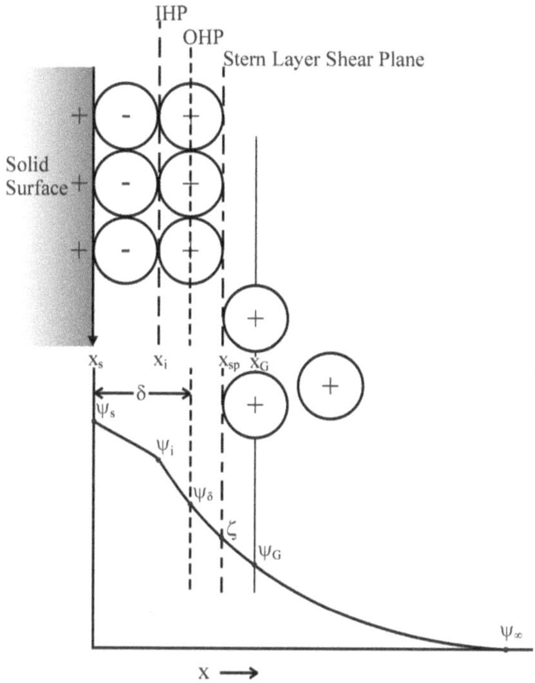

Figure 2–5b *Double layer potentials showing the Helmholtz planes and their potentials. IHP = inner Helmholtz plane (x_i). OHP = outer Helmholtz plane ($x = \delta$)*

The length of the exponential electrical field decay (from the shear plane to the bulk fluid) is known as the *Debye length* $(1/k)$. For example, if the plates of a capacitor have equal charge densities, the zeta potential is the potential difference from the center of the separation to one of the plates:

$$\frac{1}{\kappa} = x_o - x_s = \frac{\varepsilon\varepsilon_o \psi_s}{\sigma_c} = \left(\frac{\varepsilon\varepsilon_o kT}{\rho_i z^2 e^2}\right)^{\frac{1}{2}} \quad \left[\frac{C^2}{Jm}\frac{J}{C}\frac{m^2}{C} = m\right] \tag{2.34}$$

where ρ_i is the number density of ions in the solution; Eq. 2.34 also shows that the charge density of the surface (σ_c) is proportional to the surface potential (ψ_s).

With respect to an ionic solution, the Debye length is the distance from the shear plane of the Stern layer to the bulk fluid. The Debye length depends on the specific properties of the ionic solution. For aqueous solutions:

$$\frac{1}{\kappa} = \frac{B}{\sqrt{M}} \tag{2.35}$$

where B is a constant specific to the type of electrolyte. B is equal to 0.304 for monovalent cations and anions (NaCl); 0.176 where either the cation or the anion has a valency of two ($CaCl_2$ or Na_2CO_3); and 0.152 when both ions have a valency equal to two ($CaCO_3$). M is the molarity of the solution (Ex. 2.4).

The composition of the Stern layer varies with respect to the nature of the surface charge and ionic constituents of the electrolyte (Castellan, 1971):

1. The double layer may be entirely diffuse (no Stern layer) if ions are not adsorbed on the solid surface, Fig. 2–6. In this case the Stern layer does not exist and the potential difference declines exponentially from the solid surface to the bulk solution.

2. If the concentration of ions in the electrolyte is sufficient to exactly balance the surface charges of the solid, the potential will decrease linearly within the Stern layer to zero at the shear plane. Thus the zeta potential is zero (equal to the potential of the bulk fluid).

3. If the adsorption of ions does not completely balance the surface charge density, the zeta potential has a finite value with respect to the bulk fluid.

4. If the surface charge is very strong, the Stern layer may contain an excess of ions from the electrolyte and the zeta potential will have a charge opposite of the surface charge.

The zeta potential of mineral surfaces in contact with aqueous solutions is a function of pH. In general, acidic solutions promote positive charges at the surface with an attendant positive zeta potential and

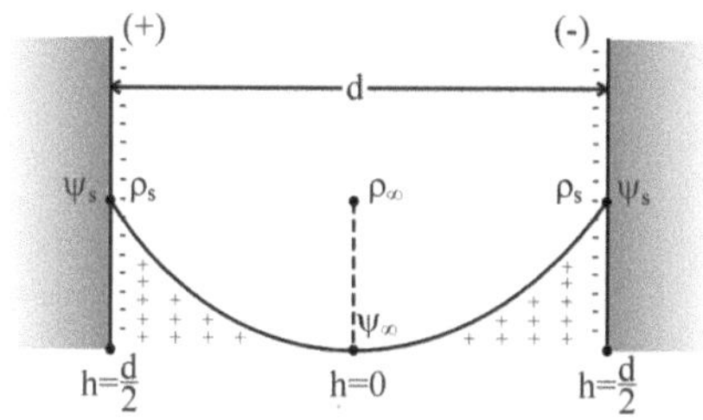

Figure 2–6 *Two charged surfaces separated by distance d with a fluid between. The film thickness on each surface is h = d/2. The number density of the sounter ions at the surface is ρ_s and the center is designated ρ_∞ which is taken as zero at the reference point in the center. The electric field, which is independent of distance is equal to the electric charge density, D, divided by the dielectric permittivity, $E_s = D/\varepsilon\varepsilon_0$.*

basic solutions produce an excess of negative charges at the surface from an increase of the hydroxide ion. The pH at which the zeta potential is equal to zero is defined as the zero point charge (zpc). When the negative and positive charges of ions in a solution are equally balanced, the solution is electrically neutral and this condition is defined as the isoelectric point (iep). Thompson and Pownall (1989) observed an approximate linear trend of the zeta potential with respect to pH for calcite in dilute solutions of sodium chloride and a mixed solution of sodium chloride and sodium bicarbonate, where $\varsigma = -6.67*\text{pH} + 40$. The zero point charge occurred at a pH of 6 and increased negative values occurred for pH > 6. Sharma et al. (1987) report inverted S-shaped trends where $\varsigma \approx -20*\text{pH} + 100$ (zpc at pH ≈ 5) for Berea cores and dilute sodium chloride solutions.

When an electrolyte is passed through a porous material (rock, glass capillaries, etc.), a potential difference develops across the ends of the core that is generally recognized as the *streaming potential.* Kruyt (1952) developed an expression for the streaming potential in terms of the zeta-potential and the combined resistivities of the electrolyte and solid (Ex. 2.5):

$$\frac{d\psi}{dP} = \frac{\varepsilon\zeta Ro}{4\pi\mu}\left[\left(\frac{C^2}{Nm^2}\right)(V)\left(\frac{1}{Pa\cdot s}\right)\left(\frac{Vsm}{C}\right) = \frac{V}{Pa}\right] \tag{2.36}$$

Considering that the total flow rate, q, in a porous medium is composed of the flow rate where there is no electrical potential effect, q_n, and an osmotic, countercurrent flow, q_{os}:

$$q = q_n - q_{os} = \frac{k_n dP}{\mu dx} - k'_{os}\left(\frac{d\psi}{dx}\right) \tag{2.37}$$

where k_n is the permeability in the absence of electrical phenomena and k'_{os} is the transport coefficient resulting from the streaming potential, ψ.

The coefficient, k'_{os} is obtained from the Helmholtz equation for the velocity of electroosmotic flow in a tortuous capillary (Adamson, 1960; Scheidegger, 1974):

$$u_{os} = \frac{e\phi\zeta d\psi}{4\pi\mu\tau^2 dx} \tag{2.38}$$

Combining Eqs. 2.36 and 2.38 into Eq. 2.37 yields the fluid flow equation that includes the effect of electroosmotic flow:

$$q = \left(\frac{k_n}{\mu} - \frac{e^2\phi\zeta^2 R_o}{(4\pi)^2\mu\tau^2}\right)\frac{dP}{dx} = \left(\frac{k_n}{\mu} - \frac{k_{os}}{\mu}\right)\frac{dP}{dx} \tag{2.39}$$

Eq. 2.39 shows that if the permeability of a dry core is measured by gas flow (when electroosmotic permeability is zero) the measured permeability, k, is equal to the permeability in the absence of electrical surface phenomena. When the permeability is measured with an electrolytic solution, however, the osmotic permeability causes the measured (apparent) permeability to be lower than the permeability obtained in the absence of surface phenomena ($k = k_n - k_{os}$).

When the permeability is measured with a low concentration electrolyte solution and then measured again with a solution containing a higher concentration of salts, the permeability will be higher for the solution with the higher salt concentration (assuming that migration of fines does not occur within the core). This occurs because when the ionic strength of the electrolyte is increased, the conductance of the solution increases and the zeta potential decreases. Therefore, the value of k_{os} decreases resulting in an increase of k (Scheidegger, 1974).

If the fluid flow velocity is increased, part of the diffuse layer is destroyed, and the zeta potential decreases with a corresponding decrease of k_{os} resulting in an increase of k.

Helander (1983) pointed out that the influence of the streaming potential increases as the pore size decreases. For example, since the Stern layer is approximately 10^{-8} m in thickness, one can compare that to the diminishing size of pore radii (for a 1.0 μm pore: 1.0 μm radius/0.01 μm layer = 100/1; and for a pore 0.1 μm: 0.1 μm radius/0.01 μm layer = 10/1). Thus the importance of the streaming potential must be greater in the smaller pores. Although the streaming potential may be considered negligible for general well-log interpretation addressing higher porosity/permeability formations, the streaming potential will become an important factor for "tight" rocks ($k < 5$ md) which have a relatively small pore-size distribution.

2.8 DLVO (Derjaguin, Landau, Verwey, and Overbeek) Theory

The DLVO theory (Derjaguin and Landau, 1941; Verwey and Overbeek, 1948) is the analysis of the competitive interactions of double-layer repulsive forces and van der Waals attractive forces as they affect the stability of colloidal suspensions of particles in electrolytic solutions. Anions and cations may be adsorbed on surfaces by van der Waals attractive forces which are not affected by variations of pH or the concentration of electrolytes. Curve 1, Fig. 2–7, describes the variation of the interactive potential energy that occurs from complete repulsive energy (in the absence of van der Waals attractive forces); this occurs for a relatively thick double layer in a solution of high electrolyte concentration.

The van der Waals forces become stronger (more negative) following an exponential curve ($U_{(r)} = -r^{n}$) as the separation decreases and hence van der Waals forces are negligible at long distances. At short distances from the wall, however, the van der Waals forces exceed the repulsive double-layer forces and the result is strong attraction of the surfaces (curve 4, Fig. 2–7).

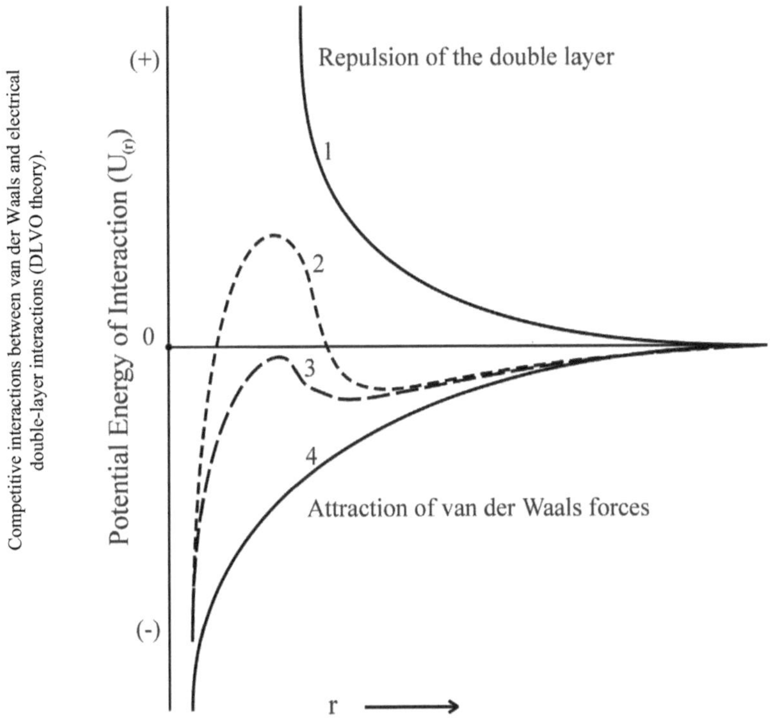

Figure 2–7 *Curve 1: Repulsion of a surface caused by double-layer Coulombic charge in the absence of van der Waals attraction. This occurs in high electrolyte concentrations. Curves 2 to 4: The increasing effect of van der Waals forces as the thickness of the double layer decreases; this results from decreasing electrolyte concentration. Curve 4: van der Waals attraction in the absence of double-layer repulsion.*

Between the two extremes, as particle surfaces approach, there is an initial attractive minimum followed by a repulsive maximum that can be too great to allow actual contact of the particle surfaces (curve 2, Fig. 2–7) and the particle will remain dispersed in the electrolyte. If the surfaces have a low charge density and the repulsive maximum of the potential energy is zero or negative (curve 3), the particles will coagulate because the forces are entirely attractive.

2.9 Hamaker Constant

The Hamaker constant (A) is a coefficient that relates the interactive van der Waals energy (E_{vdW}) to the distance of separation between two molecules where the interactive force is pair-wise additive and independent of the intervening media. Hamaker (1937) used Eq. 2.33 to evaluate the interaction between molecules based on composition and structure of the particles. Lifshiz (1956) developed a theoretical relationship for the collective interactive forces between macroscopic particles from quantum field theory that relates the interactive energy to distance of separation of particles. The interactions between the particles are relative to the macroscopic properties: the dielectric constant, ε, and the refractive index, n. The Hamaker constant for a film liquid (3, Fig. 2–8) between a gas or immiscible liquid (2), and a solid (1) is (Israelachvili, 1997):

$$A \approx \frac{3}{4}kT\left(\frac{\varepsilon_1-\varepsilon_3}{\varepsilon_1+\varepsilon_3}\right)\left(\frac{\varepsilon_2-\varepsilon_3}{\varepsilon_2+\varepsilon_3}\right)+\frac{(0.265h\nu)\left(n_2^2-n_3^2\right)\left(n_2^2-n_3^2\right)}{\left\{\left(n_1^2+n_3^2\right)^{\frac{1}{2}}\left(n_2^2+n_3^2\right)^{\frac{1}{2}}\left(n_1^2+n_3^2\right)^{\frac{1}{2}}\right\}+\left\{\left(n_2^2+n_3^2\right)^{\frac{1}{2}}\right\}} \tag{2.40}$$

The first term depends on temperature and results from the Keesom and Debye polar molecular forces, while the second term is a function of the London dispersion forces that depend on the orbiting electron frequency, ν, and the refractive index, n, of the liquid, both of which are independent of temperature (Ex. 2.6). When the forces are attractive, A is negative. It follows that the interactive forces are repulsive if the Hamaker constant is positive.

Hamaker constants for phases 1 and 2 interacting across a third medium (A_{132}) for materials whose properties are unknown may be estimated using mixing rules where:

$$\begin{aligned} A_{12} &= \sqrt{A_{11}A_{22}} \\ A_{131} &= \left(\sqrt{A_{11}}-\sqrt{A_{33}}\right)^2 \\ A_{132} &= \left(\sqrt{A_{11}}-\sqrt{A_{33}}\right)\left(\sqrt{A_{22}}-\sqrt{A_{33}}\right) \end{aligned} \tag{2.41}$$

A_{11}, A_{22}, and A_{33} are the Hamaker constant for identical phases interacting across a vacuum. A_{12} is the constant for dissimilar phases 1 and 2 interacting across a vacuum. A_{131} represents two layers of phase 1 interacting across phase 3, and A_{132} represents two dissimilar phases, 1 and 2, interacting across phase 3 (Hirasaki, 1991).

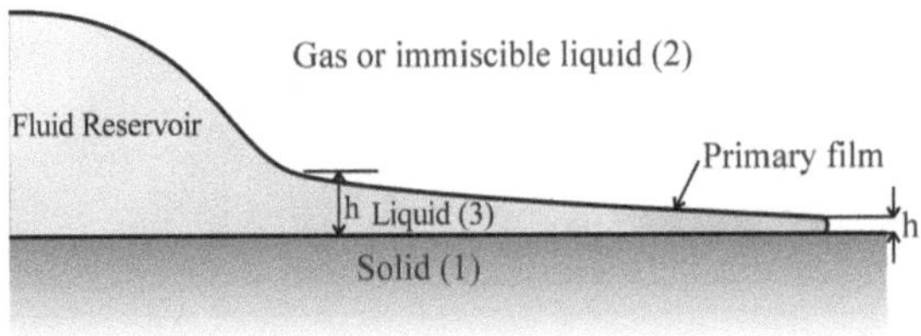

Figure 2–8 *Nonvolatile fluids spreading on a flat, horizontal surface. A thin primary flow may spread ahead of the advancing fluid before the film is established on the surface with thickness, h, which is generally < 3mm.*

Hirasaki points out that changes of the film thickness of two films on a solid surface can change the sign of the Hamaker constant because of shifting interactions between the two bulk phases. This would change the interaction from attractive to repulsive, or vice versa. Assume that the bulk liquid phases are a solid (1) and a gas (2) with an adsorbed liquid layer with thickness δ on the solid (4) and a liquid film (3) with thickness h resting on the adsorbed layer; in this case, the interacting potential for a system of materials whose Hamaker constant is estimated from Eq. 2.40 is:

$$W = \frac{\left(A_{11}^{1/2} - A_{44}^{1/2}\right)\left(A_{22}^{1/2} - A_{33}^{1/2}\right)}{12\pi(\delta + h)^2} + \frac{\left(A_{44}^{1/2} - A_{33}^{1/2}\right)\left(A_{22}^{1/2} - A_{33}^{1/2}\right)}{12\pi h^2} \tag{2.42}$$

If δ is insignificantly thin with respect to h, the interaction approaches that of a single layer of liquid, 3, between the solid, 1, and the gas, 2. On the other hand, if δ is large, the interaction potential changes to interaction between phases 2 and 4 across the film, 3. The latter change of interaction may result in a change of sign because of the overall change of the interacting relationships of the four phases which would translate to a change of wettability for the system.

For liquids having molecules that interact only by van der Waals forces, the Hamaker constant may be related to the interfacial tension by equating the work of cohesion between two surfaces separated by intermolecular distance, δ. Israelachvili has determined that the average distance, 0.165 nm, may be used for these liquids, hence:

$$W_c = 2\sigma = \frac{A_{11}}{12\pi\delta^2}$$

$$A_{11} = 2.1 * 10^{-21}\sigma \quad \left[\frac{m^2 J}{m^2} = J\right] \tag{2.43}$$

2.10 Disjoining Pressure

In a gas-liquid-solid system, if the liquid is allowed to spread from a reservoir over a flat surface, a precursor film of the liquid will advance from the reservoir and spread over the horizontal solid surface as illustrated in Fig. 2–8. The spreading is controlled by fluid properties (viscosity, density, volatility), van der Waals forces and molecular electrical interacting forces. The precursor film is usually less than one micrometer in thickness with multi-molecular layers developing more slowly on top of the precursor film. The final established thickness of the film is a function of the total interacting fluid and solid forces acting on the system. The pressure required to oppose the fluid/solid attractive forces and lift the film from the solid is the *disjoining pressure* which is composed of: (1) the van der Waals forces that are always attractive, (2) the electrostatic forces that can be either attractive or repulsive, and (3) hydration forces that become dominant if the interfaces approach within a few molecular diameters of each other.

The spontaneous spreading of a film on a flat, horizontal surface or the rise of a fluid on the walls of a capillary is the result of an imbalance of surface free energy which is driving the fluid toward thermodynamic equilibrium. For a liquid film rising on the side of a tube to a height, H, with respect to the base of the liquid, the change in free energy with respect to film thickness is zero at equilibrium:

$$\frac{\delta G}{\delta h} = 0 = \frac{A}{6\pi h^3} + \rho g H = -P_{(h)} \tag{2.44}$$

Near the base of the meniscus where the film thickness, h, is large, the van der Waals pressure reduces to zero and Eq. 2.44 becomes the familiar capillary pressure equation for rise of a fluid. On a flat surface, however, where the film thickness is relatively thin and the interfaces are parallel, the gravity term reduces to zero and the remaining negative pressure is the repulsive van der Waals component of the disjoining pressure:

$$\Pi_{vdW(h)} = -\frac{A_{12}}{6\pi h^3} \tag{2.45}$$

A_{12} is the Hamaker constant for a solid and gas interacting through a thin liquid film.

2.10.1 Electrostatic Forces

The chemical potential of a solute in a dilute solution is described thermodynamically as:

$$\mu = \mu_o + RTLn(n) \tag{2.46a}$$

$$\mu = ze\psi + kTLn(\rho_n) \tag{2.46b}$$

The chemical potential (Eq. 2.46b) is thus composed of two terms: the first, μ_o, (and its electrical equivalent $ze\psi$) is independent of concentration, but the second term is a function of the molar concentration (or the number density of ions in the case of the electrostatic system). At equilibrium the chemical potential is uniform throughout and consequently μ is equal to zero. Solving Eq. 2.46b for the number density yields the *Boltzmann distribution* for counterions in a solution with respect to the distance from a charged surface, Fig. 2–6:

$$\rho_{n(x)} = \rho_{n,\infty} e^{-ze\psi/kT} \tag{2.47}$$

If only the reference potential affects the distribution of ions, substituting $\psi_?$ for the free energy of interaction:

$$Ln\left(\frac{\rho_{n(x)}}{\rho_{n,\infty}}\right) = -\frac{ze^2}{4\pi\varepsilon\varepsilon_o xkT} \tag{2.48}$$

This relationship may be used to estimate the relative number density of ions, with respect to the average number of ions in the bulk liquid, per unit volume at a distance, x, from the oppositely charged surface, Ex. 2.47.

2.10.2 Poisson-Boltzman Equation

Introducing Poisson's equation that describes the net excess charge density at any point, x, for static conditions:

$$\nabla^2\psi = -\frac{ze\rho}{\varepsilon\varepsilon_o} \tag{2.49}$$

Combining Poisson's equation with Boltzmann's distribution (Eq. 2.47) yields the *Poisson-Boltzmann equation*:

$$\frac{d^2\psi}{dx^2} = -\frac{ze\rho}{\varepsilon\varepsilon_o} = \left(\frac{ze\rho_\infty}{\varepsilon\varepsilon_o}\right)e^{-ze\psi/kT} \tag{2.50}$$

The charge density between the surfaces (Fig. 2–6) varies with respect to distance as follows:

$$\sigma_c = -\int_0^x (xe\rho)dx = +\varepsilon\varepsilon_o\int\left(\frac{d^2\psi}{dx^2}\right)dx = \varepsilon\varepsilon_o\left(\frac{d\psi}{dx}\right)_x = \varepsilon\varepsilon_o E_x \tag{2.51}$$

Eq. 2.51 relates the electric field ($E = (d\psi)/(dx)$) to the surface charge density on each surface when x is equal to $d/2$.

2.10.3 Counterion Density with Respect to Distance (Diffuse Double Layer)

Returning to Eq. 2.47 and differentiating yields an expression of the number density with respect to distance as a function of the electric field:

$$\frac{d\rho_n}{dx} = -\left(\frac{ze\rho_\infty}{kT} e^{-ze\psi/kT}\right)\frac{d\psi}{dx} \tag{2.52}$$

Combining Eq. 2.51 with 2.52 and integrating yields the relationship for ρ_n (at any point x) in terms of ρ_∞ and the electric field:

$$\frac{d\rho}{dx} = \frac{\varepsilon\varepsilon_o}{kT}\left(\frac{d^2\psi}{dx^2}\right)\left(\frac{d\psi}{dx}\right) \tag{2.53}$$

Using the relationship:

$$\frac{d\psi}{dx}\left(\frac{d^2\psi}{dx^2}\right) = 2\frac{d}{dx}\left(\frac{d\psi}{dx}\right)\left(\frac{d\psi}{dx}\right) = 2\frac{d}{dx}\left(\frac{d\psi}{dx}\right)^2 \tag{2.54}$$

$$\rho_n - \rho_\infty = \int d\rho = \frac{\varepsilon\varepsilon_o}{2kT}\int_0^x d\left(\frac{d\psi}{dx}\right)^2 = \frac{\varepsilon\varepsilon_o}{2kT}\left(\frac{d\psi}{dx}\right)^2 \tag{2.55}$$

Introducing Eq. 2.51, shows that the concentration of counterions depends on the charge density of the surface charge and extends from the surface as a diffuse concentration to the bulk phase density, ρ_∞ which constitutes the diffuse double layer (Figs. 2–5a and 2–5b):

$$\rho_s = \rho_\infty + \frac{\sigma_c^2}{2\varepsilon\varepsilon_o kT} \tag{2.56}$$

2.10.4 Osmotic Repulsion Pressure

The repulsive pressure between two surfaces (because of electric double layers associated with each surface, Fig. 2–6) is the sum of the pressure resulting from the electric force (from the electric field) and an osmotic pressure because of the ion concentration difference between the surfaces and the midplane:

$$\frac{dP}{dx} = \rho\frac{d\mu}{dx} = ze\rho\frac{d\psi}{dx} + \frac{kTd\rho}{dx} \tag{2.57}$$

Combining Eq. 2.50 and the relationship expressed in Eq. 2.54 into the first term and integrating between $x = \infty$ (center plane of Fig. 2–6) and $x = d/2$ (the solid surface):

$$P_{x(d)} - P_{x,\infty} = \frac{\varepsilon\varepsilon_o}{2}\left(\frac{d\psi}{dx}\right)^2 - \frac{\varepsilon\varepsilon_o}{2}\left(\frac{d\psi}{dx}\right)^2 + kT\rho_{x(d)} - kT\rho_{x,\infty} \tag{2.58}$$

At $x = \infty$, the electric field and the density of counterions is equal to zero; therefore Eq. 2.61 reduces to:

$$P_{x(d)} - P_{x,\infty} = -\frac{\varepsilon\varepsilon_o}{2}\left(\frac{d\psi}{dx}\right)^2 + kT\rho_{x,d} \tag{2.59}$$

The first term is the attractive pressure exerted by the electric field and the second term is the repulsive osmotic pressure. Substituting $\rho_{x(d)}$ from Eq. 2.55 into 2.59, the attractive pressure term drops out and the pressure between the surface and center plane of two approaching bodies (Fig. 2–6) becomes the expression for the osmotic pressure as a function of the counterion concentration:

$$P_{x(d)} = kT\rho_\infty \tag{2.60}$$

Eq. 2.63 expresses the change of osmotic pressure with respect to distance between either surface and the center plane.

Substituting ρ_∞ from the Debye length ($1/\kappa$) into Eq. 2.60 and noting the Debye length approaches $\pi/2$ at large distances yields the Langmuir equation that describes wetting films of water:

$$P_{(d)} = 2\varepsilon\varepsilon_o \left(\frac{\pi kT}{ze}\right)^2 / d^2 \tag{2.61}$$

The film thickness between the surfaces and the midplane (Fig. 2–6) is equal to *d/2*. Substituting this into Eq. 2.61 yields the repulsive electrostatic pressure contribution of the *disjoining pressure* for thick liquid films (Adamson and Gast, 1997; Israelachvili, 1997):

$$\Pi_{(elec)} = \varepsilon\varepsilon_o \left(\frac{\pi kT}{ze}\right)^2 / d^2 \tag{2.62}$$

2.10.5 Hydration Forces

Charged surfaces in contact with water and ions in aqueous solutions have water molecules strongly associated with them through hydrogen bonding, thus forming hydrate complexes. The association can produce several layers of water on the surface or surrounding the solute ion. Strongly hydrated surfaces will exert a repulsion force between them because of the energy required for perturbation of the ordered structure of the bound layers of water molecules. This has been recognized as the hydration force. Hence, if there is no change in the structure of the bound water molecules as another surface approaches, there will be no hydration, or structural, interactive force. In addition, the hydration forces are absent in acidic solutions since only protons are adsorbed on the surfaces.

The hydration or structural force is repulsive between hydrophilic surfaces (generally surfaces with a net electronegative charge) and zwitterions that undergo hydrogen bonding. The strength of the hydration force depends on the energy required to perturb the hydrogen bonding network of the hydrated surface.

The repulsive hydration force between two hydrophilic surfaces apparently decays exponentially in the range of about 1 nm according to an empirical relationship:

$$\Pi_{(hy)} = Se^{-x/\delta} \ [mJ/m^3 = mNm^{-2} = Pa^{-3}] \tag{2.63}$$

where $S = 3 - 30$ mJ/m^3 and $\delta = 0.6 - 1.1$ nm.

Water cannot bind to hydrophobic surfaces, but such surfaces are strongly attracted to each other, which is the reason for rapid coagulation in excess water (Buckley, 1989; Hirasaki, 1991; Adamson and Gast, 1997; Busireddly and Rao, 2004).

2.10.6 Disjoining Pressure and the Canonical Potential

The disjoining pressure is the result of three distinct contributions: (1) the van der Waals forces that are inversely proportional to the third power of the film thickness, making them effective only at very short distances, (2) the electrostatic double-layer ionic osmotic pressure, and (3) structural forces arising from the association (hydration) of water molecules to hydrophilic surfaces:

$$\Pi_{(dis)} = \Pi_{(vdW)} + \Pi_{(elec)} + \Pi_{(hy)} = \frac{A}{6\pi h^3} + \varepsilon\varepsilon_o\left(\frac{\pi kT}{ze}\right)/\,2h^2 + Se^{-x/\delta} \tag{2.64}$$

The canonical potential, Ω, is the ratio of the van der Waals pressure to the double-layer electrical pressure. Substituting $\delta = h\kappa$ into the van der Waals disjoining pressure term yields the dimensionless canonical potential:

$$\Omega = \frac{A\kappa^3}{12\pi\rho_n kT\delta^3} \qquad \left[\frac{Nm^{-2}}{Jm^{-3}} = \frac{Jm^{-3}}{Jm^{-3}}\right] \tag{2.65}$$

where $\delta = h\kappa$ (dimensionless):

$$\Omega = A\kappa^3 / 12\pi\rho kT \tag{2.66}$$

2.11 Hydrogen Bonding

Ions in aqueous solutions are associated with water molecules through hydrogen bonding. The analysis using the dielectric constants is not sufficient to describe the electrostatic interactions. Water is polarized such that the hydrogen atoms each have a small positive charge associated with the electron configuration and those charges are balanced by the negative charges in the oxygen atom's electron cloud. When organic polar molecules (such as alcohols) dissolved in water approach the polar water molecules they become electrostatically associated, as shown in Fig. 2–9. The hydrogen, with its electron skewed toward the oxygen atom, yields a small atom that can be approached very closely by negative groups such as $O^{(-)}$ and $N^{(-)}$, to form the strong short-separation electrostatic bonds. Using Eq. 2.8 with ε = 80 and r = 0.17 nm, the strength of such a bond is 10.34 kJ/mole: for comparison, a covalent bond with ε = 4 and r = 0.1 nm is 347 kJ/mole. This bonding strength, and its higher dielectric constant, gives water unique properties in spite of its low molecular weight: high boiling point, high heat of vaporization, excellent solubility, and low compressibility, etc.

The unique properties occur because hydrogen bonding in a liquid produces a polymer-like structure that has some covalent characteristics. The bonds are weaker because of the slightly greater distance of separation (0.176 nm for hydrogen bonds, compared to 0.1 nm, or less, for covalent bonds). Their polymer-like hydrogen bonding structures of a few molecules are illustrated in Fig. 2–9 as strong directional dipole-dipole interactions. Dispersion, induction, and transfer of charges also play a role in hydrogen bonding.

Nonpolar molecules do not form hydrogen bonds and are thus excluded from the hydrogen bonding molecules forming a second, nonpolar phase. A small amount of water placed on an inert surface will form a sphere exhibiting a large, nonwetting contact angle. Conversely, molecules that have electronegative atoms in their structure that can react with hydrogen bonds are soluble in water to varying degrees depending on the size of the nonpolar portion of the molecule. Some molecules such as surfactant that have an electronegative group $(-SO_3^-, -COO^-, -N(CH)_3, -NH_2)$ attached to a large nonpolar hydrocarbon group migrate to the interface of water where they lower the interfacial tension by exposing the non-polar group at the interface. These types of molecules remain within the bulk water phase associated with hydrogen bonding (Castellan, 1971; Barrow, 1973; and Stone, 1996). Nonpolar hydrocarbon sol-

Figure 2–9 *Electrostatic polymer-like hydrogen bonding of (a) water, (b) a mixture of water and amine, and (c) an organic acid forming a dimer. Linear diagrams are shown for clarity of the concept.*

vents cannot hold water or salts in solution because the nonpolar molecules are only weakly attracted to each other by van der Waals forces that cannot overcome the strong hydrogen bonding of water molecules or salt crystals.

2.12 Summary of Interactive Forces

All of the atomic and molecular interactions are manifestations of electrostatic attractive and repulsive reactions. They are the result of quantum mechanical interactions of the electronic wave functions (constrained by the Pauli exclusion principle). For completeness of the concepts, the interactive energies should be described in terms of quantum mechanics, orbital theory, wave functions, and probability distributions. In the preceding discussions, the specific electron density (or electron cloud) was approximated by localized point charges such as ions, dipoles, induced dipoles, hydrogen bonding, etc., in order to make approximate calculations of the energies associated with each to give a better sense of the intermolecular energies associated with wettability.

Covalent bonds at very short separations are a function of e^{-r} and are the strongest bonds (where r is the distance of separation between atoms or molecules); ionic bonds are longer range functions of r^{-1}, and are very strong; dipole-dipole interactions act at shorter ranges as a function of r^{-3}, and form relatively strong bonds; the dipole-induced-dipole interactions are short range, r^{-6}, relatively weak bonds; and dispersion interactions also interact at short range as functions of r^{-6}, but are very weak.

2.13 Lewis Acid/Base Reactions

Spectrographic analyses have shown that the electrons are arranged in shells and subshells in an ordered sequence which is illustrated in Tables 2–5 and 2–6. Each circle represents an atomic orbital that can contain either one or two electrons with opposite spins in accord with Pauli's exclusion principle: Two electrons in an atomic orbital may not have the same quantum numbers, and if two electrons occupy the same orbital, they have opposite spins ($m_s = +\frac{1}{2}, -\frac{1}{2}$).

The electron orbits are described as a set of four quantum numbers: (1) n is the principal quantum number (1, 2, 3, ...) that represents the energy of the orbital and thus describes the average distance of the electron from the nucleus, (2) l is the azimuthal number (0, 1, 2, ... n–1) that describes the shape of the orbital, (3) m represents the magnetic orientation of the charge cloud with respect to the three axes (x, y and z), and (4) m_s which defines the electron spin , $+\frac{1}{2}$, $-\frac{1}{2}$ (Castellan, 1971).

The energy levels (average distances of the electron clouds from the nucleus; lower levels have the lowest energy) are described as virtual shells (*K, L, M, ...*) and subshells (*s, p, d, f, ...*) that have the same value as the azimuthal numbers, as illustrated in Table 2–7. There are $2l + 1$ values of m, but the electrons may have two values of m_s; therefore each subshell can contain a maximum of $2(2l + 1)$ electrons. The first shell ($n = 1$, $l = 0$) always has one orbital (1*s*), the

Table 2–5 Electron configurations of some elements. The asterisk indicates that the element has completed 8-electron outer obitals (Barrow, 1973; Castellan, 1971).

Z	Element	n (Shell)	1 (K)	2 (L)		3 (M)			4 (N)			
		Sub-shell:	1s	2s	2p	3s	3p	3d	4s	4p	4d	4f
1	H		1									
2	He*	(inert)	2									
3	Li		2	1								
4	Be		2	2								
5	B		2	2	1							
6	C		2	2	2							
7	N		2	2	3							
8	O		2	2	4							
9	F		2	2	5							
10	Ne*	(inert)	2	2	6							
11	Na		2	2	6	1						
12	Mg		2	2	6	2						
13	Al		2	2	6	2	1					
14	Si		2	2	6	2	2					
15	P		2	2	6	2	3					
16	S		2	2	6	2	4					
17	Cl		2	2	6	2	5					
18	Ar*	(inert)	2	2	6	2	6					
19	K		2	2	6	2	6		1			
20	Ca		2	2	6	2	6		2			
21	Sc		2	2	6	2	6	1	2			
22	Ti		2	2	6	2	6	2	2			
23	V		2	2	6	2	6	3	2			

Table 2–5 Electron configurations of some elements. The asterisk indicates that the element has completed 8-electron outer obitals (Barrow, 1973; Castellan, 1971). (cont'd)

Z	Element	n (Shell)	1 (K)	2 (L)		3 (M)			4 (N)			
		Sub-shell:	1s	2s	2p	3s	3p	3d	4s	4p	4d	4f
24	Cr		2	2	6	2	6	4	1			
25	Mn		2	2	6	2	6	5	2			
26	Fe		2	2	6	2	6	6	2			
27	Co		7	2	6	2	6	7	2			
28	Ni		2	2	6	2	6	8	2			
29	Cu		2	2	6	2	6	10	1			
30	Zn		2	2	6	2	6	10	2			
31	Ga		2	2	6	2	6	10	2	1		
32	Ge		2	2	6	2	6	10	2	1		
33	As		2	2	6	2	6	10	2	3		
34	Se		2	2	6	2	6	10	2	4		
35	Br		2	2	6	2	6	10	2	5		
36	Kr*	(inert)	2	2	6	2	6	10	2	6		

second shell ($n = 2$, $l = 1$) has 2 sublevels (2*s* and 2*p*), the third shell ($n = 3$, $l = 2$) has three sublevels (3*s*, 3*p*, and 4*d*), etc.

The sublevels (*s, p, d, f,* etc.) have a specific number of orbitals as illustrated in Figs. 2–5a and 2–5b. The *s*-sublevel always has one orbital, the *p*-sublevel has three orbitals, *d* has five, *f* has seven orbitals, etc. The *s*-orbital is a spherical shape around the central nucleus that defines the probable positions of one or two electrons. A maximum of two electrons having opposite spins can occupy the *s*-orbital. Electrons in *p*-orbitals occupy opposite spherical spaces on the sides of the nucleus akin to a dumbbell. The *d*-orbitals are visualized as four orbitals symmetrically spaced in a polar configuration about the nucleus. The *f*-orbitals form complex orbitals about the nucleus. Furthermore, some overlapping or interchange of orbitals occurs in the outer, higher

Table 2–6 The maximum permissible electrons in each sub-shell illustrated as orbitals that can only contain one or two, electrons; n is the principal quantum number (the letters K, L... are the shell designations), and l is the azimuthal quantum number. Each sub-shell can only contain a maximum of 2(2l+l) electrons.

n (shell)	l	Maximum permissible electrons in filled sub-shells
1 (K)	0	1s 
2 (L)	1	2s 2p 
3 (M)	2	3s 3p 3d 
4 (N)	3	4s 4p 4d 4f 

energy orbitals. Hence the valence electrons (or valence bonds) of most of the metals often involve the two outer sublevels giving these atoms bonding properties that cannot be explained by the original Lewis acid/base theory where bonding is restricted to completion of the outer sublevel by eight electrons. Instead one must consider only the completion of orbitals by two electrons.

Using the schematic illustration of Table 2–5, the electron configuration of the elements is shown in Table 2–6. With respect to interactions (bonding) of the atoms and molecules, it is very important to note that the electrons enter each orbital in the sublevels in sequence (one at a time) until all of the orbitals contain one electron; added electrons begin completing the two electron configuration of each orbital only after each orbital has one electron.

Bonding, by sharing the outer shell electrons, can be visualized using dots to represent the outer electrons associated with chemical bonds.

Table 2–7 Size and electrical charge of common elements found in minerals. Note that anions are two to three times larger than cations.

Element	Size, nm	Ionic Valence Charge
Aluminum	0.050	+3
Boron	0.031	+3
Calcium	0.097	+2
Chlorine	0.181	–1
Fluorine	0.136	–1
Iron	0.076	+2
Lithium	0.068	+1
Magnesium	0.065	+2
Oxygen	0.140	–2
Potassium	0.133	+1
Silicon	0.045	+4
Sodium	0.095	+1
Sulfur	0.184	–2

The inner shell electrons and nucleus are represented by the element symbol, Fig. 2–10. The original Lewis bonding theory is based on the premise that the sharing of the outer shell electrons by atoms results in an outer shell electron configuration of the inert gases: either 2 or 8 electrons in the outer shell. In the case of hydrogen, it should have two shared electrons. Sharing for most of the other atoms should result in a shared configuration of eight electrons, Fig. 2–11. The symbols thus show the paired (2 electrons) and unpaired orbitals (Lewis, 1923).

Fig. 2–11 is an idealized illustration of the valence shells of the atoms which explain many of the bonding characteristics of the elements (Castellan, 1971; Burrow, 1973). Hydrogen only has one valence orbital; therefore, it can form only one bond. The elements in the row from lithium to neon, however, have four valence orbital in their valence shell (L) that are available for bonding. The number of pos-

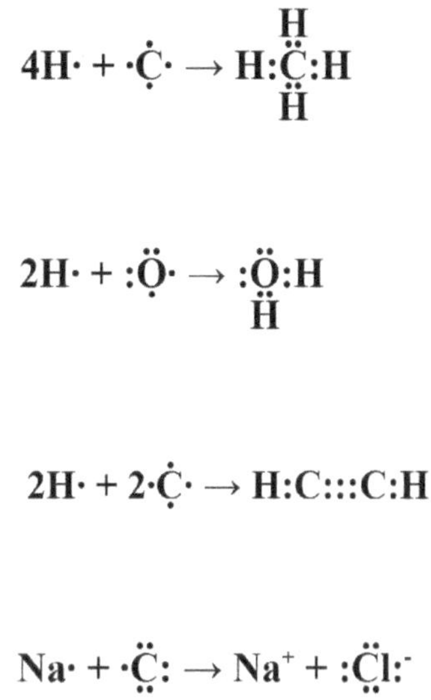

Figure 2–10 *Lewis acid/base bonding.*

sible bonds for this row of elements is thus restricted to a total of four: two electrons in the 2*s* suborbital and one each in the 2*p* suborbitals [$2s^2$, $2p_x$, $2p_y$, $2p_z$], as illustrated for carbon in Fig. 2–11d.

Lithium, like hydrogen, only has one valence electron. Beryllium has two electrons in the $2s^2$ orbital that separate in an excited state and provide two bonding electrons by raising one of the electrons into the empty $2p_x$ orbital, giving it a valence of 2. Boron has two electrons in the 2*s* orbital and one in the $2p_x$, and hence a valence of 3. Carbon, Fig. 2–11d, exhibits a valence of four because, in its excited state, an electron is moved from the $2s^2$ orbital to the empty $2p_z$ orbital of the *L*-valence shell.

The original rule of valence shell completion by eight electrons, proposed by Lewis, is only valid for the elements of the second period (lithium to neon). The number of electrons in the second period ranges from one for lithium to eight in the completed outer shell of neon where four orbitals make this element inert. The elements in the third period have the empty *d*-orbitals in the *M* valence shell making up a total of nine orbitals (one *s*-, three *p*- and 5 *d*-orbitals). This complicates the bonding characteristics because it is apparent that nine bonds to other atoms are theoretically possible.

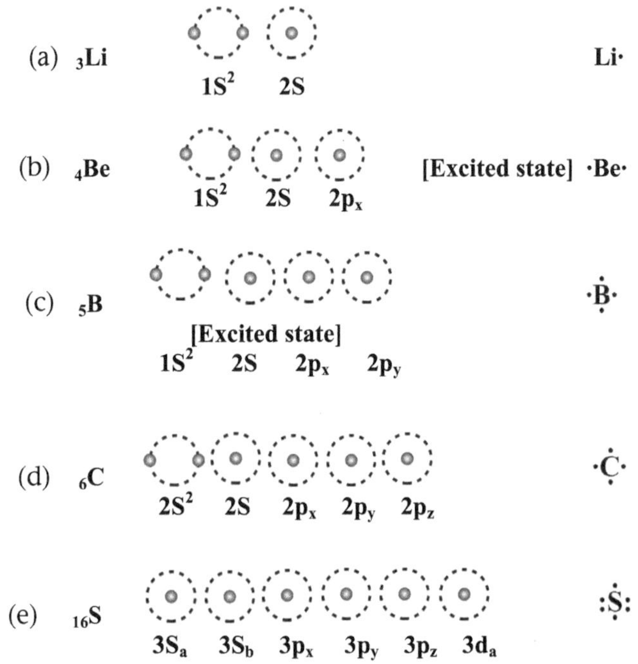

Figure 2–11 *Illustration of the virtual configuration of electrons in the valence shells. (a) Lithium has only one electron in the L-valence shell and therefore is restricted to a valence of one. (b) Beryllium has two electrons in the L-valence shell giving it a valence of two. In an excited state one electron is raised from the $2S^2$ orbital to the empty $2p_x$. This yields two orbitals with single electrons that are available for bonding: valence = 2. (c) Boron has three electrons in the L-valence shell giving it a valence of 3; however, in an excited state one of the $2S^2$-orbital electrons is shifted to an empty p-orbital. (d) Carbon has a valence of four (the maximum for the L-shell) because one of the $2S^2$ electrons can be raised to the empty $2P_z$ orbital. (e) Elements in the second period have the empty d-orbitals in their valence shell (M). Sulfur illustrates the idealized configuration, where only the M valence shell is shown. The excited state is shown where one electron is raised out of the $3S^2$ sub-orbits resulting in activation of the $3d_a$ orbital.*

The bonding properties of the fourth period (potassium, K, to krypton, Kr) are even more complex because three *d*-electrons in the *M*-shell can shift into the empty orbitals of the *N*-shell. Thus cobalt has nine valence electrons (six of which are paired), Table 2–7.

The high energy electrons in the outer shells (the valence electrons) can thus enter into chemical reactions forming:

1. Ions, where electrons are removed to complete the orbitals of another atom (Fig. 2–12a). Ionic bonds are important with respect to wettability because petroleum reservoir waters contain a host of salts in various concentrations (Collins, 1975). Jadhunandan and Morrow (1995) found that sensitivity of wettability to brine composition decreases as the aging time of cores saturated with brine and crude oil is increased, and that preferential oil-wetting is greater when the brine contains a high concentration of calcium. They concluded that one of the parameters affecting wettability is brine composition.

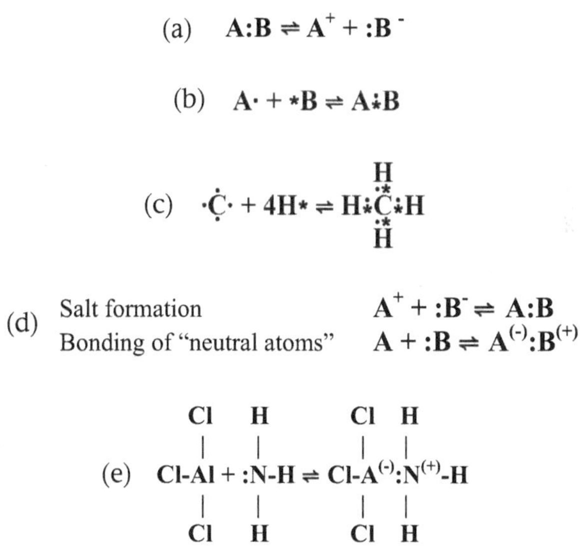

Figure 2–12 *Formation of ions: Lewis acid/base reactions. (a) Electrons are removed to complete the orbitals of another atom. (b) Covalent bond: each atom furnishes one of the electrons to complete the orbital. (c) Covalent bonding of carbon: completion of an octet of electrons. (d) Coordinate-covalent bonding: sharing of paired electrons. (e) Aluminum chloride's behavior as an acid accepting a pair of electrons to complete its outer four orbitals.*

2. Covalent bonds, where each atom furnishes one electron to complete the orbital (Fig. 2–12b). The very strong covalent bonds are formed when two atoms or molecules each have an incomplete orbital and they join together sharing the electrons between them. The resulting bonds establish neutral molecules when all of the orbitals are satisfied. Carbon, for example, has four valence electrons that can be shared with other atoms to form four covalent bonds completing an octet of electrons for carbon. When hydrogen is the other atom, the hydrogen atom is completely satisfied by completion of the $1s^2$ orbital with two shared electrons (Fig. 2–12c). Covalent bonding at an interface can immobilize the molecules involved and drastically change the chemical and physical properties of the interface. Reactions such as this that destroy the interface and/or change the chemical characteristics of the bulk phases are not considered in wettability. Preferential wettability develops where reactions at the liquid/liquid and liquid/solid interfaces do not change the bulk properties of the phases involved.

3. Coordinate-covalent bonds occur where pairs of electrons are shared, as in the formation of a salt with ionic reactions, or in the weak bonding of two neutral molecules. The molecule accepting the electrons acquires a slight negative charge while the donor molecule is left with a slight positive charge resulting in a dipolar complex (Fig. 2–12d).

Aluminum chloride is a Lewis acid that can accept two electrons from a base to complete its outer four orbitals (Fig. 2–11e). In terms of orbital theory, a Lewis acid is any substance that can accept an electron density donated by a base, and a base is a substance capable of donating an electron density to be shared. An acid is an acceptor and a base is a donor of an electron cloud, resulting in a compound made up of a simple addition of an electron acceptor and an electron donor ($A + B \leftrightarrow A{:}B$). The acid/base compound ($A{:}B$) may then go on to perform as an acid or a base (depending on its electron cloud configuration) and thus form larger add-groups [$A{:}B + A' \leftrightarrow A{:}B{:}A'$ *or* $A{:}B + B' \leftrightarrow B'{:}A{:}B$]. For example, water has two acid hydrogens attached to a basic oxygen atom, and these are linked in a 3-D tetrahedral structure in liquid water, Fig. 2–13.

The Lewis acid/base adducts can participate in displacement reactions in which the acid/base groups acts as a unit in a substitution reaction

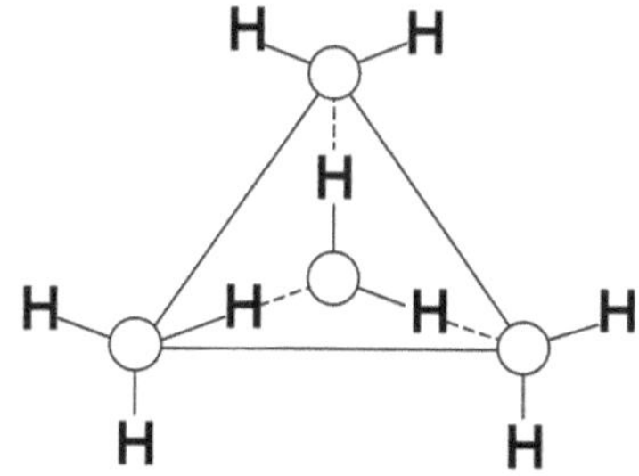

Figure 2–13 *Tetrahedral water hydrogen bonding.*

(where one of the original adducts leaves the group), Fig. 2–14. The behavior of the adduct complex, as an acid or base functional group, is not always obvious from its structure. Most substances are not completely inert because they are bifunctional to some degree. Thus, water can behave as an acid or a base: the oxygen in acetone is an electron donor (base), but it also can act as a weak acedic functional group in a basic environment. There is no doubt that phenol is a strong acid, but the pie-electron cloud of the aromatic ring exhibits some basic functionality. This bifunctional behavior allows self-association that contributes to the adhesive attributes of liquids (their Hamaker constant).

The self-associated liquids are not miscible with inert solvents, such as saturated hydrocarbons, because the inert solvents cannot separate the acid/base complexes, which is required for solution.

In the general description of electron orbits of molecules, the orbitals of the molecule (composed of two or more atoms) are combined as an electron cloud containing the complete valence electrons of the molecule. In some cases the electron density may be greater at one part of the molecule (accumulated around a particular atom) because of a difference in the overall energy distribution, but the individual electrons are not confined to a single atom in the complex. The analysis of the bonding in terms of the Lewis acid/base and acceptor/donor diagrams of electrons as point charges is nevertheless valid for depicting the behavior of reactants. In terms of molecular orbitals, an acid is defined as a particle with empty orbitals (or as having its lowest unoccupied molecular orbital) which are available for completion by transfer or by sharing of electrons. A base is depicted as having electrons that are

$$\text{(a)} \quad \mathbf{A{:}B + A' \rightleftharpoons A{:}B{:}A'}$$

$$\text{(b)} \quad \mathbf{B' + A{:}B \rightleftharpoons B'{:}A{:}B}$$

$$\text{(c)} \quad \mathbf{A{:}B + A'{:}B' \rightleftharpoons A{:}B : A'{:}B'}$$

$$\text{(d)} \quad \mathbf{A'{:}B' + A{:}B \rightleftharpoons A'{:}B'{:}A{:}B}$$

$$\text{(e)} \quad :\underset{\cdot\cdot}{\ddot{\mathbf{I}}}^{+} + :\underset{\cdot\cdot}{\ddot{\mathbf{I}}}:^{-} \rightleftharpoons :\underset{\cdot\cdot}{\ddot{\mathbf{I}}}:\underset{\cdot\cdot}{\ddot{\mathbf{I}}}:$$

$$\text{(f)} \quad \mathbf{AH^{+} + {:}NH_3 \rightleftharpoons NH_4^{\;+}}$$

Figure 2–14 *Acid-base adduct. (a) Acid adduct. An acid (electron acceptor) shares an electron pair of the base portion of an acid-base compound. (b) Base adduct. A base sharing a pair of its electrons with an acid-base compound. (c) Two compounds form an adduct complex molecule. (d) The inverse of (c). (e) Charge transfer to form a neutral salt. (f) Adduct resulting in a complex ion with a charge.*

usually in the highest occupied orbitals of the molecule and are available for initiation of reactions (Luder and Zuffanti, 1961; Jensen, 1980).

The acid/base reactions are principally determined by the perturbations of the outer orbitals as described above, but other ortibals in the molecule also may be perturbed when the reaction takes place. In addition, the acceptor/donor orbitals are not necessarily parts of a single atom or localized between two atoms, as shown by the Lewis dot-structures. Electron donation may range from zero to complete transfer of electrons from a donor to an acceptor; *A* and *B* may represent neutral molecules or ions and the *A-B* bonded molecule is the representation after reaction of the idealized shared electron bonding. The relative strengths of acids, bases, and the bonded products depend on the electronic structure of the atoms and molecules and the local environment (for example, the presence of a solvent); stearic hindrance can become a factor. Thus the nature of the reactions depends on the specific species of particles involved and the nature of the environmental system. Classifications of reactivity are convenient for special descriptions of types of reactions, but have no absolute significance.

2.14 Solvent Behavior

In the condensed state, water forms a self-associated, three-dimensional structure because of hydrogen bonding. Nonaqueous, inert molecules such as hydrogen-saturated hydrocarbons form a loose association through van der Waals attractive forces. If a solute is introduced to a solvent, in order to form a solution the solute must break the self-associated bonding of the solvent by interacting to form discrete ions associated with the solvent molecules (Luder and Zuffanti, 1961).

The cations and anions of amphoteric solutes can act as either an acid (accepting a pair of electrons) or a base (donating a pair of electrons), and will normally be surrounded by solvent molecules. The result is dissociation of the ions in the solvent, Fig. 2–15a.

(a) $$A{:}B + xH_2O \rightleftharpoons [A\cdots H_2O_x]^+ + [B\cdots H_2O_x]^-$$

(b)
$$HCl + 2H_2O \rightleftharpoons [HO\cdot\cdot HCl]^- + H_3O^+$$
$$SO_3 + 2H_2O \rightleftharpoons [HO\cdots SO_3]^- + H_3O^+$$
$$NaOH + H_2O \rightleftharpoons [NaOH\cdots H]^+ + OH^-$$
$$NH_3 + H_2O \rightleftharpoons Na_4^+ + OH^-$$

(c)

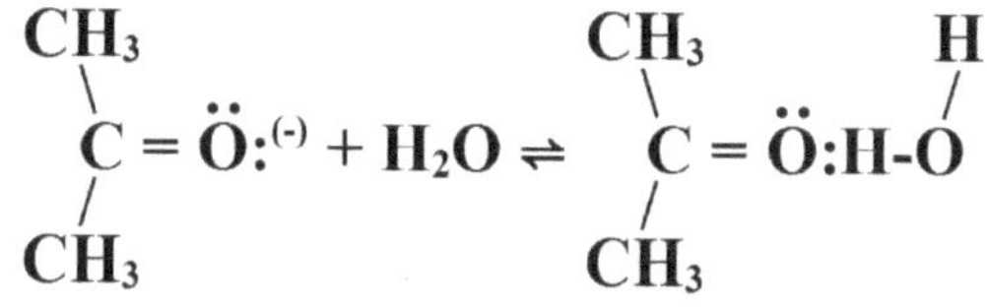

(d)
$$Al^{(+)}(OH)_3^{(-)} + 3H_2O\cdot + 3Cl^- \rightleftharpoons Al^{3+} + 6H_2O + 3Cl^-$$
$$Al^{(+)}(OH)_3^{(-)} + {:}OH^- + K^+ \rightleftharpoons [HO{:}Al(Cl)_3]^- + K^+$$

Figure 2–15 *Solute-solvent behavior. (a) Dissociation of ions in a solvent. (b) The shift in the cation/anion balance of a solution: increase of hydronium or hydoxyl ion concentrate. (c) Formation of adduct complex with water. (d) Amphoteric behavior of a solute.*

The solution of an acid or base in water causes a shift of the cation/anion balance: an acid will increase the hydronium ion concentration, whereas a base dissolved in water increases the hydroxyl ion concentration, Fig. 2–15b. Nonelectrolyte solutes with small charge dipoles and weak acids and bases form simple adduct complexes with the water molecules and are solvated by surrounding water molecules, Fig. 2–15c. Amphoteric behavior of a solute occurs when the molecule can accept or donate a pair of electrons depending on whether the water solvent is acetic or basic, Fig. 2–15d.

Alkenes, aromatic compounds, and other molecules where resonance occurs between electrons associated with double bonds are soluble in water to some degree, ranging from completely inert compounds that have zero solubility to amphoteric solutes with complete solubility in water. The motions of the electrons produce a charge distribution that can enter into hydrogen bonding with water, Fig. 2–16.

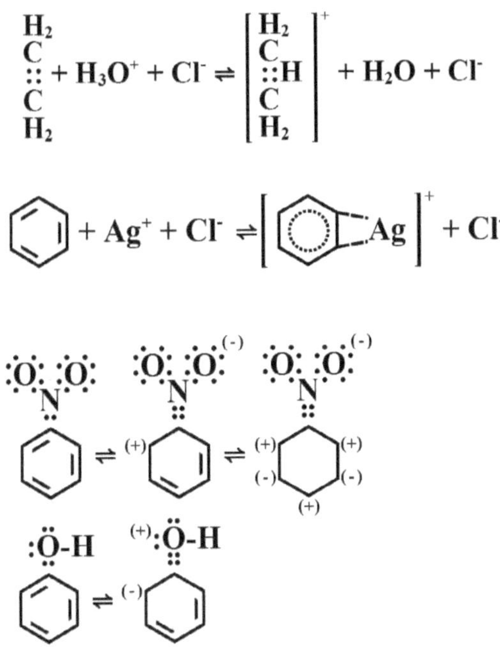

Figure 2–16 *The motions of the electrons in molecules with double bonds (alkenes and aromatics) produce electrical charge distributions that can interact with hydrogen bonds in water and thus promote limited solubility in water.*

Inert (insoluble in water) compounds do not have sufficient charge capacity to overcome the self-association of water molecules. Hence they are excluded from solution, forming a second phase that separates in accord with their density differences. Thus weakly associated molecules depending only on van der Waals forces for cohesion will not be soluble in a strongly associated, amphoteric solvent such as water but they will be soluble in other inert solvents.

2.15 Solid Surface Interactions with Solutions

Because of the strong coulombic adsorption of counterions on the surface, and hydrogen bonding by water and proton donating dissolved molecules, the intrinsic surface charge of the solid is neutralized and the wetting behavior is a relationship between the Stern layer charge and constituents in the bulk fluid. Rocks derived from granite have net negative surface charges (Lewis bases: electron donors), but in contact with water their behavior is that of Lewis acids because of hydrogen bonding. Quartz, sodium and potassium feldspars, and mica exhibit characteristic acetic properties in contact with water. Whereas, rocks derived from extrusive magma generally exhibit basic properties in contact with water, for example, andesite, basalt, olivine, pyroxene, and calcium feldspar. Amphibolite is intermediate between the two groups, exhibiting amphoteric acid/base properties depending on the specific properties of ions in solution and pH (Tiab and Donaldson, 2004).

Multivalent cations adsorb preferentially at the surface with respect to monovalent cations, bonding strongly with negatively charged surfaces. The surfaces may acquire considerably higher concentrations of multivalent cations even when the bulk solution is very dilute. Metals, silicates, and other oxide surfaces in contact with water may be described as hydroxylated oxide surfaces. The resulting hydroxide groups covering the surface will behave like Lewis acids or Lewis bases depending on the properties of the aqueous solution and the strength of the electrical surface charge. Water can interact with the hydroxyl groups at a pH greater than 7, producing a negatively charged surface through nonassociative adsorption of hydroxyl ions, Fig. 2–17:

$$\begin{aligned} &MOH + H_2O \leftrightarrow MO^- + H_3O^+ \\ &MOH + OH^- \leftrightarrow MOH \cdots OH^- \end{aligned} \quad (\text{pH} > 7) \tag{2.67}$$

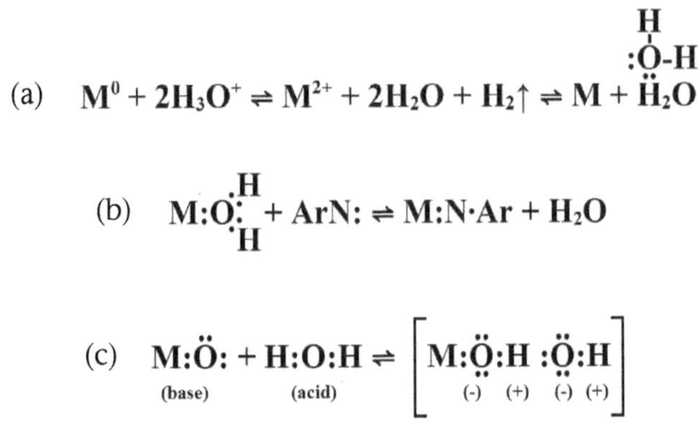

Figure 2–17 *Behavior of metals and electrolytic solutes. (a) Hydration of a metal. (b) Displacement of a water adduct by a base. (c) Formation of hydroxyl groups by metal oxides.*

Reactions with water in acetic solutions result in solid surfaces that are positively charged:

$$\begin{array}{l} MOH + H_2O \leftrightarrow MOH_2^+ + OH^- \\ MOH + H_3O^+ \leftrightarrow MOH \cdots H_3O^+ \end{array} \quad (\text{pH} < 7) \tag{2.68}$$

At a pH where the negative and positive surface charges are exactly balanced, the zeta-potential is zero, and is known as the isoelectric point.

Interaction of the hydroxylated surfaces results in retention of several molecular layers of hydrogen-bonded water molecules. The presence of this bonded water reduces the Lewis acid or base reactivity toward polar organic compounds, such as NSO compounds in crude oils. Thus these surfaces are preferentially water-wet in contrast to acetic or basic aqueous solutions.

Throughout the world, oil is found primarily in sedimentary rocks. There are a few reservoirs in highly fractured igneous rocks but the quantity of oil from these is insignificant compared to the total amount extracted from sedimentary rocks originating from the mechanical and chemical weathering of igneous and metamorphic rocks and the biogenic accumulation of carbonated rocks. Rocks are then the cohesive aggregates of one or more types of minerals. Therefore, it is the surface atomic structure and chemical and physical properties that are of utmost importance to wettability beginning with a general division into those with acetic and basic surface properties.

Minerals crystallize from molten magma of igneous rocks as a function of temperature, as the magma cools. Minerals that have basic properties ($M^{+}{\cdot}OH^{-}$, olivine, pyroxene, and calcium feldspar) solidify first, followed by minerals whose surface properties are amphoteric (having both negative and positive electrical sites) such as amphibole and biotite mica. As the temperature decreases the last minerals that crystallize exhibit acetic properties ($M^{-}{\cdot}H^{+}$, quartz, muscovite mica and, sodium and potassium feldspar) following Bowen's crystallization series (Tiab and Donaldson, 2004). Intrusive rocks crystallize from magma in a subsurface environment and over a longer period of time as the magma cools slowly surrounded by preexisting rocks. This results in vast expanses of coarse-grained granite, diorite, and gabbro that have decreasing amounts of silica and range from acetic (granite) to basic (gabbro) in contact with water.

Extrusive rocks are composed of fine-grained crystals as a result of rapid crystallization on the surface. Particle size and chemical composition are used for their classification. Fine-grained basalt exhibits basic chemical properties, andesite is amphoteric, and rhyolite is an acetic lava.

Metamorphic rocks form at high temperatures and pressure in the subsurface, changing from granite to gneiss, basalt to schist, limestone to marble, etc. (Tiab and Donaldson, 2004). The metamorphic rocks will retain the acid/base properties of their original mineral constituents because most (quartz, feldspars, micas, olivines, etc.) are stable at high temperatures. Foliation of metamorphic rocks produces layers that can have opposing electrical surface properties. Foliation occurs when the tectonic pressure is exerted in a single direction on recrystallizing rocks; this causes long crystals of minerals to accumulate perpendicular to the direction of the imposed pressure (foliation does not occur if the forces are exerted from all sides).

The electrical surface properties of the mineral are a consequence of the exposed elements, their size, and ionic characteristics. Cations are smaller than anions (Table 2–8); hence, steric hindrance from the anions can play a role in surface electrical properties that affect wettability.

Table 2–8 General wettability of carbonate and sandstone rocks reported by numerous authors.

Rock type	Percent water-wet	Percent neutral	Percent oil-wet	Total studied	Reference
Carbonate	8	8	84	25	Trieber et al. (1972)
Sandstone	43	7	50	30	
Carbonate	—	—	—	—	Lorenz et al. (1974)
Sandstone	100	—	—	5	
Carbonate	8	64	28	25	Morrow et al. (1976)
Sandstone	40	33	27	30	
Carbonate	8	12	80	16	Chilingar et. al (1983)
Sandstone	—	—	—		

The sedimentary formations that are oil reservoirs are principally composed of either silicate or carbonate rocks. The silicate surfaces and clay inclusions are primarily acetic, while the carbonates are predominantly basic because of the positive charges of exposed calcium and magnesium ions. Numerous studies of wettability have shown that silicate reservoirs exhibit a range of wettability from neutral to strongly water-wet while the carbonates are neutral to strongly oil-wet. Table 2–9 is a list of wettability measurements of carbonates and sandstones by various authors. The majority of carbonate rocks have been found to be oil-wet and sandstones water-wet. These data, however, must be considered with respect to the properties of the oils that were used in making the measurements because wettability is influenced by constitutents in the oils; even so, the predominant characteristics of the rock surfaces are evident (McGhee et al., 1979).

2.15.1 Clay Swelling and Particle Movement

The planar particles of clay are held together by a combination of (1) negative electric charges generally satisfied by hydrogen and sodium cations at pH ≤ 7, and multivalent cations can bridge between sites on adjacent clay plates inhibiting swelling by displacement of cations and hydration of the surfaces; (2) van der Waals attractive forces, and (3) adhesion between adsorbed organic compounds. The

Table 2–9 Ion exchange properties of rocks (Crocker et al., 1983).

Cations displaced, mg/kg	Bandera	Berea	Coffeyville	Cottage Grove	Noxie	Torpedo
Water Dispalcement						
Calcium	0.85	0.44	0.87	0.97	0.39	1.16
Magnesium	0.09	0.82	0	0.07	0.21	1.00
Potassium	0	0.18	0	0.19	0.06	0.28
Sodium	0.04	0.09	0	0.03	0	0
Iron	0.10	0	0	0	0	0
Lanthnum Displacement						
Calcium	8.96	2.30	2.32	16.40	9.06	19.69
Magnesium	3.03	2.80	0	1.40	0.95	9.58
Potassium	0	0.01	0	0.06	0	0
Sodium	0	0.17	0.72	0.10	0	0
Cation Exchange Capacity						
meq/kg —	11.99	5.28	23.92	17.96	10.01	29.27

dispersion of clay particles can be attributed to reversal of the forces causing adhesion: (1) removal of the cations and subsequent hydration by introduction of dimineralized water or increase of pH with a base, (2) introduction of hydration forces that overcome van der Waals forces, and (3) desorption of bonding organic compounds by hydration. High rates of fluid flow (near a wellbore, for example) can mechanically dislodge clay particles resulting in permeability decrease as the moving particles bridge large pores and block pore throats of small pores (Donaldson et al., 1977).

Veley (1969) found that the amount of permeability damage in cores containing water-sensitive clays becomes progressively less as the valance of cations added to water increases from monovalent to trivalent cations. Monovalent cations are not held as firmly in the Stern double electric layer of negatively charged surfaces as multivalent cations. Furthermore, monovalent cations are readily exchanged by the introduction of multivalent cations.

Ion exchange is a stoichiometric diffusion-controlled process. Electroneutrality of the exchange surface is maintained because every ion removed from the surface is replaced by an equivalent amount of another ion having the same charge (cations replace cations, and anions replace anions). This differs from adsorption where an ion or molecule is attached to an adsorption site without replacing another species. Selectivity for the exchange of one ion at the expense of another is governed by three principal conditions: (1) the size and valence of the ions; ions with the greatest valence will replace those having a lower valence, thus calcium ions will replace hydrogen and sodium ions on a surface; (2) steric hindrance; large ions may be excluded from the surface; and (3) chemical reactions of the ions with the surface.

The ion-exchange capacity of a given surface is a constant for the particular sample because of the stoichiometric relationship that exists between displaced and displacing ions. The value may differ from one test to another if steric hindrance is a factor or if the environmental conditions (pH, etc.) of the test are changed. This is a valuable number when determined under specifically controlled conditions because it gives a direct measure of the number of exchangeable sites and can be used for comparisons of samples. Methods of defining ion-exchange capacity differ among authors; the ion-exchange capacities presented in Table 2–9 are expressed as miliequivalents of exchangeable ions per kilogram of dry solid. The samples were first flooded with water and the effluent was analyzed showing that only a very small amount of monovalent cations (calcium and magnesium) were displaced by the demineralized water. When cation exchange was conducted using trivalent lanthanum, a considerable amount of calcium and magnesium were removed from the rock surface. The Bandera rock is cemented by siderite, but an insignificant amount of iron appeared in the water effluent and none appeared when the core was flooded with lanthanum chloride. The amount of cations displaced can be used to calculate the charge (C/m^2) on the surface of the rocks, Ex. 2.8.

Veley (1969) reported only a 2-3% decline in the permeability of Berea cores when they were contacted with calcium chloride brine followed by sodium chloride brine and distilled water. When the rock was contacted by sodium chloride brine and followed by distilled water, a permeability decrease of approximately 90% occurred because of dispersion of clay particles.

Multivalent cations are hydrolyzed by water at pH > 7. The resulting complex hydrate ionizes with an increase of hydrogen ion concentration that lowers the pH of the solution toward neutral:

$$M^{+y} + xH_2O \rightleftharpoons [M(H_2O)_x]^{+y} \rightleftharpoons \left[M(OH)_2(H_2O)_{x-y}\right] + yH_3O^+ \quad (2.69)$$

The hydrates remain stable in a basic environment; an increase of pH, however, will increase the size of the hydrates until precipitation of the metal hydrates takes place. A decrease of pH will destroy the hydrates and stabilize the clays; however, strong acid may dissolve grain-cementing agents resulting in the entrainment of particles in a flowing fluid causing subsequent blocking of pores.

2.15.2 Crude Oil/Solid Surface Interactions

The classification of naturally occurring petroleum is very vague and may depend only on the general characteristics of a product from a specific location. The following terms are drawn from discussions at various conferences, among them the UNITAR Conference, Bowman (1981): Light oils: API° > 20; heavy oils: 10 < API° ≤ 20 and μ < 10 Pa-s (10,000 cP) at 15°C; Bitumen: API° < 10, μ > 10 Pa-s at 15°C. Furthermore, petroleum from different locations and variable mixtures of classes of compounds: paraffins or saturates (hydrogen-saturated carbon compounds), naphtenes (hydrogen-saturated cyclic carbon compounds), aromatics (partially unsaturated cyclic compounds), and asphaltenes or polar compounds (polynuclear aromatic compounds containing variable amounts of nitrogen, sulfur, and oxygen, or NSO compounds). The asphaltenes are further defined as a fraction of petroleum that precipitates upon dilution with light alkanes such as *n*-pentane. The mixed asphaltenes range in molecular weight from 10^3 to more than 10^6.

The oil-wetting properties of petroleum are attributed to (1) precipitation as asphaltenes because of poor solubility, followed by adsorption or electrostatic reaction with the rock surface, and (2) electrostatic reactions of NSO compounds at the oil-water or oil-solid interfaces with electrostatic charges on the rock surfaces. The tendency of polar compounds to migrate to the interfaces of oil depends on the aliphatic/aromatic ratio of the oil; thus the oil-wetting tendencies of a specific oil is a function of both the insolubility and concentration of polar compounds in the oil (Donaldson and Crocker, 1980; Morrow, et al., 1994; Buckley and Liu, 1998; and Legens, et al., 1999).

The Lewis acid/base interactions, which are pH dependent, are the principle reactions occurring between a water film and the oil-water-rock interfaces. As long as a water film remains between the oil and rock, the rock surface will remain water-wet, but if the film collapses (because of diffusion of polar compounds from the oil-water interface through the water film to the rock surface, or the influence of DLVO forces) polar compounds can adhere to the rock by physical adsorption and/or coulombic reactions leading to a change of wettability from water-wet to oil-wet. The presence of multivalent cations in the water film can promote oil-wetting by partial reaction with polar compounds at the oil-water interface: polar$^{(-)}$-M$^{(++)}$-rock$^{(-)}$. This reaction and diffusion is time dependent and thus responsible for the gradual rate of attaining a stable wetting condition after a rock sample is saturated with water and oil (aging time).

Limestone and dolomite rocks have both negative and positive surface charges because of random exposure of oxygen (with negative orbital clouds) and calcium/magnesium atoms with positive charges. If the surface is in contact with water, hydronium ions are attached to the negative charges and the hydroxyl ion to the positive metallic ions producing a water-wet surface. When an acetic asphaltene compound arrives at the site, its chemical potential for coulombic reaction is greater than the associated water molecules and displacement takes place. The van der Waals and electrostatic forces are both negative, therefore the approach of a basic polar compound to the surface is increasingly attractive without an energy barrier to overcome for a reaction to take place.

As a general rule, oils with an acid/base ratio greater than 1.0 will exhibit preferentially water-wet conditions on silicate-base rocks because the abundance of negative ions (on the organic polar ion) at the oil interface will cause repulsion at the oil-rock interface, resulting in contact angles less than 90° and USBM wettability indices greater than 0.0. If the acid/base ratio is less than 1.0, there will be more positive charges at the oil interface (on the organic polar ion) and thus the oil will exhibit more oil-wetting tendencies on rocks with negative surfaces, resulting in contact angles greater than 90° and USBM wettability indices less than 0.0. Buckley and Liu (1998) found that the reactivity is pH controlled: Silicate-based rocks are negative when the pH is greater than 2, and thus basic; positively charged NSO compounds will readily interact with the surfaces; carbonate rocks are positively charged below a pH of 9.5 and thus the acetic, negatively charged NSO compounds will react with the surface. (Pashley and Israelachvilli, 1984; Basu and Sharma, 1996; Dubey and Doe, 1993; Skauge et al., 1999; Austad and Standes, 2003; and Alkafeef and Smith, 2005).

The number and complexity of the water-oil-rock interactions is so great that it is more probable that many reactions are involved simultaneously at multiple sites on rock surfaces to produce the overall average wettability character of the particular water-oil-rock system (Amott and USBM wettability indices provide the average preferential wettability of the fluid-rock systems). The aliphatic/aromatic ratio of oils controls their solvent properties toward polar compounds, and any change of temperature and pressure (near a wellbore, for example) can change the ratio and induce precipitation of both asphaltenes and high molecular weight paraffins. The overall chemical properties of the polar compounds and brines control the oil-water and oil-rock electrostatic interactions. Direct oil-rock contact, leading to adsorption and chemical interaction, is not an important phenomenon because of the ubiquitous presence of water.

2.16 Example Problems

Example 2–1

Derive Bolzmann's constant and calculate the kinetic energy of an ideal gas at 25°C.

One mole of ideal gas occupies a volume of 22.414 liters (0.022414 m^3) at one atmosphere of pressure (101.325 kPa) and 0°C. (273.15 K) Therefore, the gas constant is:

$$R = \frac{PV}{nT} = \left(\frac{101315}{m^2}\right)\left(\frac{0.022414\ m^3}{(1)(273.25\ K)}\right) = 8.3143\frac{Nm}{mol\ K} = 8.3143\ J\,mol^{-1}\,K^{-1}$$

The average kinetic energy of a mole of an ideal gas is $\left(\overline{KE} = \frac{3}{2}\frac{RT}{n}\right)$, and Boltzmann's constant is the average kinetic energy of a single molecule with respect to temperature:

$$k_B = \frac{3}{2}\frac{R}{nK} = \left(\frac{8.3141\,J}{mol\,K}\right)\frac{1}{(6.022*10^{23}\ molecules/mole)} = 1.3806*10^{-23}\ JK^{-1}$$

The average kinetic energy of a single molecule at 25°C (298 K) is:

$$\overline{KE} = \frac{3}{2}k_B T = \left(\frac{3}{2}\right)(1.3806*10^{-23})(298) = 4.11*10^{-21}\ J$$

Example 2-2

In a vacuum (dielectric constant = 1.0) and in water (dielectric constant = 78.5):

a. Calculate the binding interactive force between ions of a monoionic molecule if the separation between the ions is equal to 0.4 nm. From Eq. 2.7:

$$F(r) = \frac{z^+ z^- e^2}{(4\pi \epsilon_o)\epsilon r^2} = \frac{(+1)(-1)(1.602*10^{-19})^2}{1.1126*10^{-10}(1)(0.4*10^{-9})^2}$$
$$= -1.44*10^{-9}\ N/molecule\quad [in\ vacuum]$$
$$= -1.86*10^{-11} N/molecule\quad [in\ water]$$

b. Calculate the amount of energy required to separate the ionic molecule in a vacuum and in water. From Eq. 2.12:

$$E_{(r)} = \frac{z^+ z^- e^2}{4\pi \epsilon \epsilon_o r} = \frac{(+1)(-1)(1.602*10^{-19})^2}{(1.1126*10^{-10})(1)(0.4*10^{-9})}$$
$$= -5.77*10^{-19}\ J/molecule\ = -348\,kJ/mol\quad [in\ vacuum]$$

$$= -7.21*10^{-21}\ J/molecule\ = -4.34\ kJ/mol\quad [in\ water]$$
$$\{where\ Avogadro's\ Number = 6.022*10^{23}\ molecules/mol\}$$

Example 2–3

Using Eq. 2.12, calculate the first Bohr radius of an electron where the interaction energy is equal to $2h\nu$:

$$r_b = \frac{e^2}{2(4\pi\varepsilon_o\varepsilon_r)h\nu} = \frac{(1.602*10^{-19})^2}{25.12\left(8.0854*10^{-12}\frac{C^2}{Jm}\right)\left(60626*10^{-34}Js\right)\left(3.3*10^{15}/s\right)}$$

$$= 00053*10^{-9} = 0.053\ nm$$

Example 2–4

Calculate the Debye length for a 0.2M solution of calcium chloride:

$$\frac{1}{\kappa} = \frac{B}{M} = \frac{0.152}{\sqrt{0.2}} = 3.4\ nm$$

Example 2–5

Determine the zeta potential for a rock saturated with brine where: $R_o = 1.5$ ohm-m, $\Delta P = 70*10^3\ Pa$, $\mu = 8*10^{-4}\ Pa{\cdot}s$, $\varepsilon_r = 70$, and $\Delta\psi = 36\ \mu V$.

$$\zeta = \frac{4\pi\left(8*10^{-4}\right)\left(36*10^{-6}\right)}{\left(70*8.854*10^{-12}\right)(1.5)\left(70*10^{3}\right)} = 5.6mV$$

Example 2–6

Calculate the Hamaker constant for two bodies of dodecane (1 and 2) interacting across water (3). Eq. 2.40, reduces to the following for interaction of two identical phases across a third phase because $\varepsilon_1 = \varepsilon_2$ and $n_1 = n_2$.

$$A = \frac{3}{4}kT\left(\frac{\varepsilon_1 - \varepsilon_3}{\varepsilon_1 + \varepsilon_3}\right)^2 + 0.133h\upsilon\frac{\left(n_1^2 - n_3^2\right)^2}{\left(n_1^2 + n_3^2\right)^{\frac{3}{2}}}$$

Where: $\varepsilon_1 = 2.01$, $\varepsilon_3 = 80$, $n_1 = 1.411$, $n_3 = 1.333$, $h = 6.626*10^{-21}$, $\nu = 3.0*10^{-21}$

$$A = 2.79*10^{-21} + 1.65*10^{-21} = 4.45*10^{-21}$$

Example 2-7

a. Determine the ratio of the number density (ions/m^3) of ions with respect to the average density in the bulk liquid as a function of distance from an oppositely charged stationary surface.

$$Ln\left(\frac{\rho_{n(x)}}{\rho_{n,\infty}}\right) = -\frac{ze^2}{4\pi\varepsilon\varepsilon_o xkT}$$

$$z = -1;\ e = 1.602(10^{-19})C;\ \varepsilon = 10;\ \varepsilon_o = 8.854(10^{-12})CJ^{-1}m^{-1}$$
$$kT = 4.114(10^{-21})J;\ x(meters)$$

$$\frac{Ln\left(\rho_{n(x)} / \rho_{n,\infty}\right)}{x(10^{-9})m}\begin{bmatrix}5.61\\ \hline 1\end{bmatrix}\begin{bmatrix}1.40\\ \hline 4\end{bmatrix}\begin{bmatrix}0.7\\ \hline 8\end{bmatrix}\begin{bmatrix}0.47\\ \hline 12\end{bmatrix}\begin{bmatrix}0.28\\ \hline 20\end{bmatrix}\begin{bmatrix}0.19\\ \hline 30\end{bmatrix}\begin{bmatrix}0.11\\ \hline 50\end{bmatrix}$$

b. Calculate the calcium ion concentration at a surface where the surface potential is 200 mV, temperature is 298K, and bulk concentration of calcium ions is 0.01 mol/dm^3.

From Eq. 2.50, at the surface where $x = 0$, the surface values of $\rho_{n(x=0)}$ and ψ are related to the bulk ionic concentration if $\rho_{n,\infty} = [Ca^{++}]_\infty$; thus:

$$\left[Ca^{++}\right]_s = \left[Ca^{++}\right]_\infty e^{(+z\psi/25.69)}$$

where kT/e = 25.69 mV at 298 K

$$\left[Ca^{++}\right]_s = \left(\frac{0.01\ mol}{dm^3}\right)e^{\left(\frac{2*200}{25.69}\right)} = 2.75\left(\frac{0.01}{0.001}\right) \approx 0.03\ mol/dm^3$$

$$\rho_s = \frac{0.03 mol}{dm^3} = \frac{(0.03\ mol)(6.022*10^{23}\ ions/mol)}{0.001\ m^3} \approx 1.8*20^{25}\ ions{\bullet}m^{-3}$$

Example 2-8

Assume that the amount of calcium displaced from a rock by ion exchange is 10 meq/kg. Calculate the surface charge on a rock that has a surface area equal to 2 m^3g^{-1}.

An electrical charge is equal to $1.602*10^{-19}$C. One charge per square meter is then $0.16*10^{-18}$ C/m^2; or C/m^2 = charge/0.16 nm^2 = $6.25*10^{18}$ charges/m^2.

$$\frac{mg\,Ca^{++}}{kg} = \frac{0.56g}{kg} = \frac{kg}{2000m^2} = \frac{2.8*10^{-4}g\,Ca^{++}}{m^2}$$

$$\left(\frac{2.8*10^{-4}g}{m^2}\right)\left(\frac{mol}{112\,g}\right) = \left(\frac{2.5*10^{-6}mol}{m^2}\right)\left(\frac{6.022*10^{23}\,ions}{mol}\right)\left(\frac{2\,\text{Charges}}{ion}\right) = \frac{3*10^{18}\,\text{Charges}}{m^2}$$

$$\sigma_c = \left(\frac{3*10^{18}\,\text{Charges}}{m^2}\right)\left(\frac{m^2}{6.25*10^{18}}\right) = 0.48 Cm^2$$

CHAPTER 3

Wettability and Production

3.1 Introduction. Origin and Accumulation of Hydrocarbons

The influence of the relative wetting of water and hydrocarbons on geologic materials begins with the migration of hydrocarbons from source-rocks to stratigraphic and structural traps. Therefore, a brief review of the theories that have been developed to explain the genesis of hydrocarbons and their accumulation in geologic structures is presented as an introduction to the effect of wettability on production of oil.

The origin of petroleum begins with the accumulation of organic matter in lacustrine and shallow marine environments together with fine-grained sediments transported by surface waters and winds. In marine environments, the deposition of carbonate sediments can occur along with the organic matter and algae. Oxygen is depleted in the lower zones where anaerobic bacteria begin the conversion of the organic matter into precursors of petroleum hydrocarbons. If the region of accumulation is subjected to tectonic forces that produce subsidence and burial of the organic mass, the gradual increase of temperature and pressure transforms the material into kerogen. Kerogen is a waxy, bituminous material that yields petroleum when it is heated.

It is generally conceded that the formation of hydrocarbons is the result of thermal alteration of plant and animal organic matter in a reducing aquatic environment (principally shallow marine and lacustrine environments with fine-grain sediments) and microbial conversion of organic substances (kerogen) found in source rocks, usually shales and carbonates. Carbonate source rocks originate (1) in lacustrine environments where they are associated with shale, (2) in marine environments where some of the calcareous sediments have undergone dolomitization, and (3) are associated with large influxes of clastic sediments. Any oxygen present in the environment would be consumed by oxidation of hydrocarbons to water and carbon dioxide. In carbonates, algae are important contributors to the organic content of the source rocks. The algae provide sapropelic (oil-forming) organic matter, and bacteria help maintain axoic conditions in the depositional environments. The pore size distribution and heterogeneity is so great in many carbonate reservoirs that oil recovery is principally controlled by the variations of overall permeability. The difference in the fluid-flow characteristics between carbonates and sandstones is largely because of permeability heterogeneity and the preferential oil wetting tendencies of carbonates.

During the early stages of diagenesis, bacteria present in the shallow sediments produce methane and small amounts of light hydrocarbons. In time the mixture of fine sediments and organic material is gradually buried where compaction and increasing subsurface temperature change the mass into a kerogen/clay organic mixture that has no porosity or water. Catalytic reactions (by minerals in the sediments), pressure, and thermal stress from tectonic events and deeper burial convert the kerogen into gas and liquids that subsequently accumulate in hydrocarbon traps. The rate of transformation from kerogen to petroleum increases as the environmental temperature increases during burial of the sediments. The transformation to petroleum begins to take place at a depth of about 3,500 meters (about 95°C). Expansion of the oil as maturation takes place causes expulsion of oil from the source rocks into migrating pathways to hydrocarbon traps.

3.2 Petroleum Traps

There are two general classifications of petroleum traps: structural and stratigraphic. Structural traps are formed when porous sedimentary layers are altered by tectonic movements that produce (1) folding of the sediments, (2) up-dip sealing of anticlines by faults, and (3) intrusion of salt diapers that cause inclination of the porous sediments and sealing of the up-dip region by the penetrating salt (Fig. 3–1). These are the principal structural traps, but variations in the composition of the sediments and the forces acting on them produce a large number of more complex structures that serve as petroleum reservoirs (Link, 1982; and Chapman, 1983).

Stratigraphic traps develop because of their unique depositional properties: A porous sedimentary accumulation can be deposited followed by (1) cessation of the accumulation of sediments, (2) burial because of subsidence or other local environmental changes, and (3) uplift and erosion of a porous sediment followed by deposition of a different, impermeable sediment on top resulting in an unconformity. These occur for lacustrine deposits, ancient river channels, marine shoreline sand bars, shallow marine coral reefs, and precipitation of carbonates. Thus there are many types of stratigraphic traps. Some of the principal traps are illustrated in Fig. 3–2.

The majority of petroleum reservoirs are found in porous clastic sediments and carbonate structures that were originally deposited as horizontal or gradually inclined layers, forming a depression or zone of massive deposition of clastic materials carried by rivers and winds.

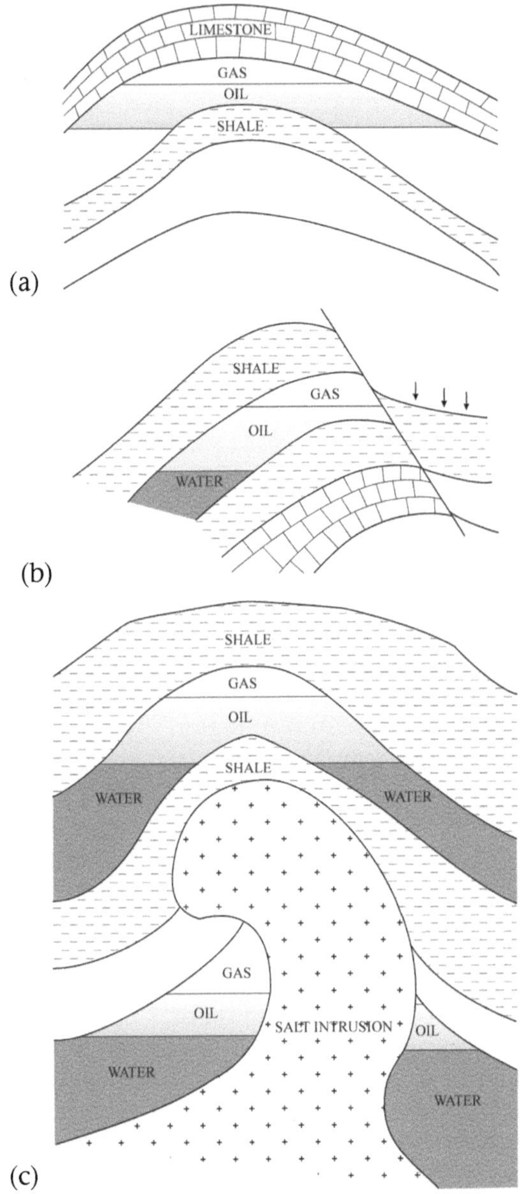

Figure 3–1 *(a) Folding of layered sediment caused by tectonic compression forming an anticline petroleum structural trap. (b) Up-dip sealing of an anticline by a fault to form a structural trap for accumulation of oil. (c) Intrusion of salt diapers produce structural, folding, traps for accumulation of oil.*

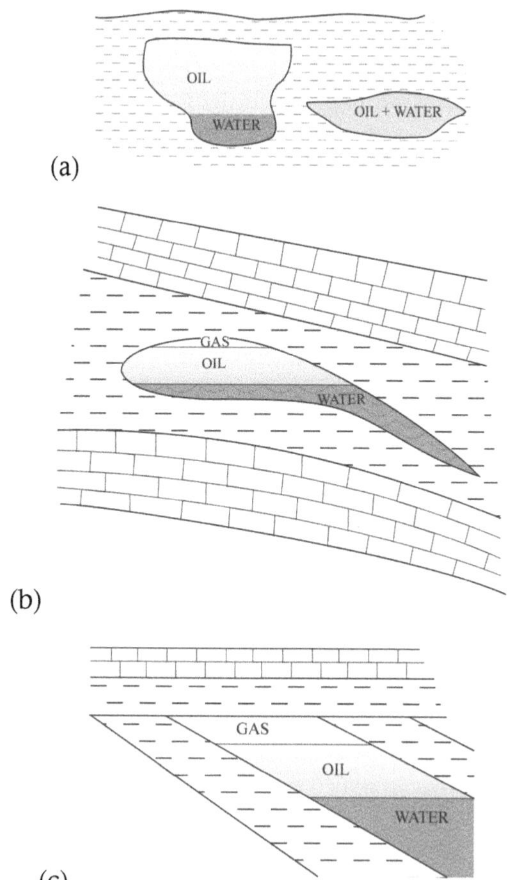

Figure 3–2 *(a) Cross section of a channel-fill stratigraphic trap formed by filling of a stream bed with porous clastic sediments which are later surrounded by impermeable sediments represented as shale. (b) Pinch-out of sand with a mass of shale forming a lense-type stratigraphic for gas and oil. (c) Unconformity stratigraphic trap formed by erosion of an anticline followed by subsidence and subsequent deposition of impermeable sediments.*

The earth surface, however, is a dynamic system that is gradually moving and undergoing deformation in response to tectonic stresses and varying climatic changes. Simultaneously, the sediments undergo diagenetic changes such as cementation of grains by clays and precipitation of minerals and other materials between the grains. The grains are compacted by burial under other accumulating sediments some of which form impermeable layers of shales, carbonates, and other impermeable materials. While these changes are taking place, the dynamic stresses of tectonic motions, overburden pressures, and internal fluid pressures all contribute to the deformation of the sedimentary layers from movement and compaction of individual grains to folding and faulting of the sediments over hundreds of kilometers. Structural traps originate from the faulting of large blocks that provide seals along the fault plane, folding of formations that provide zones for accumulation of hydrocarbons at the crests of anticlines, and the intrusion of salt domes that occurs when overburden pressures on a large deposit of plastic salt are great enough to push diapers of salt upward into the overlying sedimentary layers, providing anticline-like structures and faults sealed against the salt dome. Stratigraphic traps are formed when the permeability of overlying and adjacent geologic structures undergo changes that destroy their permeability. Sand lenses (deltaic and lacustrine sand bodies) and reefs can become surrounded by shale-forming deposits or biologically produced impermeable layers (Link, 1982; Chilingarian et al., 1995; Lyons and Plisga, 2005).

3.3 Primary Migration

The generation of gas and liquid hydrocarbons within the kerogen-clay mass increases the volume, and thus the internal pressure acting on the fluids (together with the temperature and overburden stress) create fractures in the matrix that initiate primary migration of the fluids to adjacent porous structures. The source beds form in low energy depositional environments with little or no movement of fluids during the deposition of organic matter, thus the source rocks are usually composed of fines, silt, and clay. Therefore, the source beds have a very small pore size distribution and consequently the fluids moving out are probably in a dispersed state and must accumulate into a continuous phase for movement out of its initial storage adjacent to the source. Continuous phase flow is the principal flow mechanism of oil out of the source rock driven by a high pressure difference brought about by compaction and expansion of the volume of kerogen during maturation as it is transformed into oil; this transformation takes place for oil between 100°C and 150°C and gas formation occurs from 150°C to 200°C. During secondary migration from the source rock to traps, the hydrocarbons become associated with brine and wettability. Capillary pressure and relative permeability influences become important aspects of the secondary migration to confining reservoirs (Jones, 1980; Tissot, 1984).

3.4 Secondary Migration

Continued changes of fluid pressure and fracturing of the host sediments by mechanical stress from tectonic changes, together with the formation of porous and permeable pathways for movement of the water and hydrocarbons through an inclined porous structure, result in secondary migration. The fluids move from the source rocks to geologic traps and areas at or near the surface. Once hydrodynamic equilibrium is established by confinement in a reservoir or pressure release near the surface, migration stops, resulting in an accumulation recognized as a gas and/or petroleum reservoir. Some of the accumulations reach the surface and become static tar pits, while others have sufficient energy to allow slow seepage forming tar pits; however, any hydrocarbons exposed to the atmosphere will be lost over time by oxidation and microbial action unless subsidence and reburial of the oil sediment occurs.

The oil found in the Little Buffalo Basin, Wyo. (U.S.) has been referred to as *recycle oil* because the oil trap was uplifted to the surface where it underwent rapid erosion with redeposition of the eroded sediments, associated water, and oil into the basin. While the oil was briefly exposed to the surface during erosion of the original reservoir, bacteria and evaporation of light hydrocarbons removed paraffins and low molecular weight fractions, leaving an asphalt-base, 20°API oil (Emmet et al., 1969). The reservoir that developed is laminated and cross-bedded with a wide grain size distribution. Secondary cementation or pore-filling occurred in the upper portion of the reservoir from infiltration of carbonates and anhydrite.

The movement of water along with the hydrocarbons during the course of migration will generally be segregated by interfacial tension and density differences. But water and the effects of wettability are present throughout the formation during migration and accumulation of gas and oil. Several mechanisms are responsible for the transport of hydrocarbons from source rocks to traps. Diffusion of hydrocarbons composed of fewer than seven carbon atoms has been shown to occur, resulting in accumulation and loss of the hydrocarbons if they eventually diffuse through the overburdened structures to the surface. Solutions of hydrocarbons, especially aromatics such as benzene and toluene, in moving groundwater can occur until the hydrocarbons are released as the fluid pressure diminishes or accumulation takes place at a structural or stratigraphic trap. Expansion of dissolved gas can furnish the required energy for transport if the migration path is tilted to provide gravity and pressure differences.

Oil also may travel along the interface between an impermeable shale and sandstone that becomes tilted sufficiently for the isolated oil globules to join and form long filaments that dominate the migration (Baker, 1959; Gussow, 1968; Hunt, 1984).

Secondary migration probably begins when a sediment containing hydrocarbons is tilted, setting into motion the various mechanisms for transport up-dip, such as differential stress gradients from differences in elevation, gas pressure, and buoyancy of oil which are sufficient to overcome the pore-throat capillary pressure retaining the oil, etc. When a trap is encountered, if free gas is associated with the moving hydrocarbons it will accumulate at the top of the reservoir, displacing water until an irreducible saturation of water is attained. Oil accumulation below the gas cap occurs with the displacement of water until the trap is filled with hydrocarbons and additional oil arriving from the migrating path displaces oil at an up-dip spill-point of the trap or continues along the bottom past the spill-point to enter a trap further up-dip. Migration may occur along several narrow paths joining to form larger ones in major pathways to up-dip traps. As the traps fill, hydrocarbons will continue up-dip until they are all accumulated in traps or passed on to an outcrop. Many combinations of migration and accumulation patterns occur, filling adjacent traps as well as those that are along the up-dip path of migration (Gussow, 1968).

Oil may enter a trap before the arrival of gas, either from release of dissolved gas or from later gas migration arriving from another source. If gas displaces oil from such a gas cap, a residual oil saturation will remain in the gas cap. This sequence of events could result in an oil-wet zone in the gas cap from adsorption of the oil component on the rock matrix as the gas displaces water and the lighter components of the oil. If the reservoir is buried deeper and the gas is forced into solution in the oil by the increasing fluid pressure, oil will once more occupy the former gas cap which has become an oil-wet zone.

3.4.1 Buoyant Forces

The transition zone (from S_{wi} to S_{or}, Figs. 3–5) of a reservoir can vary in wettability from the oil-wet zone in the former gas cap to neutral or water-wet as the free water level (FWL) is approached. The combined theories of accumulation of oil in traps without a gas cap for carbonate reservoirs is applicable to any geologic system where upward migration of oil depends principally on buoyant forces. In some cases, as oil begins to accumulate, it enters the larger pores of the structure first, invading smaller pores as the accumulation of oil and buoyancy increases. At this point, the pore-size distribution has an important role in establishing the final preferential wettability of the reservoir. As the larger pores become more oil-wet from adsorption of surface active compounds or asphaltene-type compounds of the walls, they expel more water from the pores by imbibition of oil which is facilitated as the pores become more oil-wet. Thus the large pores become oil-wet and filled with oil.

Entry of oil into smaller pores requires greater buoyant pressure leading to partial displacement of the water originally filling the pores. This leads to conditions of fractional and mixed wettability. Finally, at an even smaller pore size the threshold capillary pressure (required for drainage of water) is too great for entry of oil; all pores with entry sizes less than this critical size for oil entry will remain filled with water. The nature of the pore-size distribution in the rocks where oil is accumulating thus exerts a strong influence on the state of wettability that develops within the oil trap. A large number of small water-filled pores distributed throughout the matrix of larger pores can be surrounded by oil-wet regions, resulting in macroscopic mixed wettability of the reservoir. In addition, water is not present as a continuous phase at the top of the reservoir and is thus immobile. Moving down from the top of the reservoir, the amount of water in isolated pockets increases until it becomes mobile within the transition zone. Therefore, the final preferential wettability of the reservoir can vary from strongly oil-wet at the top, to neutral-wet in the transition zone grading, to water-wet at the bottom, Fig. 3–3 (Morris and Wieland, 1963).

The mineral composition of the reservoirs has a vital role in establishing the preferred wettability of the reservoir because the surfaces of sandstones exhibit wettability tendencies ranging from neutral to strongly water-wet, and carbonates have been found to range from neutral to strongly oil-wet. Thus the grading of wettability shown in Fig. 3–3 will be more complex, but in general the upper portions of thick reservoir will be more oil-wet grading to a water-wet zone near the water-oil contact.

Jerauld and Rathmell (1997) found that the Prudhoe Bay reservoir in Alaska varies in preferential wettability from mixed-wet near the top of the structure to water-wet near the water-oil contact. The mixed wettability had areas that were oil-wet and others that were water-wet

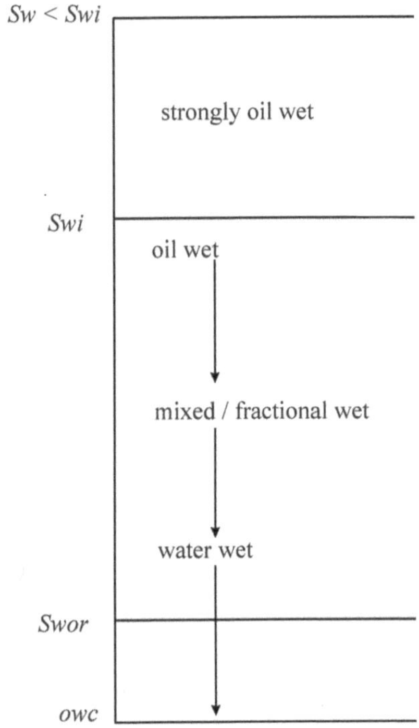

Figure 3–3 *Gradation of wettability that has been found in many sandstone and carbonate reservoirs.*

within the pores and films of oil coating connecting portions of multiple pores. They found that kaolinite clay, siderite ($FeCO_3$), and pyrite (FeS_2) present at the pore surfaces were all preferentially oil-wet and contributed to the mixed wetting character of the reservoir by their distribution within the matrix.

Crude oil–containing polar compounds and high molecular weight polynuclear aromatic compounds, such as asphaltenes, can exert a controlling influence by deposition of films of organic materials on pore walls, rendering the surfaces oil-wet. Reisberg and Doscher (1956) observed the deposition of thick films on glass and quartz surfaces in contact with oil from the Ventura oilfield in California.

Thus the reservoir could have vertically spaced dual wettability at discovery, and production from the two zones would behave quite different from each other. Water breakthrough will occur earlier in the oil-wet zone and the oil production rate will be low; in the water-wet zone, water breakthrough will occur after considerable production of oil at a rapid rate, but then the rate of production will rapidly decrease along with an increase in the water/oil production ratio.

Many changes in the composition of gas and oil, pressure and temperature of the reservoir, depth of burial, etc., can occur after accumulation in a trap as a result of the continual churning of sediments in the geologic time-frame because of tectonic motions. These changing situations, as well as diagenetic changes within the sediments, result in various degrees of preferential wetting being established over the long geologic time periods that follow the origin, migration, and accumulation of petroleum reservoirs (Chilingarian and Yen, 1978; Chapman, 1983; Hunt, 1984).

3.4.2 Capillary Forces

There is a large influence of capillary pressure over the migration of oil. Consider the influence of capillary pressure on an isolated drop of oil in a pore with an opening at the bottom of 50 μm and the top equal to 10 μm (Ex. 3.1). The capillary pressure largely exceeds the buoyant pressure acting on the drop of oil; thus the capillary forces will hold isolated drops of oil in place. The problem that arises in laboratory experiments is that thousands of pore volumes of water can be used for waterflood displacements of oil, but still there will remain a considerable residual oil saturation. Continued flooding with brine displaces oil at a diminishing rate in minute quantities by a viscous drag process recognized as film drainage. The oil relative permeability curves indicate displacement may continue toward zero oil saturation if the passage of water continues.

Secondary migration often occurs over long distances (25–200 km) leaving only trace amounts of hydrocarbons in the porous conduit. The movement of water in the conduit is slow and is certainly not sufficient for the passage of many pore volumes of water for displacement of oil. This leaves the problem of determining the mechanism of oil migration that moves the oil at almost 100% displacement efficiency. Displacement by gas will leave an even larger practical residual oil. The conduit for transport of the oil from the source rock is not flooded by the passage or transport of large volumes of gas or water. The mechanisms of transport are apparently only differential gas pressure, hydraulic pressure differences developed from tectonic movement, and buoyancy, but these will leave a residual oil saturation in the porous conduit along the path from the source to the reservoir. Buoyant forces are insufficient to move residual oil because capillary pressure phenomena will trap petroleum in small pores where extremely large pressure differences are necessary to displace the petroleum. Hydrodynamic forces are insufficient because there is not enough water passing through the conduit. Salathiel (1973) plotted the diminishing residual oil saturation of an East Texas field core as a function of pore volumes of water injected. He discovered that the relationship described a straight line on log-log paper for injection of almost 5,000 pore volumes, expressed as:

$$\bar{S}_o = 42 * W_i^{-0.183} \tag{3.1}$$

By making the broad assumption that Eq. 3.1 remains stable until all of the oil has been produced, a gross estimate of the number of pore volumes required for complete removal of the residual oil can be obtained by taking the logarithm of both sides and solving for the limit of W_i as $\bar{S}_o$ approaches zero. The enormous amount of injected water required is approximately 900 million pore volumes. The linear flow of this much water through a permeable conduit from a source rock to a hydrocarbon trap is clearly not possible. Other methods for transport that have been proposed are: (1) movement of oil through fractures along the passage from the source rock to the reservoir, but this would still leave behind residual oil in the blocks of the matrix; (2) transport in a dispersed phase in water or in solution, but this leaves the question of how a large volume of water is moving through the conduit carrying small quantities of hydrocarbons that eventually accumulate at some stagnant zone; (3) movement through a static inclined path such as an anticline; and (4) there are other theories of secondary transport but none of them address the problem of a remaining residual oil saturation (Cartmill and Dickey, 1970; Bonham, 1978; Welte and Yukler, 1981; Bishop et al., 1983). It is quite possible that the energy required for large scale movement of hydrocarbons is another source that has not been considered: the diurnal motion of subsurface fluids by earth tides.

3.4.3 Earth Tides and Secondary Migration

Earth tides activate the motion of fluids in subsurface formation by continual expansion and contraction as simple harmonic oscillations resulting from diurnal tides. This motion is detectable as two measurable modes in wells connected to subsurface reservoirs as: (1) water level variations and (2) a phase lag between the tidal force and the hydraulic motion in the subsurface (Morland and Donaldson, 1984). The periodic fluctuation of fluid level in the well is out of phase with the changes in gravity because of a number of drag forces: fluid viscosity, resistance to fluid flow, density, temperature, overburden pressure, compressibility, and shear modulus. The amplitude of the water fluctuations in the well is a function of permeability and the reciprocal of the porosity. Although the relative gravity variation in earth tides is very small ($< 10^{-7}$), it has a significant effect on the motion of subsurface fluids (Bredehoeft, 1967). This effect was even greater earlier in geologic history because the orbit of the moon has changed considerably; the moon was much closer to earth (it is currently moving away from the earth at about 3 cm per year). Therefore, millions of years earlier when hydrocarbons were moving through porous conduits from source rocks to reservoirs, the motions of subsurface fluids because of earth tides were more rapid and energetic (greater dilation and compression of the formations). If the porous conduit containing the water and oil is inclined, the back and forth movement of the fluids will produce segregated upward movement of oil by buoyant drag forces that produce film drainage. As the pores expand, water and hydrocarbons are pulled into the pores, and when the pore volume decreases (as the earth tide is relaxed) the overburdened mass compresses the pores expelling fluid. An elastic change of porosity equal to 0.1% in a 10-m thick reservoir over a radius of 100 m would result in an oscillatory motion of 47 m^3 (294 bbl) of fluid. Over geologic time periods, these motions could deplete the oil saturation to zero, leaving only a trace amount of oil in the pathway. Wettability effects can retard the movement of oil, but even the oil-wet conduits would eventually empty as indicated by the continuing trend of the oil relative permeability toward zero saturation with the passage of sufficient water that would occur in the geologic time period (Robinson and Bell, 1977; Arditty and Ramey, 1978; Moreland, 1977, 1978, 1992).

3.4.4 Accumulation of Hydrocarbons in Traps

For accumulation of hydrocarbons to form a petroleum reservoir, a geologic trap must be connected to the path of migration from the source rock. The matrix of the trap must be porous, permeable, and sealed on top with impermeable sediments to contain the accumulating gas and oil. In some cases the seals above the traps have fractures and other porous paths, resulting in the loss of small amounts of the light components (especially C_1-methane to C_5-pentane, and others) that can migrate to the surface and escape. Some prospecting by searching for surface emission of hydrocarbons has been used with limited success.

The loss of lighter components from a trap over a long period of time can leave behind a viscous, gas-depleted oil if the reservoir pressure and permeability of the porous channels are great enough.

3.5 Subsurface Environment

The average lithostatic (overburden) pressure (P_{ob}) composed of the combined pressure from interstitial fluids (P_f) and the grain-to-grain stress of the rock matrix (P_g), is 22.6 kPa m^{-1} (1-psi/ft). The hydrostatic pressure gradient is an average value of 10.5 kPa m^{-1} (0.465 psi/ft that corresponds to water containing 80,000 ppm NaCl at 77°F). Any reservoir with a gradient greater than this value is considered to have an abnormally high pressure. A few petroleum reservoirs with pressures less than the normal hydrostatic gradient have been discovered (the eastern Colorado plateau and Oklahoma-Texas panhandle in the United States, Siberian oilfields of Russia, central Alberta, Canada, and regions of the Middle East [Chilingarian et al., 2002]). Subnormal formation pressures have been artificially induced by water and hydrocarbon production, resulting in surface subsidence that can cause destructive damage to buildings and other structures. Abnormally high pressures resulting from tectonic and diagenetic events are commonly found in hydrocarbon reservoirs where natural fluid flow within the reservoir and communicating aquifers has been restricted after hydrocarbon accumulation in the traps. Petroleum reservoir fluid pressure is also increased after accumulation of hydrocarbons by temperature increase because of deeper burial as subsidence occurs. The average geothermal gradient resulting from the heat flux is about 18°C/km (Ex. 3.2); therefore, as subsidence followed by deeper burial occurs because of dynamic motions of the earth's crust, the fluids trapped in the sediments are subjected to higher temperatures. The sediments overlying hydrocarbon traps may have lower thermal conductivities than the porous zones below that contain fluids. Compacted, fine-grained materials such as clay can act as a "blanket" that decreases the natural flow of heat transfer from the mantle (1.2 µcal/cm^2-s), resulting in abnormally high temperatures in some subsurface zones (Chilingarian et al., 1995).

3.6 Pore Surface Properties

When hydrocarbon reservoirs are discovered, the connate water saturation can be any value greater than zero. As oil begins to accumulate, it begins to interact with the rock surface and eventually establishes an equilibrium condition of wettability. If the reservoir undergoes further development by changes in the environmental conditions (temperature, pressure, water salinity, diagenetic changes in the rock, etc.), the preferential wettability also changes adjusting to another state of wettability equilibrium. Different zones within a large reservoir can therefore have very different states of preferential wettability.

The establishment of reservoir wettability is a complicated process that depends on a large number of conditions, principally:

1. The area and pore-size distribution within the matrix of the geologic structure. Small pores and crevices will retain connate water with capillary forces; thus these are strongly water-wet. The larger pores from which water was displaced by oil have continuous paths for transport of fluids. Continuous paths of preferentially oil-wet pores are required for efficient displacement of oil. How these flow paths are distributed through the rock affects the overall production characteristics of the reservoir.

2. The minerals may exhibit a variety of preferential wetting properties from strongly water-wet to strongly oil-wet.

3. Many oils contain surface active compounds that migrate to the oil-water and oil-rock interfaces and are adsorbed on the mineral surfaces.

4. Halogen salts ($NaCl$, KCl, $CaCl_2$, $MgCl_2$, etc.) and water soluble organic compounds control the pH and surface activity of the water.

The surface area for oil-rock interaction is enormous. Donaldson et al. (1975) determined the surface area of glass beads, unconsolidated building sand, and sandstones (Table 3–1). One cubic centimeter of Berea sandstone saturated with 50% water and oil has just 0.094 cm^3 of oil and an equal amount of water in a pore space with 2.27 m^2 of surface area. This large surface area in sandstones is randomly interspersed with various minerals that can react with surface active components in the oils. Crocker et al. (1983) used an electron microscope to determine the types of minerals that are exposed on the surface of rocks and an emission spectrograph to quantitatively measure the bulk mineral compositions, Table 3–2. Less silica was exposed to the surface than was present in the bulk phase, and more reactive minerals were found on the surfaces of the grains where they were available for physical and chemical reactions of adhesion (wettability

Table 3–1 Surface areas of various materials measured by nitrogen adsorption.

Material	Porosity (Fraction)	Permeability (Darcy)	Density (g/ml)	Surface Area (m^2/cm^3)	Ion Ex meq/kg
Glass Beads (100–140 mesh)	0.388	6.73	1.557	0.047	—
Glass Beads (100–250 mesh)	0.385	4.70	1.557	0.048	—
Glass Beads (50–60 mesh)	0.383	31.31	2.576	0.010	—
Building Sand	0.374	5.00	1.110	1.277	—
Building Sand (40 mesh)	0.364	4.00	1.640	2.649	—
Bandera SS	0.174	0.012	2.18	12.0	11.99
Berea SS	0.187	0.25	2.12	2.268	5.28
Coffeyville SS (Oil Reservoir)	0.217	0.06	2.12	7.399	23.92
Cottage Grove SS (Outcrop)	0.263	0.30	1.94	5.374	17.96
Noxie, course-grain SS (Outcrop)	0.258	0.35	1.86	3.385	10.01
Torpedo, fine-grain SS (Outcrop)	0.224	0.03	2.04	8.650	29.27

interactions), adsorption of active components in oils, and ion exchange. Kaolinite, chlorite, and illite clays minerals were found in all of the sandstone samples, Table 3–3. The Coffeyville oil reservoir rock has aluminum (probably bound in the clays), iron, magnesium, calcium, titanium, and potassium as surface minerals. Crocker et al. (1983) found that the physical properties and pore-size distribution measurements of the cores showed considerable differences between the samples even though the surface mineral analyses were similar in some cases. The Coffeyville and Cottage Grove oil reservoirs have two to three times the surface area of the Berea and yet the mineral anal-

Table 3–2 Bulk and surface elemental analyses of rocks. Reported as percent mineral oxides (Crocker et al., 1983).

Rock	SiO_2	Al_2O_3	Fe_2O_3	CaO	TiO_2	K_2O	Mn_2O_3	SO_3
Berea	78.0	10.3	3.1	2.8	2.1	4.4	—	—
Coffeyville	65.1	15.4	10.5	2.7	3.8	2.4	—	—
Cottage	70.4	15.9	11.2	0.6	1.1	2.8	—	—
Noxie	64.4	9.5	22.5	0.7	0.6	1.5	3.1	—
Sweetwater	72.4	13.3	8.1	2.9	—	3.1	—	—
Torpedo	72.3	—	11.0	0.7	1.9	2.3	2.6	—
Oswego limestone	2.0	1.0	—	53.0	—	—	—	1.3

Table 3–3 Clays exposed on the surface of rock pores (Crocker et al., 1983; Twehyo et al., 1999 North Sea sandstones A and B).

Sandstone	Kaolinite	Clorite	Illite
Bandera	6.0	5.0	8.0
Berea	7.0	0.0	4.0
Coffeyville	4.0	4.0	6.0
Cottage	6.0	1.0	6.0
Noxie	5.0	1.0	7.0
Sweetwater	0.0	2.0	8.0
Torpedo	6.0	<0.1	7.0
North Sea A	47.2	0.2	1.5
North Sea B	1.7	42.7	2.5

yses are fairly similar. The Sweetwater (from 3,790 m depth) had a bimodal pore-size distribution that the SEM showed was because of a series of microfractures in addition to the pores; the microfractures could have resulted from expansion of the rock at the surface indicating that other physical measurements made at ambient conditions are not representative of the subsurface properties.

Because the pores are initially filled with water, when the water is displaced by oil a film of water may remain in contact with the rock pore surface. The water film must be ruptured to expose the rock surface, or reactive compounds in the oil (acidic, basic, electrostatic dipole, etc.) must diffuse through the water film to bind with minerals exposed on the surface. In the laboratory, a finite amount of *aging time* at about 75°C is required for changes of wettability to take place after a rock has been first saturated with water and then with oil. Emery et al. (1970) aged water- and oil-saturated sand packs at 71°F and 7 MPa for various periods of time and then examined the waterflood production curves. Production after 30 pore volumes of water was injected decreased as the systems became more oil-wet at longer aging times. Production was 55% of the original oil in place after aging for only 5 hours, 53% after 100 hours, 47% after 500 hours, and 46% after 1100 hours. The 9% decrease of production clearly shows that slow interaction between surface active components of the oil and rock surface takes place until an equilibrium condition of wettability is established. A decrease of the initial water saturation (lower than S_{wi}) before aging of the sample will produce a more oil-wet system, probably because oil will have access to smaller pores and thus greater overall surface area for contact with the rock. In the natural environment, time is infinite with respect to the aging times used in laboratory experiments; thus the preferential wettability of the reservoir has attained a stable equilibrium condition ranging anywhere between strongly water-wet and strongly oil-wet. In general, light oils in sandstones range from neutral wettability to water-wet and limestones have been found to range from neutral to oil-wet (Jadhunandon and Morrow, 1995).

Local differences of wettability in a reservoir may be governed by films of water left in contact with the oil surface following displacement of water by hydrocarbons. Films of water in contact with the surface are unstable in circumstances where the local capillary pressure is greater than the film disjoining pressure. The water films will reduce in thickness and desorb from the surface, allowing surface active compounds in the oil to contact the rock surface. Electrostatic forces at the rock-fluid interfaces also may increase the disjoining pressure, causing rupture of the film (Buckley, 1989 and Hirasaki, 1991). Thus it has been experimentally shown that when a rock surface is exposed to water and oil, surface active compounds in the oil can react physically and chemically with the surface rendering the contacted area more oil-wet.

If the local environmental conditions are such that the films formed between the rock surface and the invading oil remain stable, the reservoir will be one which is water-wet. Oil reservoirs are never completely water-wet; there are always some portions of the rock surface that are oil-wet indicating that polar, surface, active oil components diffuse through the water films to interact with the rock surface altering the reservoir wettability.

3.7 Multiphase Flow

When a rock is 100% saturated with a single phase and placed under a pressure differential, its laminar flow is opposed by viscous drag forces and restrictions imposed by the size distribution of the rock and Darcy's law describes the flow capacity. The conductivity of the fluid (expressed as the absolute permeability, K, in Darcy's equation, Ex. 3.3) can be determined from the flow rate and pressure drop across the porous medium. The absolute permeability is independent of fluid properties because the viscosity is incorporated in Darcy's equation and capillary pressure effects are zero for a single phase. When a second, immiscible fluid is present in the rock sample, several multifluid physical and chemical properties of the fluids influence laminar flow through the rock.

The effective permeability of each phase is a variable which is a function of saturation, wettability, and viscosity; it is expressed as the permeability relative to the single–phase, absolute permeability of the porous medium (in some cases, when applied to oil, instead of using the single-phase absolute permeability of the porous medium, the permeability to oil at irreducible water saturation is used):

$$k_{rw(S_w,\underline{W})} = \frac{k_{ew(S_w,\underline{W})}}{K} \qquad k_{ro(S_w,\underline{W})} = \frac{k_{eo(S_w,\underline{W})}}{K}$$
$$k_{rg(S_w,\underline{W})} = \frac{k_{eg(S_w,\underline{W})}}{K} \tag{3.2}$$

Laboratory procedures for determining relative permeability and several methods for its calculation are presented by Tiab and Donaldson (2004). The generally used procedure is to (1) saturate a cleaned core with brine, (2) displace the brine to S_{wi} with oil, (3) displace the oil to S_{or} with brine while making measurements of the rates of effluent brine and oil, and (4) calculate the relative permeabilities as functions of the increasing brine saturation (see *Program Listing 6.20 RELPERM.FOR*). Wettability is an intrinsic parameter that is expressed in the variation of the relative permeability curves. Figs. 3–4a and 3–4b compare relative permeability curves of water-wet and oil-wet systems. Fig. 3–4b is a plot on semi-log paper to show the characteristic decline of the curves, especially the diminishing value of oil-relative permeability that continues to approach zero by film drainage.

Under the same pressure drop used for single-phase flow when two fluids are present, each fluid will flow at a rate that is less than their single-phase flow dependent on their specific, instantaneous saturations. The effective permeability of the displacing phase increases while the displaced phase permeability decreases with respect to time as their fluid saturations change within the rock. In order to calculate the effective permeabilities of each phase, Darcy's law is assumed to apply independently to each phase. The effective permeabilities are functions not only of saturation, but also wettability, and are expressed relative to the single-phase absolute permeability of the porous medium as shown in Figs. 3–4a and 3–4b, Ex. 3.4a.

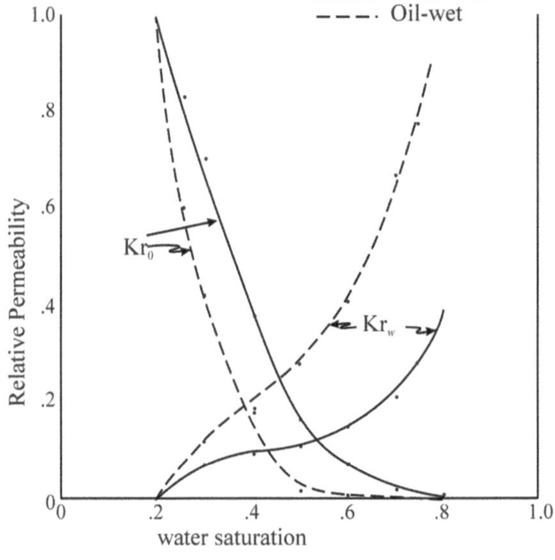

Figure 3–4a *Water-wet and oil-wet relative permeability causes the crossover point (where the relative permeabilities are equal) to have a higher value in the oil-wet systems, Ex 3.4a.*

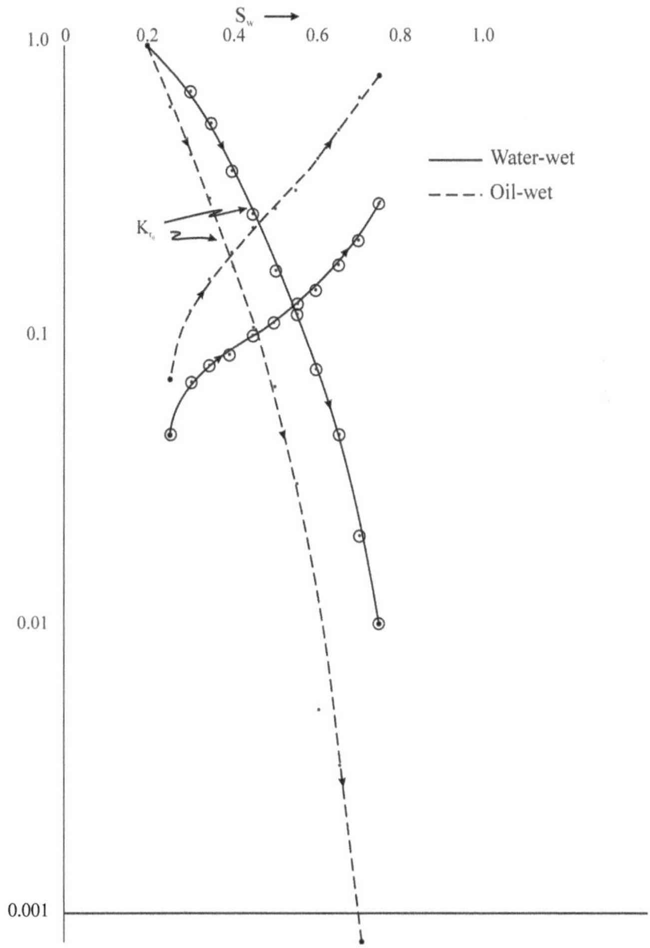

Figure 3–4b *Expanded relative permeability curves showing the diminishing value of the oil relative permeability as it continues to approach zero.*

The saturation history of the rock is important because it affects the wettability of the system. If a rock is saturated first with oil, the relative permeability curves will tend to reflect an oil-wet system and vice versa. There is considerable evidence that water was displaced by gas and oil accumulation in reservoirs. Therefore, the standard procedures for measuring relative permeabilities begins with a core 100% saturated with water which is reduced to S_{wi} before displacement with water is conducted for the k_{rw} and k_{ro} calculations. The overall wettability of the rock changes from the initial condition (after the irreducible water saturation is established) to an equilibrium state of wettability for the particular combination of porous medium, water solution, and oil that is being used. Therefore, after the irreducible saturation is established, the water-oil-rock system should be allowed to equilibrate with respect to wettability before the relative permeabilities are determined. The equilibration of wettability is accelerated by raising the temperature of the system.

3.8 Vertical Saturation Profile

The vertical saturation profiles at wells throughout the oilfield can be determined from the initial drainage capillary pressure curve (oil displacing water from $S_w = 1.0$ to S_{wi}). The saturation distribution above the free water level which occurs at zero capillary pressure can be estimated from the oil-water contact (OWC), which occurs at the threshold pressure for water-wet reservoirs and can be located on well-logs for assignment of its position with respect to the depth of the well. This depth is used as the beginning point for calculation of the vertical saturation profile. The OWC and the FWL both occur at zero capillary pressure for oil-wet cores because the water displacement begins with imbibition of oil.

Reservoirs can have a large vertical saturation distribution if the zone is thick enough (< 100 m). It can have a significant transition zone (from FWL to S_{wi}) that may vary throughout several hundred meters. This zone can have a significantly variable wettability ranging from water-wet at the FWL and becoming progressively more oil-wet as S_{wi} is approached; the geater capillary pressure of a higher oil saturation can promote contact of the pore walls by interaction of surface active compounds. Thus the wettability of a reservoir at discovery can be more oil-wet at the top of the reservoir. The initial saturation with respect to height in the transition zone is a function of the drainage (oil displacing water) capillary forces and gravity forces acting on the fluids (buoyant forces acting on the density differences). The capillary forces, however, may be modified by movement of the FWL after initial filling of the oil trap and by wettability alteration (Ingsoy et al., 1998). If the reservoir remains water-wet, there will not be any mobile oil below the FWL and the oil saturation in this zone (FWL down to the OWC) will remain constant. According to work by Jackson et al. (2005), if wettability has changed from water-wet to oil-wet, there can be mobile oil below the FWL. Jackson et al. (2005) used pore-scale modeling and reservoir simulation to study the effect of wettability change on oil mobility in the transition zone. They found that the initial water saturation in the transition is controlled by the primary drainage curve, and if the initial water saturation is changed by movement of the FWL after oil has filled the trap, oil and water flow will be different than it would be before movement of the FWL. Wettability variation may yield water-free production and greater oil production.

The capillary pressure curve should be obtained using reservoir fluids to reflect the influence of wettability; however, cleaned reservoir cores are used to obtain mercury-air capillary pressure curves for estimation of the vertical saturation profile (assuming that the pore-size distribution exerts the major influence on the capillary pressure curves) and to calculate the pore-size distribution.

Welge and Bruce (1945) derived the equation for calculation of the saturation profile by balancing the vertical gravity and buoyant forces in a capillary. The vertical force exerted by the interfacial tension acting upward is equated to the difference between the gravity and buoyant forces:

$$\pi r^2 hg\rho_w - \pi r^2 hg\rho_o = 2\pi r\sigma_{wo} Cos\theta_{wo} \tag{3.3}$$

$$h = \frac{2\sigma Cos\theta}{rg(\rho_w - \rho_o)} = \frac{P_c}{g(\rho_w - \rho_o)} \left[\left(\frac{kg}{s^2 \bullet m}\right)\left(\frac{1}{9.81\ m/s^2}\right)\left(\frac{1}{kg/m^3}\right)\right] = \frac{0.102P_c}{\rho_w - \rho_o}\ [m] \tag{3.4}$$

Fig. 3–5 illustrates the calculation of a vertical saturation profile. The OWC occurs at 0.115 m above the FWL. The residual oil saturation occurs at 0.169 m, thus the oil below this height is at residual oil saturation and cannot be produced by waterflood. Oil occurring between 0.169 m (S_{or}) and 1.15 m (S_{wi}) has been labeled the transition zone where oil will be produced along with water, but any oil occurring at a height greater than 1.15 m can be produced free of water. In an oil-wet system the OWC and FWL both occur at zero capillary pressure because oil will imbibe into the water-filled pores at zero capillary pressure.

An overall average estimate of the vertical saturation profile of an oilfield can be obtained from the Leverett J-function which was developed for correlation of capillary pressure curves from a specific formation (Leverett, 1941). Refer to page 349 of Tiab and Donaldson

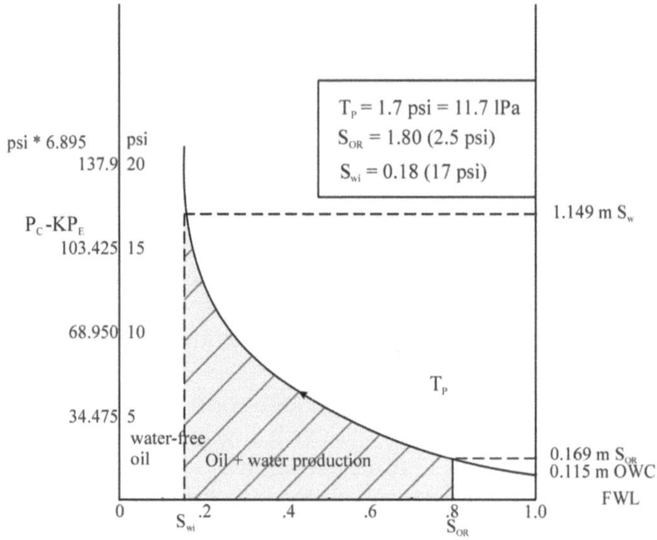

Figure 3–5 *Vertical saturation profile reservoir calculated from a drainage capillary pressure curve.*

(2004) for a complete explanation and sample calculation using the J-function.

Wettability throughout the vertical height above the water-oil contact (WOC) may vary because of variations of: (1) the relationships between gravity and capillary forces and the chemistry of the fluid/rock systems, (2) the vertical brine/oil/rock chemical compositions, or (3) changes in the pore-size distribution with respect to height above the WOC. Initial pore water is displaced from the larger pores by buoyant forces (differences in densities) as oil begins to accumulate at the top of the structure. As the oil accumulates the pressure becomes great enough to break films of water covering the pore walls, oil contacts the walls and begins to displace water from smaller pores (Hirasaki, 1991). Hence, the top portion of the reservoir can go through a change from water-wet to mixed wet (large pores containing oil and smaller pores containing water), and if the oil contains sufficient polar components, the change can form an oil-wet zone at the top of the reservoir. Lower in the structure as the oil saturation decreases, the system can gradually become less oil-wet until it eventually is strongly water-wet near the WOC, if the pore-size distribution and the chemistry of the fluids remains constant with depth.

Tang and Morrow (1997) found that differences of brine and oil compositions can have a governing influence on wettability; therefore, a vertical change in the amount and chemical composition of the polar compounds can produce a profound change of wettability with respect to height above the WOC.

Hamon (2000) measured drainage and imbibition capillary pressure curves of 26 samples taken from 4–175 meters above the WOC of an oilfield and concluded that the structural position and absolute permeabilities of the samples controlled the vertical variation of wettability. Measurements of the wettability index with respect to height above the WOC indicated a gradual change of wettability from strongly water-wet to mixed wettability. They also confirmed the fact that greater recovery is obtained (lower S_{or}) from samples that exhibit neutral or mixed wettability.

3.9 Fractured Reservoirs

In fractured reservoirs wettability has a large influence on production because oil must be displaced from the matrix blocks into the fractures communicating with the wells. Thus imbibition of water into the blocks accompanied by displacement of oil (from blocks that are preferentially water-wet) is one of the principal mechanisms for recovery of oil from fractured reservoirs. The top portion of the capillary in Fig. 3–6a for a water-wet system shows that water will imbibe until the capillary pressure becomes zero; some gravity-assisted forced entry of water also will occur if the height of the water column is sufficient. After imbibition stops, the lower portion of Fig. 3–6a shows that oil displacement occurs from the combination of pressure exerted by gravity and pressure induced by water injection.

A fractured reservoir which is preferentially oil-wet, especially one with low rock permeability, will be difficult to produce because the capillary forces inhibit entry of water by imbibition (Fig. 3–6b). In the absence of imbibition, recovery of oil by forced displacement with injected water is not practical because the injected water flows rapidly through the highly permeability network of fractures causing early water breakthrough and an uneconomical water/oil-producing ratio.

If the fracture blocks are large and exhibit large surface area, countercurrent imbibition is the most probable mechanism because each phase develops its own series of capillaries for flow of fluids. On the other hand, if the reservoir is highly fractured, the injected water will tend to channel readily from the injector to the producing wells;

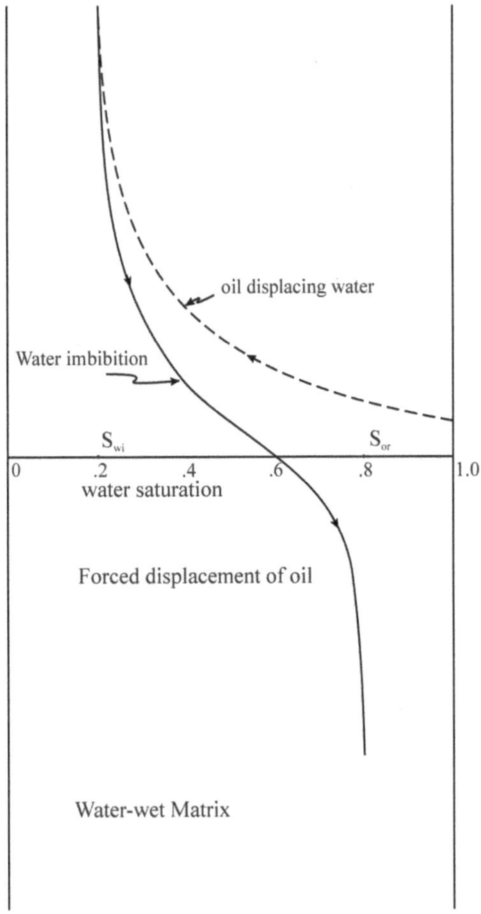

Figure 3–6a *Capillary pressure curve for water-wet matrix. The threshold pressure for water displacement of oil is zero; therefore, water spontaneously imbibes displacing oil to a water saturation equal to 60%. Subsequently, water must be forced into the matrix to displace oil to the residual oil saturation.*

therefore, the injected water will sweep a much smaller volume of the reservoir. Channeling of the injected water results in high flow rates through fractures imposing high pressure gradients across blocks that have low flow in fractures on one side; the pressure difference encourages co-current imbibition, but the produced oil is bypassed in the low flow rate channels.

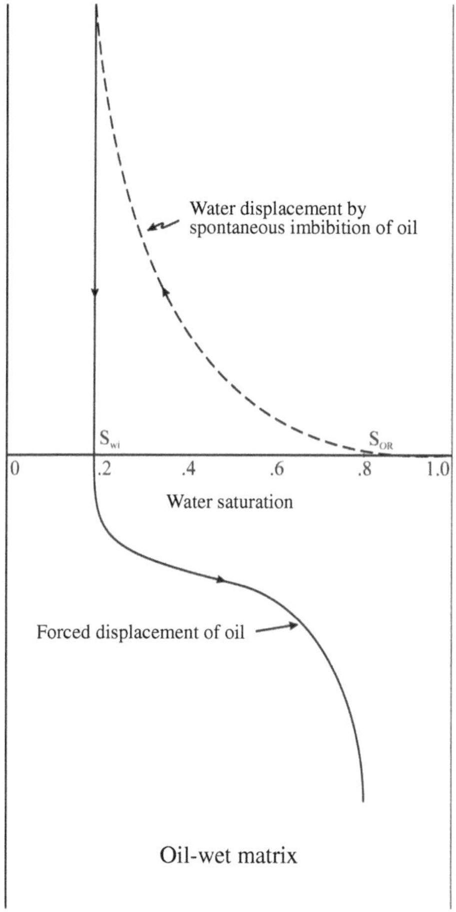

Figure 3–6b *Capillary pressure curve for an oil-wet matrix. The threshold pressure for initial entry. Oil is zero; therefore, imbibes spontaneously into oil-wet matrix which is saturation of 100% water.*

The fluid within the fractures between the blocks actually controls the imbibition and forced displacement processes. In a water-wet matrix imbibition of water will occur and oil will be displaced into the fracture channels. In addition, the height of the water within the fractures provides a gravitational pressure for invasion of water into the matrix blocks:

$$P_{(gravity)} = gh(\rho_w - \rho_o) \quad \left[(9.81m/s^2)(m)(kg/m^3)\right] = 9.81 \quad [Pa/m] \qquad (3.5)$$

If the matrix is oil-wet, water cannot imbibe into the matrix and production of oil from the blocks can only occur if water pressure from gravity forces and water injection is high enough to overcome the water threshold capillary pressure for forced displacement of oil from the blocks, Fig. 3–6b. The threshold pressure is inversely proportional to the pore-size distribution; thus a matrix with large pore sizes will have a lower water threshold pressure. On the other hand, the threshold pressure is directly proportional to the water-oil interfacial tension; therefore, artificially lowering the IFT by adding a surfactant or caustic to the injection water can enhance displacement of oil from the matrix into the fracture channels.

A low flow rate of injected water results in greater sweep efficiency and greater exposure of the reservoir for countercurrent imbibition to take place, resulting in greater oil recovery at a lower water/oil-production ratio (Minour and Knapp, 1980).

The network of fractures in naturally fractured reservoirs is randomly distributed, and blocks between the fractures vary in size and shape. Parts of the reservoir may have large areas that are not fractured, thus the position of wells can be a critical component for oilfield development. Communicating wells within a large unfractured area will behave entirely different from those in the fractured zone. The unfractured area will require considerably higher injection pressure for the same rate of injection and water breakthrough will occur much later; direct displacement of oil takes place while the fractured portion depends on imbibition for oil displacement.

About half of all reservoirs are carbonates. They usually exhibit low oil recovery (< 30% OOIP) because a large number of the reservoirs are naturally fractured. The overall permeability can range from 1–10 mD, hence water rapidly displaces the oil in the fractures and early water breakthrough occurs. In water-wet and mixed-wet fractured reservoirs, the second phase of production by waterflood depends on spontaneous imbibition of water into the fracture blocks. Høgnesen et al. (2005) found that sulfate ions in the injected water can alter the preferential wettability of carbonate reservoirs from oil-wet to water-wet by suppressing the positive charge potential in the brine and rock surface. The rate of spontaneous imbibition of water increased as the concentration of sulfate ions in the injected water was increased. They concluded that sea water has enough sulfate ions to change the wettability of carbonates from oil-wet to water-wet.

Hirasaki et al. (2004) used an anionic surfactant and sodium carbonate/sodium bicarbonate to promote spontaneous imbibition in carbonate rocks by changing the wettability. Sodium carbonate was best because it suppressed the positive calcium ions in the brine, saponified organic acids in the crude oil, and inhibited adsorption of the anionic surfactant. They concluded that oil displacement from fractured carbonate blocks is assisted by buoyancy if a surfactant is used along with an alkali to reduce the IFT and alter the wettability.

Graue and Bogno (1999) discussed the results of waterflood experiments conducted using fractured blocks (5 * 12 * 20 cm) of outcrop chalk at strongly water-wet and moderately water-wet conditions. The initial wettability of the chalks was strongly water-wet (I_A = 1.0) and could only be changed to moderately water-wet conditions (I_A = 0.3 to 0.5) by aging at S_{wi} with crude oil for 83 days at 90°C. Repeated wettability imbibition tests using the same cores showed that the altered wettability produced by aging was permanent.

Overall waterfloods with fractured and unfractured blocks showed that oil recovery was only slightly lower for the fractured blocks. This is the significant point of these experiments and correlates with results reported by Lorenz et al. (1974), Amott (1959), and Kennedy et al. (1955). Although oil recovery was higher for the moderately water-wet system, the pressure was higher for any given flow rate. They also observed that the system with the moderately water-wet condition had a lower drainage threshold pressure for entry of oil into the 100% water-saturated core, and the rates of spontaneous water imbibition (from S_{wi} to S_{or}) reduced as the wettability was changed toward the moderately water-wet condition. The influence of the closed fractures on the flow patterns of water during the waterflood was less (compared to nonfractured rocks) for the moderately water-wet than for the strongly water-wet rocks. The authors believe that two mechanisms account for this observation: (1) the moderately water-wet rocks exhibited less capillary end effects (increased water saturation at the fluid exit from the fracture), and (2) water droplet formation at the water exit face of a fracture was greater for the moderately water-wet rocks; the droplets formed water-bridges across the fractures enhancing the transfer of injection water between fracture blocks. Water movement was monitored using radioactive $^{22}NaCl$ in the artificial brine (5% $^{22}NaCl$ and 5% $CaCO_2$). All of the observations supported the theoretical behavior of water-wet versus oil-wet fractured reservoirs rocks with intermediate wettability, with (1) lower drainage threshold pressure and higher required injection pressure for a given water injection rate, (2) lower rates of imbibition, and (3) had greater overall oil recovery (lower S_{or}).

3.10 Mobility Ratio

The mobility ratio is defined as the mobility of the displacing fluid at the average water saturation behind the advancing front of displaced oil ($S_{w\alpha}$), divided by the mobility of the oil at the average saturation in the advancing oil bank ($S_{w\beta}$), Fig. 3–7:

$$M = \left(\frac{k_{rw}}{\mu_w}\right)_{S_{w\alpha}} \left(\frac{\mu_o}{k_{ro}}\right)_{S_{w\beta}} \tag{3.6}$$

The mobility ratio is a single term that describes the rate and efficiency of oil displacement by other immiscible fluids. Diminishing values of M (less than one) are generally considered as favorable values, and increasing values (greater than one) are considered as unfavorable values. For a waterflood to be economically successful, M should have a value equal to or less than 10. Enhanced oil recovery processes are designed to decrease the value of this single parameter.

The relative permeabilities are intrinsic functions of wettability; therefore, chemical solutions of surfactants and caustics are used to change reservoir wettability in EOR processes and decrease the mobility ratio. Polymer solutions are used to increase the water viscosity and decrease M, while thermal processes decrease M by decreasing the oil viscosity.

Miscible displacement EOR processes that use solvents such as light petroleum fractions, condensed gases, carbon dioxide, and alcohol to dissolve oil at the advancing front eliminate capillary forces by reducing the interfacial tension to zero. As the miscible front expands away from the injecting well, more and more oil is dissolved into the solvent until the solvent is no longer effective and thus making the displacement less efficient than theoretical predictions from linear laboratory core floods.

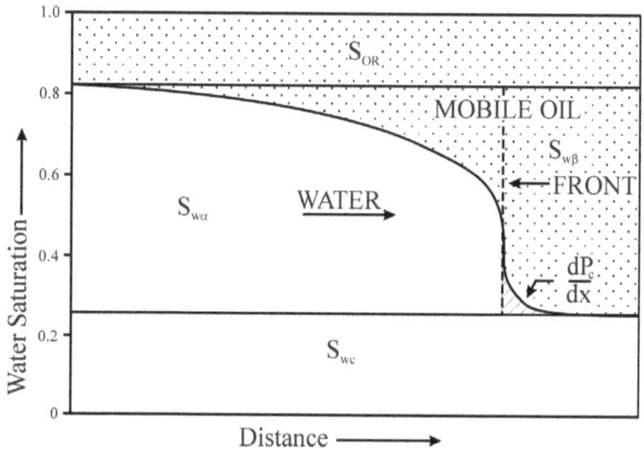

Figure 3–7 *Displacement of oil by water in a water-wet system. A bank of mobile oil develops ahead of the advancing water with an average water saturation, S_{WB}, and an oil saturation quotient develops behind the front from 1-S_{WC} to S_{OR} at the point of injection; the average water saturation behind the front is $S_{w\alpha}$.*

The displacement of oil by water initiated by a pressure difference between points of water injection and oil production will result in displacement of oil from its connate (or initial) saturation (S_{wc}) to an economical (or practical) residual oil saturation. If the mobility ratio is small enough, a definite front of mobile oil develops in front of the advancing water. The displacement is not an exact, piston-like process; instead, an oil saturation gradient exists from $S_o = 1 - S_{wc}$ in the mobile oil to S_{or} at the zone of injection where a large volume of water has contacted the porous medium, shown in Fig. 3–7. If the system is water-wet, water will imbibe ahead of the front as a function of the imbibition capillaries' pressure relationship with respect to distance (dP_c/dx). This produces an initial rapid production of oil that was calculated as oil relative permeability greater than 1.0 and interpreted at the time as an effect caused by capillary end effects (Donaldson et al., 1966). In an oil-wet reservoir, there will be an opposite effect as some oil trails behind the front because of the threshold capillary pressure required for entry of water into the pores.

3.11 Fractional Flow Equation

The fractional flow equation was developed by Leverett (1941) to describe the flow of immiscible fluids:

$$f_w(S_w) = \frac{1 + \dfrac{k_o}{(q_t / A)\mu_o}\left[\dfrac{\delta P_c}{\delta x} - \left(\dfrac{g}{g_c}\right)(\Delta\rho)(\sin\alpha)\right]}{1 + \dfrac{\mu_w}{\mu_o}\dfrac{k_o}{k_w}} \tag{3.7}$$

where units are: Darcy, atm, cm, cP and:

q_t/A [cm/s] = flow rates of water and oil ($q_w + q_o$)

α = angle of inclination of the porous system

k_o/k_w = the ratio of the effective (or relative permeabilities)

In the fractional flow equation, the ratio of the relative permeabilities is a single function of specific water saturation in the porous system. This is different from the definition of the mobility ratio that is defined with respect to the average saturation of the advancing water behind the mobile oil front and the saturation in the mobile oil bank.

Eq. 3.7 may be separated into three distinct terms that express the properties of any two flowing fluids. The effective permeabilities have been replaced by relative permeabilities:

$$f_w = \frac{1}{1 + \dfrac{k_{ro}}{k_{rw}}\dfrac{\mu_w}{\mu_o}} + \frac{\dfrac{Kk_{ro}}{\mu_o q_T}\dfrac{\delta P_c}{\delta x}}{1 + \dfrac{k_{ro}}{k_{rw}}\dfrac{\mu_w}{\mu_o}} - \frac{\dfrac{k_{ro}}{\mu_o q_T}\dfrac{g}{g_c}\Delta P \sin\alpha}{1 + \dfrac{k_{ro}}{k_{rw}}\dfrac{\mu_w}{\mu_o}} \tag{3.8}$$

The first term in Eq. 3.8 is the *fluid displacement term* that is used to calculate the fraction of water flowing as a function of saturation. The second term is the *capillary pressure term* that is generally neglected because it is a low potential (<30 kPa) that occurs over a long distance (150 m or more) and is difficult to estimate with any degree of accuracy for field cases. The capillary pressure gradient, however, can be separated and expressed as a function of water saturation:

$$\frac{\delta P_c}{\delta x} = \frac{\delta P_c}{\delta S_w}\frac{\delta S_w}{\delta x} \tag{3.9}$$

The change of saturation with respect to length can be evaluated by methods for imaging dynamic flow in cores, Combes et al. (1989). Eq. 3.9 also leads to qualitative evaluation of the influence of the term on the linear flow of fluids in a forced oil displacement waterflood. The capillary pressure curve that applies is the negative curve (curve 3, Fig. 1–8) describing increasing water saturation from S_{wi} to S_{or}. Thus the change of capillary pressure with respect to water saturation is negative and the change of saturation in the porous medium with respect to length also is negative because the water saturation gradient in the direction of flow decreases with respect to length. Thus the capillary pressure gradient $((\delta P_c)/(\delta x))$ is a positive term and its effect is to increase the fractional flow of water. Therefore, neglecting the capillary pressure gradient actually yields a slightly lower relative flow of water as a function of saturation.

If flow through the porous medium is horizontal, the inclination expressed in the sine of the angle of inclination from a horizontal plane is zero and Eq. 3.8 reduces to:

$$f_w(S_w) = \frac{1}{1+\dfrac{\mu_w}{\mu_o}\dfrac{k_{ro}}{k_{rw}}} = \frac{1}{1+\left(\dfrac{1-f_w}{f_w}\right)} \tag{3.10}$$

The fractional flow of water (f_w) is the ratio of the flow rate of water to the total flow of both fluids at any specific saturation; f_w is a function of saturation because the effective permeabilities are functions of saturation (see *Program Listing 6.6 FRAFLOW.FOR*). The effective permeabilities in Eq. 3.8 appear as a ratio and therefore the relative permeabilities can be used in the simplified version, Eq. 3.10. Although

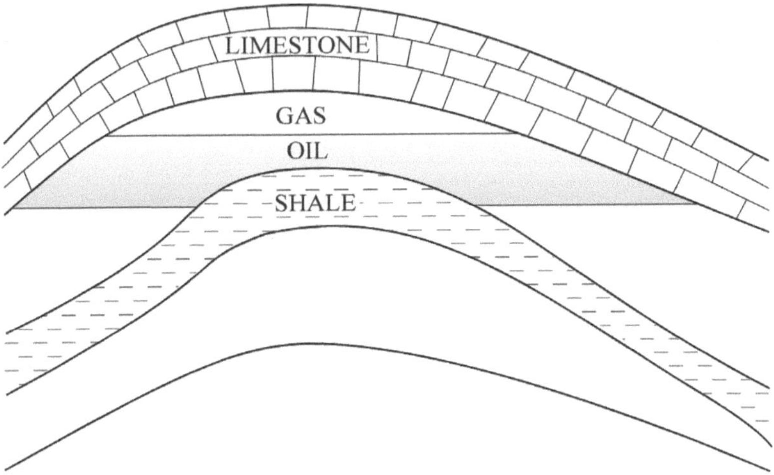

Folding of layered sediment caused by tectonic compression forming an anticline petroleum structural trap.

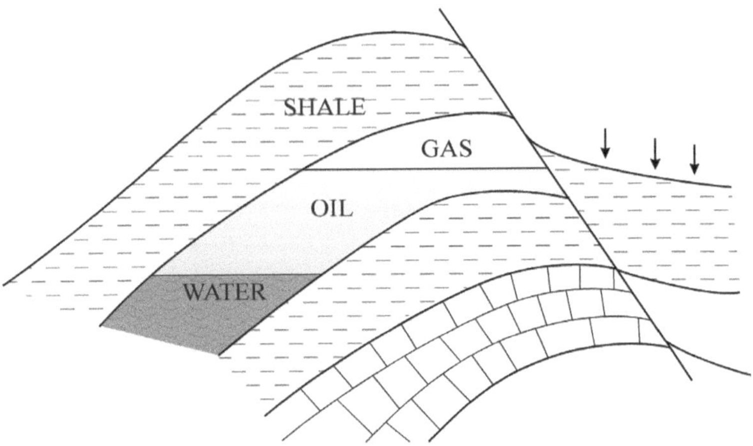

Up-dip sealing of an anticline by a fault to form a structural trap for accumulation of oil.

Figure 3–8 *A porous medium with a positive tilt up-dip with a flow potential tilt to the pressure drop minus gravitational drainage potential* ($\Pi = \Delta P - pgh$).

the capillary pressure term was neglected, the influence of wettability on the fractional flows of water and oil is expressed because the overall wettability of the water-oil-rock system is a property of relative permeability. Figure 3–9 illustrates the effect of relative permeability

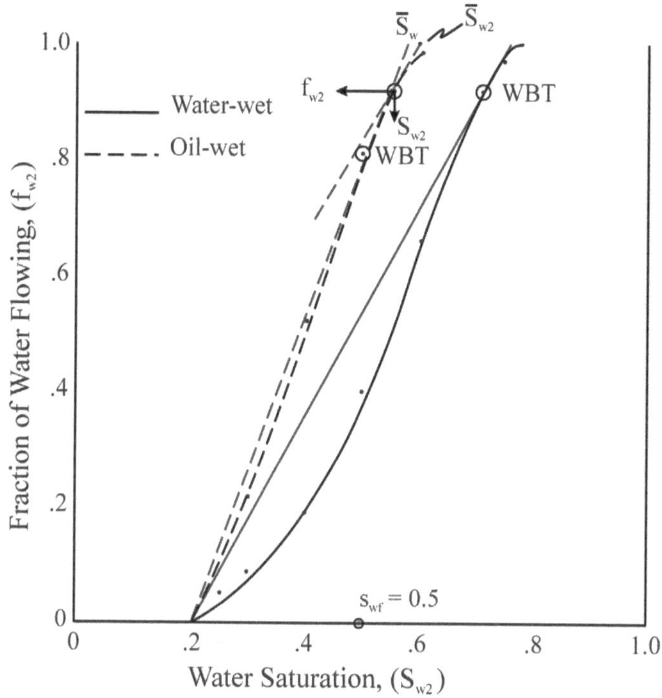

Figure 3–9 *Fractional curve for water-wet and oil-wet systems calculated from data Ex. 3.3 using a viscosity ratio equal to 1.0. Water breakthrough (WBT) occurs at tangent to the curve drawn from S_{wi} = 0.2. The fraction of water flowing (f_{w2}) and average saturation (S_{w2}) are found from subsequent tangents to the curve. S_{wf} is the saturation at the displacement front, Ex. 3.4b.*

(and hence wettability) on water-wet and oil-wet systems where the viscosity ratio is constant and equal to 1.0 (Ex. 3.4b). The curves show that a more oil-wet system has less efficient production than a water-wet system at any saturation, and the fraction of water flowing is considerably greater for the oil-wet system. Thus one of the objectives of EOR processes is to change the reservoir wettability from an oil-wet system to one that is more water-wet in character and thus improve overall recovery and decrease the producing water/oil ratio.

An increase of oil viscosity from 1 to 5 cP (with wettability assumed to be the same and constant for both conditions), will result in a less favorable mobility ratio and an increase of the fraction of water flowing at any saturation, Fig. 3–10. This is the same effect on multiphase flow

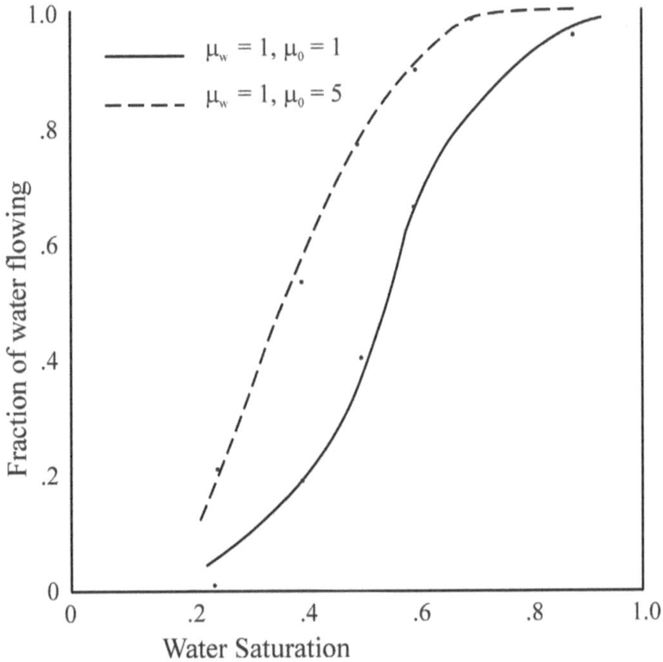

Figure 3–10 *Fractional flow curves for a water-wet system showing the effect of a change of oil viscosity from 1.0 to 5.0 cP. An inverse of* f_w *occurs which the same effect results from a change of wettability from water-wet to oil-wet (data from Ex 3.3)*

that would be caused by a change of wettability toward a more oil-wet system.

Analyses of the fractional flow curves (Fig. 3–9) are of considerable practical value because a tangent line drawn from the connate water saturation yields the water/oil production ratio that occurs soon after water breakthrough of a water-flood (S_{wbt}). Production of water-free oil stops and the producing fraction of water rises to the value indicated by the tangent line (0.70 for the water-wet rocks and 0.50 for the oil-wet rocks). Furthermore, the average saturation behind the water-flood front is given by the intersection of the tangent line at $f_w = 1.0$, and tangents above water breakthrough give the fraction of producing water after the flood-front has passed by the well (or end of the core for laboratory samples). Fig. 3–9 shows that the average saturation behind the front is considerably higher for the oil-wet system because less oil has been produced when water breakthrough occurs. The saturation change from connate water saturation to the much higher saturation at water breakthrough does not occur as an abrupt change. The change is gradual because it is moderated by capillary pressure effects (the neglected term in Eq. 3.7).

For the oil-wet systems, the pore volume of water injected when water breakthrough occurs is:

$$Q_{ibt} = \frac{W_i}{V_p} = S_{wbt} - S_{wi} = 0.57 - 0.20 = 0.37 \tag{3.11}$$

The value of Q_{ibt} at breakthrough also may be computed from graphical analysis of an expanded plot of the fractional flow curve (Fig. 3–11) where:

$$Q_{ibt} = \frac{\bar{S}_{wbt} - S_{wbt}}{1.0 - f_{wbt}} = \frac{0.57 - 0.5}{1.0 - 0.812} = 0.37 \tag{3.12}$$

The same analysis for the water-wet system produced the following results:

$$\begin{aligned} Q_{ibt} &= 0.75 - 0.20 = 0.55 \\ Q_{ibt}(graphical) &= \frac{0.75 - 0.70}{1.0 - 0.915} = 0.59 \end{aligned} \tag{3.13}$$

The small error is because of difficulty in obtaining accurate values from a graph.

The fractional flow equation presents a representative model of the flow of two fluids in a homogenous porous medium; however, reservoirs and large laboratory rock samples are not homogenous with respect to their physical and chemical properties. As presented earlier, a vertical saturation gradient can exist that is defined by the capillary

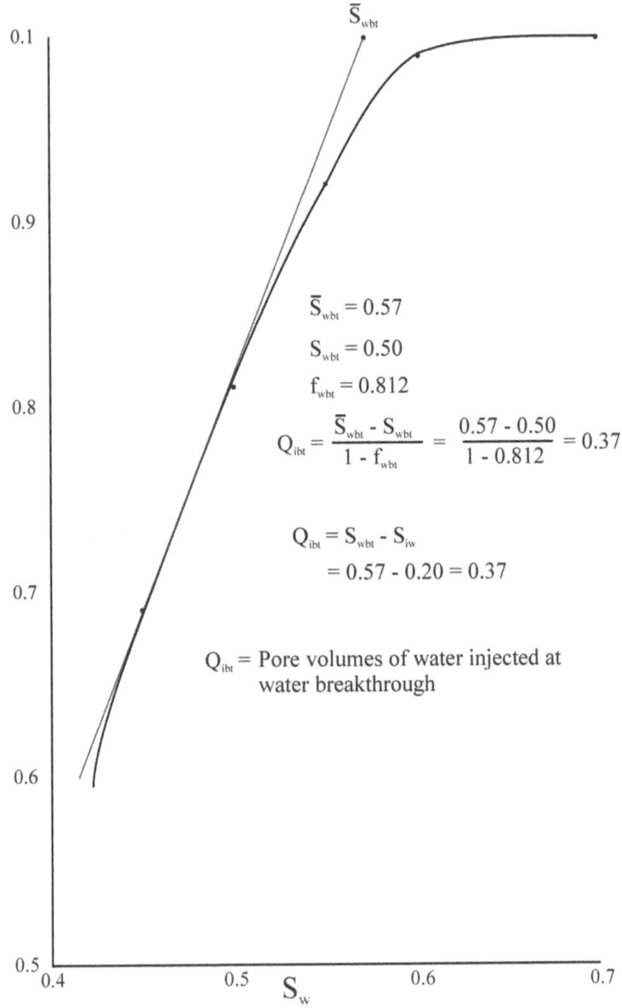

Figure 3–11 *Expanded fractional flow curve for the oil-wet system.*

pressure phenomena, and vertical and horizontal variations of porosity and permeability may be present as well as differences of wettability. Where the vertical distribution effects can be determined the reservoir is divided into parallel layers, assuming typical average properties within each layer, which are then treated separately to estimate the overall fluid flow behavior of the reservoir by applying fractional flow and simulation equations to each layer. Estimated vertical and volumetric fluid flow efficiencies are incorporated by method clearly described by Craig Jr. (1971) and later developments for numerical simulators in the literature.

3.12 Frontal Advance Equation

The frontal advance equation, developed by Buckley and Leverett (1942), predicts the velocity of a moving displacement front. A detailed derivation of the equation is presented by Craig Jr. (1971) and Willhite (1986).

$$\left(\frac{dx}{dt}\right)_{S_w} = \frac{q_T}{A\phi}\left(\frac{df_w}{dS_w}\right) \tag{3.14}$$

$$\frac{x}{L} = \frac{q_T t}{A\phi}\left(\frac{df_w}{dS_w}\right) \tag{3.15}$$

Eq. 3.14 gives the velocity of the displacing fluid at any saturation within the porous medium. Eq. 3.15 yields the fractional length of the porous medium from the inlet, which a specific *saturation* has moved after water breakthrough. In addition, the theory yields: (1) the saturation distribution within the system under waterflood, (2) the time after initiation that water breakthrough will occur, (3) the producing water/oil ratio as a function of time, and (4) the amount of oil recovery that can be expected.

For saturations less than S_{wbt} but greater than S_{iw}, the location of a specific saturation may be computed using the value of the slope of the fractional flow curve at water breakthrough saturation (Eq. 3.16). If the slopes of the fractional flow curve at saturations less than breakthrough saturation are used, two different saturations having the same velocity (or distance from the inlet) will appear to exist. Terwilliger et al. (1951) found that saturations less than the breakthrough saturation move at the same velocity, therefore they labeled this zone ($S_{w\alpha}$ behind the advancing front, Fig. 3–7) as the *stabilized flow zone*. Furthermore, they discovered that saturations ahead of the advancing front ($S_{w\beta}$, Fig. 3–7) move at variable velocities and thus this area was labeled the *nonstabilized zone*. For this calculation, Eq. 3.16 is expressed as follows and an example calculation is presented in Ex. 3.4c:

$$\left(\frac{x}{L}\right)_{S_w} = \frac{q_T t}{A\phi}\left(\frac{\delta f_w}{\delta S_w}\right)_{S_{wbt}} \qquad [S_{iw} < S_w \leq S_{wbt}] \tag{3.16}$$

$$\left(\frac{x}{L}\right)_{S_w} = \frac{q_T t}{A\phi}\left(\frac{\delta f_w}{\delta S_w}\right)_{S_w} \qquad [S_{wbt} \leq S_w \leq (1 - S_{or})] \tag{3.17}$$

Table 3–4 is a list of results from calculations using the fractional flow and frontal advance equations for analyses of the water-wet and oil-wet systems (Ex. 3.4c). The results show a clear distinction between the characteristic fluid flow behaviors of waterfloods of cores that are water-wet and oil-wet with both floods beginning at an initial water saturation (S_{iw}) equal to 0.20. When water breakthrough occurs, the water saturation at the front and the average saturation in the rock are greater for the water-wet system. The pore volumes of water injected (Q_{ibt}) and the amounts of oil produced (N_{pbt}) also are greater at breakthrough for the water-wet system. The fraction of water flowing at breakthrough, and thus the producing water/oil ratio, is greater for the water-wet system. The time required for water breakthrough after initiation of the waterflood is greater for the water-wet system because a more uniform displacement front travels through the rock; whereas, in the oil-wet system, water fingers develop ahead of a diffused advancing front and cause earlier water breakthrough. Thus the fractional flow and frontal advance theories furnish an excellent tool for understanding the different fluid flow properties of water-wet and oil-wet systems.

Tang and Morrow (1997) found that oil recovery from cores is sensitive to the salinity of the connate brine. Both oil recovery at breakthrough and final recovery increased when the salinity of the connate water was reduced. Connate brine and waterflood brine with a salinity equal to 1 % of the reservoir brine resulted in recovery which was about 10% greater than the recovery for cores using the formation brine. They also found that the rate and total amount of water imbibition increased when the temperature of the system was increased from the ambient temperature of 22°C to 75°C. This occurred because the increase of temperature caused the equilibrium wettability of the rock to change toward a more water-wet system. When a refined oil (Soltrol) was used in place of the crude oil, temperature had no effect on recovery efficiency; this is attributed to the

Table 3–4 Comparison of the fluid flow properties for water-wet and oil-wet systems at water breakthrough.

	Water-wet Core	Oil-wet Core
S_{iw}	0.20	0.20
S_{wbt}	0.70	0.50
$\bar{S}_{wbt}$	0.75	0.57
Q_{ibt}	0.55	0.37
N_{pbt} (ml)	13.75	9.25
f_{wbt} = WOR	0.91	0.81
t_{bt} (s)	32.2	23.8
$(x/L)_{t=20}$	0.62	0.83
$(df_w/dS_w)_{bt}$	1.70	2.28

lack of polar organic compounds that adsorb on the rock surface and establish a dynamic equilibrium wettability for a given set of environmental conditions. Reservoir wettability is not an absolute, stable condition; instead, it is an equilibrium condition that can be changed during production by affecting changes in the environmental and chemical conditions of the fluids and rock.

Leach et al. (1962) used advancing contact angle measurement of the Muddy-J oil (Nebraska, U.S.) and artificial brine (5,500 ppm NaCl) on a quartz surface to determine wettability reversal from oil-wet to water-wet with sodium hydroxide. The water-oil contact angle was 135 (IFT – 25 mN/m). Increasing concentrations of sodium chloride in the brine decreased the contact angle and IFT; 0.1 molar NaOH produced a moderate decrease of the contact angle from 135° to 90° and a concentration of 0.4 molar NaOH (IFT = 9 mN/m) decreased the contact angle to 70°. Assuming the quartz surface offered a good representation of the reservoir rock surface, they concluded that insitu reservoir wettability reversal could occur in the reservoir if NaOH was used for water injection. Thus a portion of the oilfield with flooded-out wells was chosen for the test. At this location the depth was 1800 m and the average reservoir properties were: thickness approximately 3 m, porosity 15%, permeability 119 md, and wettability was oil-wet. A central well was treated with a 6,500 m^3 slug of 2% (0.5 M) NaOH followed by normal field brine injection. Sodium hydroxide began to appear at one of the adjacent wells within four months and oil production began in other nearby wells at increasing rates at the same time. The total production, attributed to a wettability reversal to a water-wet condition, was 1,400 m^3 (8,750 bbl). Wettability reversal from oil-wet ($\theta = 135°$) to water-wet ($\theta = 40°$) resulted from a decrease of IFT (25 to 9 mN/m) and strong interaction of sodium hydroxide with the rock producing sodium silicate which was detected in the produced water.

Contact angle measurements do not show the complete scenario because the solid surface is poorly represented when a smooth surface is used for determination of contact angles. The complexity of the pore surface is not included; however, the Muddy-J oil probably had a sufficiently large amount of active polar N, S, and O compounds to provide an overbearing influence on the wetting characteristics and reaction with sodium hydroxide to provide definitive results of the wettability behavior using a quartz surface. More precise and dependable data of wettability characterization of a reservoir could be obtained using spontaneous imbibition analyses or the Amott and USBM wettability indices.

Duquerroiz et al. (1990) investigated the kinetics of silicon dissolution by sodium carbonate and sodium hydroxide. The alkaline reaction with sandstones is complicated by the multitude of components in the rock matrix that can react with alkaline solutions. Therefore, their work was restricted to quartz grains. A 0.1 pore volume slug of NaOH (0.4 M solution) and 0.2 M Na_2CO_3 were injected into sand packs (2 cm diameter by 30 cm long), and sand kaolinite mixtures using 8% clay in the sand. The quartz grains exhibited the highest reactivity with the alkaline solutions; one of their conclusions was that the amount and rate of dissolution of silicate is highly dependent on the pH of the alkaline solution. The reactions of sodium hydroxide with other constituents in the rock and formation brine must be evaluated in the laboratory before field injection. Calcium sulfate, in particular, will consume sodium hydroxide and form a precipitate of calcium hydroxide; other divalent cations also will have the same effect. Some clay minerals will dilute the sodium hydroxide by exchanging hydrogen ions for the sodium and thus transform the alkali to water. If the conditions are favorable—oil-wet reservoir, lack of divalent cations, and absence of cation exchange properties—a caustic flood can produce additional oil that would otherwise be lost to abandonment of the field.

The clay/sand mixture had lower reactivity than pure siliceous sand because of inhibition of the silica reaction by the clay-alkaline reactions that precipitated products on the sand surface.

A slug of NaOH will not be consumed by silica-clay reactions because precipitation of reaction products on the surface inhibits further reaction; therefore, one may expect arrival of some caustic and silica compounds at the production wells along with increased production if a wettability change from oil-wet to water-wet can take place in specific reservoir being used.

Furthermore, the capillary number (Eq. 3.18) must be greater than 10^{-4} to enhance the production of oil beyond the practical residual oil left at the end of a water flood (Tiab and Donaldson, 2004). Reservoir fluids should be used to evaluate the stability of N_c for selection of the optimum concentration of sodium hydroxide to overcome losses from chemical reactions.

$$N_c = \frac{\mu_w u_w}{\phi \sigma_{ow}} \tag{3.18}$$

3.13 Impact of Wettability on Waterfloods

Reservoirs can have zones with very different wettability conditions, and different states of wettability have definite consequences for waterflood behavior. McDougall and Sorbie (1995) used pore network simulation to study the effects of fractionally wet and mixed-wet conditions on the efficiencies of waterfloods. In strongly water-wet reservoirs the rock surfaces are wet by water, and at the end of the waterflood oil is trapped as isolated globules (on ganglia) in the larger pores. When residual oil is attained the oil is no longer connected to the outlet face by oil films or filaments, and because there are no oil-wet pores, the movement of oil through surface films is not possible. Mixed wettability means that the small pores are water-wet and filled with water, whereas the larger pores are oil-wet and filled with oil. This may occur when oil migrates into a rock which is initially water-wet, and oil preferentially fills the larger pores. The wettability is then altered by deposition and adsorption of polar organic compounds making the larger pores oil-wet. Fractional wettability is related to differences of surface chemistry of the rock pores. Crude oil components may be adsorbed on the pore walls and the core may contain areas which are water-wet and others that are oil-wet.

McDougall and Sorbie (1995) found that with fractional wettability a larger volume of water is imbibed than for the water-wet case. At the end of imbibition, oil is saturating pores of all sizes. Invading water fills almost all of the water-wet pores (displacing oil from these pores). When imbibition ends and forced oil displacement begins, water is forced into the larger (oil-wet) pores. Some oil is displaced through remaining pathways filled with oil and films of oil. In fact, most of the oil is displaced by the mechanism of film drainage if the waterflood is continued for a very long time.

For fractionally wet systems, as the overall fraction of oil-wet pores becomes greater the total volume of imbibed water decreases until it stops when the whole system is oil-wet because fewer paths for movement of water are available. Eventually a definite threshold pressure for forced entry becomes necessary as the system becomes strongly oil-wet.

For mixed wettability systems, an increase in the fraction of oil-wet pores results in a decrease in the total volume of imbibed water and thus the amount of water imbibed is an indication of wettability. They found that the greatest amount of production at water breakthrough for various states of wettability occurs when the rock is at neutral wettability, or 50% wet by water. For both mixed and fractionally wet rocks at intermediate wettability, oil recovery continues even after the economical residual oil saturation is passed.

The influence of wettability on production can be summarized as having three distinctly different properties, Fig. 3–12. Assuming that each one has the same water/oil viscosity ratio, water-wet systems exhibit piston-like displacement until water breakthrough occurs at or just after one pore volume of water has been injected. Neutral-wet systems exhibit earlier water-breakthrough because of fingers of water advancing ahead of the main production front, but production continues at increasing water/oil ratios to a practical residual oil saturation that is less than can be attained from water-wet systems. Water breakthrough occurs much earlier for oil-wet systems and production proceeds at a lower rate to a practical residual oil saturation that is greater than the others.

At high flow rates the injected phase will move rapidly through the larger pores, much like fingering through a viscous fluid, and produce early water breakthrough. The flood will behave as if the oil viscosity is very high or the system is definitely oil-wet.

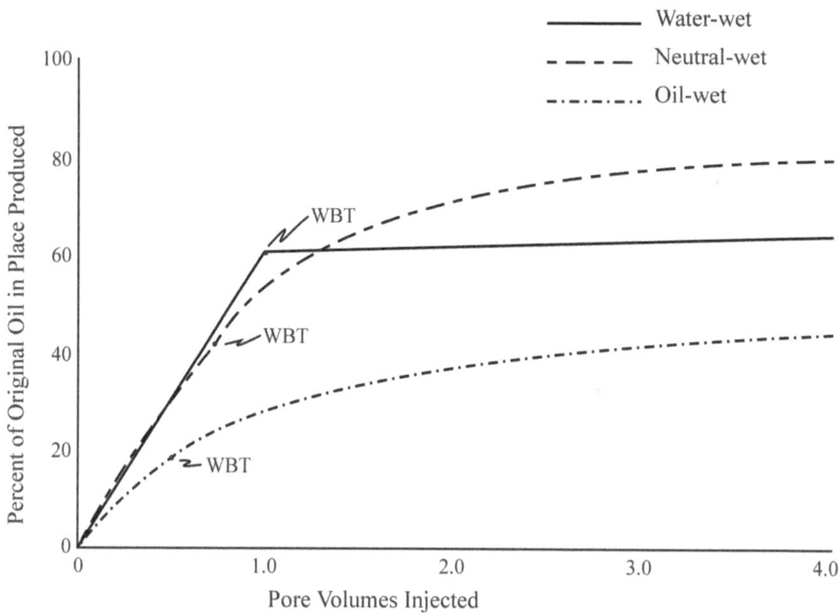

Figure 3–12 *Idealized production from three conditions of wettability.*

3.14 Example Problems

Example 3–1

The migration of oil as drops is opposed by capillary pressure which is related to the pore entry and exit sizes, interfacial tension, and the cosine of the contact angle:

$$P_{c(leading)} - P_{c(trailing)} = \left(\frac{2\sigma Cos\theta}{r}\right)_{leading} - \left(\frac{2\sigma Cos\theta}{r}\right)_{trailing}$$

Assume: $\sigma = 30\ N(10^{-3})/m$; $Cos\theta = 0.5$; $r_L = 10\ \mu m$; $r_T = 50\ \mu m$.

$$\Delta P_c = 2\sigma Cos\theta\left[\frac{1}{r_L} - \frac{1}{r_T}\right] = 2 * 30(10^{-3}) * 0.5\left[\frac{1}{10(10^{-6})} - \frac{1}{50(10^{-6})}\right]$$
$$= 2400\ N/m^2 = 2400\ Pa\ (0.348\ psi)$$

The buoyant pressure acting on the drop is:

$$P_B = zg_c(\rho_w - \rho_o)$$

Assume: Vertical length of the drop of oil (z) = 100 µm; ρ_w = 1.050; ρ_o = 0.850.

$$P_B = 100(10^{-6})(9.8)\left[(1.050 - 0.850)10^3\right]\left[m \cdot \frac{m}{s^2} \cdot \frac{kg}{m^3}\right] = 0.196\ Pa$$

Example 3–2

$$G_T = \frac{T_D - T_s}{D} = \frac{65^0 C - 25^0 C}{2.2 km} = 18.2^0 C\ km^{-1} = \frac{152^0 F - 77^0 F}{6688\ ft} = 1.1^0 F(100\,ft)^{-1}$$

Example 3–3

This example is listed to define the units used in three of the principal forms of Darcy's Law which are sometimes confusing.

a. SI-units.

$$q\left[\frac{m^3}{s}\right] = \frac{KA}{\mu L}(\Delta P - 9.81\rho z)\ \left[\left(\frac{(m^2)(m^2)}{(Pa \cdot s)(m)}\right)\left(Pa - 9.81\left(\frac{m}{s^2}\right)\left(\frac{kg}{m^3}\right)(m)\right)\right]$$

$$\text{where}: \left(\frac{m}{s^2}\right)\left(\frac{kg}{m^3}\right)(m) = \frac{kg}{(m \cdot s^2)(m^2)} = \frac{N}{m^2} = Pa$$

b. Units usually used in laboratories.

$$q\left[\frac{cm^3}{s}\right] = -\left(\frac{KA}{\mu L}\right)\left(\Delta P - \rho\left(\frac{a}{g_c}\right)z\right) \left[\left(\frac{(Darcy)(cm^2)}{(cP)(cm)}\right)\left(atm - \frac{atm}{1.01325 * 10^6}\right)\right]$$

$$where: \ \rho\left(\frac{a}{g_c}\right)z = \left(\frac{g^m}{cm^3}\right)\left(\frac{cm/s^2}{\left(\frac{g^m \bullet cm}{dyne \bullet s^2}\right)}\right)(cm) = \frac{dyne}{cm^2} = \frac{atm}{1.01325 * 10^6}$$

$$cP = \frac{g}{cm \bullet s}; \qquad a = local \ \ acceleration \ \ of \ \ gravity$$

c. English engineering units:

$$q\left[\frac{bbl}{D}\right] = -1.127\left(\frac{KA}{\mu L}\right)\left(\Delta P - \rho\left(\frac{a}{g_c}\right)z\right) \ \left[\left(\frac{(Darcy)(ft^2)}{(cP)(ft)}\right)(psi - 0.433psi/ft)\right]$$

$where:$

$$\rho\left(\frac{a}{g_c}\right)z = \left(\frac{62.4 \ lb^m}{ft^3}\right)\left(\frac{xx \bullet ft/s^2}{\left(\frac{32.174 \ lb^m \bullet ft^2}{lb^F \bullet s^2}\right)}\right)(ft)$$

$$= \frac{62.4 \ lb^F}{\left(\frac{144 \ in^2}{ft^2}\right)(ft^2)} ft = 0.433psi \ / \ ft$$

$where: \ density \ \ of \ \ water = 62.4 \ lb^m \ / \ ft^3$

$local \ \ acceleration \ \ of \ \ gravity \ = xx \bullet ft \ / \ s^2$

$\mu = cP = 0.000672 \ lb \ / \ ft \bullet s = 2.42 \ lb \ / \ ft \bullet hr$

Hydrostatic pressure is frequently expressed (in English engineering units) in terms of the specific weight of water (lb^F/ft^3) and height of the water column (*z* or *h*):

$$P(\text{psi}) = \gamma_w * z$$

The specific weight of fresh water is 62.4 lb/ft^3; therefore, the hydrostatic pressure gradient is expressed as

$$(G_H((psi)/(ft)) = \gamma_w/144 = 0.433((psi)/(ft)))$$

The specific weight of water is a function of the salinity of the brine and temperature; hence the hydrostatic pressure gradient varies in different oilfields. Where the gradient is not known, it is generally estimated as 0.465 psi/ft which corresponds to water containing 80,000 ppm sodium chloride at 77°F.

Example 3–4a

Plot relative permeability curves for water-wet and oil-wet systems to show the deviation from water-wet to oil-wet curves. Water-wet relative permeability curves for water are higher than those for oil-wet system, while the oil-wet relative permeability curves are lower. The cross-over point (where the relative permeabilities are equal) has a higher value for lower water saturations than those for oil-wet systems.

CORE DATA

$K = 167$ md
$A = 11.4\ cm^2$
$L = 25$ cm
$PHI = 0.24$
$V_B = 284.88\ cm^3$
$V_P = 68.4$

Relative Permeability (Water-wet)		Oil-Wet	Water-wet	
S_w	K_{rw}	K_{ro}	K_{rw}	K_{ro}
0				
0.05				
0.1				
0.15				
0.2	0	1	0	1
0.25	0.07	0.6	0.045	0.835
0.3	0.12	0.42	0.068	0.7
0.35	0.155	0.305	0.078	0.535
0.4	0.19	0.17	0.085	0.375
0.45	0.235	0.105	0.1	0.26
0.5	0.28	0.065	0.11	0.165
0.55	0.34	0.03	0.125	0.115
0.6	0.41	0.005	0.145	0.075
0.65	0.525	0.0032	0.175	0.045
0.7	0.66	0.0008	0.215	0.02
0.75	0.78	0.0002	0.28	0.01
0.8				
0.85				
0.9				
0.95				
1				

Example 3–4b

Using the relative permeability data above, calculate the fraction of water flowing versus water saturation (Eq. 3.9) for the water-wet and oil-wet cases using a viscosity ratio equal to 1.0.

$$f_w(S_w = 0.25) = \frac{1}{1 + \dfrac{\mu_w k_{ro}}{\mu_o k_{rw}}} = \frac{1}{1 + \dfrac{0.84}{0.045}} = 0.051$$

S_w	Water-wet, f_w	Oil-wet, f_w
0.25	0.051	0.104
0.30	0.089	0.222
0.35	0.127	0.337
0.40	0.185	0.528
0.45	0.278	0.691
0.50	0.400	0.812
0.55	0.521	0.919
0.60	0.659	0.988
0.65	0.795	0.994
0.70	0.915	0.999
0.75	0.966	1.000

Example 3–4c

Estimate the derivatives of the fractional flow curves (Fig. 3–9) by taking differences of f_w and S_w.

Calculate the distance traveled by the advancing front for saturation less than S_{wbt} (Eq. 3.15) and the locations of saturations greater than S_{wbt} (Eq. 3.16) for various times (seconds) after the beginning of the waterflood.

$$\left(\frac{df_w}{dS_w}\right)_{S_w=0.30}^{Water-wet} = \frac{f_{w(0.35)} - f_{w(0.25)}}{S_{w(0.35)} - S_{w(0.25)}} = \frac{0.127 - 0.052}{0.35 - 0.25} = 0.761$$

$$\left(\frac{x}{L}\right)_{0.3}^{Water-wet} = \frac{(0.05)(10)(1.70)}{(11.4)(0.24)} = 0.31$$

$$\left(\frac{x}{L}\right)_{0.75}^{Water-wet} = \frac{(0.05)(50)(0.851)}{(11.4)(0.24)} = 0.78$$

Water-wet

	$S_{iw} < S_w \le S_{wbt}$				$S_{wbt} \le S_w \le S_{wbt}$		
S_w	f_w	df_w/dS_w	x/L t = 10	x/L t = 20	x/L t = 32.15	x/L t = 50	x/L t = 75
0.20	0.000						
0.25	0.051	0.885	0.31	0.61			
0.30	0.089	0.761	0.31	0.61			
0.35	0.127	0.962	0.31	0.61			
0.40	0.185	1.505	0.31	0.61			
0.45	0.278	2.152	0.31	0.61			
0.50	0.400	2.431	0.31	0.61			
0.55	0.521	2.591	0.31	0.61			
0.60	0.659	2.746	0.31	0.61			
0.65	0.795	2.558	0.31	0.61			
0.70	0.915	1.701	0.31	0.61	1.00		
0.75	0.966	0.851	0.31	0.61	0.50	0.78	0.85
0.80							

Oil-wet

	$S_{iw} < S_w \le S_{wbt}$			$S_{wbt} \le S_w \le (1-S_{or})$			
S_w	df_w/dS_w	x/L t=10	x/L t=20	x/L t=23.82	x/L t=30	x/L t=50	x/L t=75
0.20	0.000						
0.25	0.104	2.222	0.42				
0.30	0.222	2.325	0.42				
0.35	0.337	3.056	0.42				
0.40	0.528	3.542	0.42				
0.45	0.691	2.838	0.42				
0.50	0.812	2.277	0.42	0.99			
0.55	0.919	1.764		0.77	0.97		
0.60	0.988	0.750		0.33	0.41	0.69	
0.65	0.994	0.108		0.05	0.06	0.10	0.15
0.70	0.999	0.058		0.3	0.03	0.05	0.08
0.75	1.000						

Example 3–4d

Calculate the fraction of water flowing versus saturation (Eq. 3.10) using the relative permeability data for the water-wet system and a viscosity ratio equal to 5.0. Plot the $f_w - S_w$ curves for viscosity ratios of 1.0 and 5.0 to show the effects of viscosity on the fractional flow of water and oil (Fig. 3–9).

Table 3–5 Fraction of water flowing, as a function of water saturation, for water-wet systems with oil/water viscosity ratios of 1.0 and 5.0.

S_w	$\mu_o = 1\ f_w$	$\mu_o = 5\ f_w$
0.25	0.051	0.212
0.30	0.089	0.327
0.35	0.127	0.422
0.40	0.185	0.531
0.45	0.278	0.658
0.50	0.400	0.769
0.55	0.521	0.845
0.60	0.659	0.906
0.65	0.795	0.951
0.70	0.915	0.982
0.75	0.966	0.993

CHAPTER 4

Pore Size Effects and Wettability Alteration

4.1 Introduction

Wettability has gradually been recognized as a phenomenon that is affected by many subtle properties of the environment within the pores of rocks. The recognition and study of wettability began in the 1930's and as its effects on production became more evident, different aspects of wettability were studied using the useful variations that are evident in capillary pressure behavior. Gradually other variables were introduced as the body of literature increased. The properties of rocks, mineral, pore structure, pore geometry, pore surface chemistry, and pore size, etc., have been recognized as some of the most important criteria for recognition of wettability and its alteration for enhancement of oil recovery.

The production of petroleum from deep subsurface reservoirs and shallow accumulations of heavy oils and bitumens is the objective of laboratory and field research. This research has been generally centered on specific areas of the chemical and physical aspects of brine/oil/rock system properties and particular reservoirs. The theories and conclusions of the complex mixture of components that affect fluid flow properties are generalized to predict the behavior of fluids in reservoirs. Overall it is the interaction of the chemical and physical properties of the three phases that govern the rates and limits of economic hydrocarbon production. With respect to wettability, oil reservoirs can be classified into very broad categories: (1) silicate-based sand/sandstones that have net negative surface charges ranging from neutral to several hundred millivolts; these sediments produce reservoirs that exhibit neutral to strongly water-wet conditions; and (2) biologically deposited carbonate reservoirs that have positive surface charges and are preferentially oil-wet, ranging from neutral to strongly oil-wet. The differences of wettability between silicate and carbonate reservoirs is evident in the differences of multiple parameters: capillary pressure, contact angle, production rates, water/oil producing ratios, sweep efficiency, and residual oil remaining after the economic limit of production has been attained.

The overall production characteristics of individual reservoirs, however, are affected by the physical properties of the sedimentary basins that can be viewed, for simplicity, as forming under two different conditions, beginning with subsidence. First, if the region surrounding the depression is one of low topographical relief, biological carbonate deposition is more probable in a shallow marine environment. If the physiographic region is mountainous, clastic silicate particles derived from erosion of the mountains will accumulate in the basin. Reservoirs composed of clastic deposits vary considerably because of different environmental conditions involved in their genesis; however, they have the fact that they are fragments of eroded rocks deposited by transport to a basin, river, or marine environment by wind and moving surface water in common. Some clastic reservoirs are composed of carbonate grains (oolites) and fragments of shells deposited in place in a shallow marine environment. The process can change to a fluid regressive phase (where the volume of sediments accumulating in the basin exceeds the rate of subsidence). Carbonate depositions generally do not occur during regressive phases of basin formation. Second, where the rate and volume of subsidence is greater than the rate of accumulation of the sediments, a fluid transgressive phase controls the process. If the region surrounding the depression is one of low topographical relief, biological, carbonate deposition is more probable.

Carbonate and dolomite reservoirs exhibit two types of porosity. The primary porosity is formed during the initial stages of deposition and may be modified by diagenetic changes occurring after deposition and during the development of secondary porosity from circulating subsurface water and from tectonic activity.

Overall, the petroleum reservoirs are composed of a wide variety of rock types and geometries with porosity and permeability that are the consequence of the changing local environmental conditions during genesis and burial. In the resulting petroleum reservoirs, if there is a network of fractures, wettability is an important aspect of production because wettability controls the water volume and rate of imbibition into the fracture-blocks with subsequent oil displacement into the fracture channels that conduct fluid to production wells. A thorough knowledge of petroleum reservoir geology, sedimentary history, and petrophysics must be integrated into the analyses of the influence of wettability on a specific reservoir.

4.2 Porosity

The interactions of wettability and fluid-flow properties with multiphase fluids in porous media are influenced to a large degree by the pattern and structure of the pores. Porosity is the ratio of pore space to the volume of the rock sample which includes all of the voids whether interconnected or not. From an engineering perspective, it is the interconnected pore space which is capable of storing and conducting fluids that is of practical importance. The shapes of the pores, their individual sizes, and the pore size distribution of the pores are features that exert a large influence on the physical behavior of multiphase fluids. In very small capillaries (< 2 μm) liquids tend to be completely immobile because of strong molecular attraction between the rock and fluids and because of very high capillary pressures that are required for displacement. Thus, the effective porosity for fluid flow must consist of a distribution of pores and intersecting pore openings greater than 2 μm. The shapes of the pore throats may be spherical, cylindrical, highly irregular with sharp projections, cavernous, vuggy, etc.

The voids may be classified as *primary voids* originating from (1) intercrystalline phases that depend on the degree of heterogeneous stacking and compaction for size and shape; (2) voids between grains of minerals in sediments accumulated from the erosion of rock outcrops such as granites, stacking of oolites, and rapid deposition of detritus particles, large cavernous voids and vugs (>2 mm, irregular in shape often formed by dissolution of the matrix) from highly irregular size particles at the time of deposition; (3) layered depositions of different types of sediments during variable environmental periods of deposition, such as sand-shale sequences; and (4) voids resulting from volcanic activity, such as the formation of gas bubbles in pumice, producing high porosity accompanied by almost zero permeability, and cooling and motion fractures within bodies of lava. *Secondary voids* arising from post deposition events that modify the primary voids originate from: (1) alteration by meteoric waters moving slowly through the porous rocks carrying suspended particles that are deposited in pores; (2) dissolved minerals that precipitate between grains as cementing agents, eventually producing massive formations of consolidated porous sandstones that have surprisingly uniform characteristics over large distances; (3) voids caused by decomposition or disintegration of rocks by meteoric water; and (4) various fracture voids produced by tectonic events.

Porosity depends on grain size, shape and sorting, the mechanics of sedimentation, compaction by overburden structures, and cementation of the grains. Some processes diminish or destroy primary porosity while others enhance it. Porosity may be destroyed and reopened with a completely different interconnected pattern. Thus each sedimentary rock has a different history of deposition and genesis of porosity and permeability that depend on long periods of local changing environmental conditions.

Neasham (1977) used a scanning electron microscope (SEM) to describe the crystalline morphology of clay particles in the pores of sandstones as a result of diagenetic changes. Neasham distinguished three different types of clay morphology produced by diagenetic changes in sandstones: (1) *discrete particles* identified as crystals of kaolinite resting on the surfaces of sand grains which are undoubtedly mobile and will reduce the permeability when fluids are mobilized for production; consequently the reduction of permeability will be a function of fluid velocity and will have its greatest effect near the wellbore creating the skin-effect (permeability reduction) observed in well tests; (2) *pore lining* where clay deposits on the grains as a continuous growth-structure reducing the size of the pores and severely restricting the flow of fluids; three common types of clay were identified (illite, montmorillonite, and chlorite), but this phenomena may be attributed to many other clay types (pore lining apparently produced low porosity but exhibited relatively high permeability); and (3) *pore bridging* by fibrous crystal growth across the pores (high porosity was observed at 18%, accompanied by low permeability, 0.5 md). The primary and secondary genesis of voids in porous rocks listed above are a very general set of more prominent conditions.

Rocks are divided into principal classifications according to their mode of formation: (1) igneous intrusive rocks: granite, diorite, and gabbro; (2) igneous extrusive rocks: ryolite, andesite, and basalt; (3) metamorphic rocks: gneiss, amphibolite, schist, and slate; (4) sedimentary rocks originating from decomposition and transport of rock fragments to depositional basins; (5) chemical rocks formed from accumulations of minerals that precipitate from aqueous solutions because of changes of temperature and concentration; and (6) biochemical rocks.

The chemical compositions of these rocks reveal the relative abundance of elements (expressed as oxides). Table 4–1 shows that the major constituents are few in number (silicon, aluminum, iron, magnesium, calcium, potassium, and sodium). The rocks, however, are composed of a mixture of minerals that are responsible for their petrophysical properties and are summarized by Tiab and Donaldson (2004).

The fragmental texture (size, shape, sorting, and grain-to-grain boundaries) essentially govern porosity and permeability of detritus rocks, and the crystalline texture of biogenic and chemical rocks govern porosity and permeability. There are three principal components of textural properties: (1) particles or grains and crystals whose influence is a function of the size and shape; (2) a matrix composed of

Table 4–1 Elemental oxide compositions (percent of total) of major igneous rocks (after Daly, 1933).

Oxide	Granite	Diorite	Gabbro	Rhyolite	Andesiste	Basalt
SiO_2	70.2	56.8	48.2	72.8	59.6	49.1
Al_2O_3	14.5	16.7	17.9	13.5	17.3	15.7
Fe_2O_3	3.4	7.6	8.5	2.3	6.5	11.8
MgO	0.9	4.2	7.5	0.4	2.8	6.2
CaO	2.0	6.7	11.0	1.2	5.8	9.0
K_2O	4.1	2.1	0.9	4.5	2.0	1.5
Na_2O	3.5	3.4	2.6	3.4	3.6	3.1
TiO_2	0.4	0.8	1.0	0.3	0.8	1.4
Other	1.0	1.7	2.4	1.6	1.6	2.2

materials <1/256 mm filling interstitial areas and in some cases supporting grains; and (3) cement binding the grains and matrix. Matrix materials are more important in poorly sorted sands where they contain a greater percentage of fine material within pores and separating grain contacts. Cementing materials enter the sedimentary grains after deposition as precipitates from circulating water as local environmental changes take place.

4.2.1 Pore Structure

A large body of literature has developed discussing the morphology and pore geometry of rocks based on discoveries using X-rays, electron microscopes, nuclear magnetic resonance, etc. Many of the discoveries of pore geometry apply to petroleum reservoirs and are critical to understanding production characteristics.

Tight Gas Sands

The U.S. Bureau of Mines defined "tight gas sands" as reservoirs with a dry gas permeability equal to 100 micro-darcies. However, the industry's accepted definition is a reservoir that can produce economical quantities of gas after fracture stimulation. The porosity is generally <10% for tight gas sands and they are generally composed of layered clastic sedimentary deposits of quartz sand, mudstones, and shale. Their petrophysical properties are functions of composition and insitu stress porosities. Therefore, all petrophysical properties must be determined on cores equilibrated at simulated reservoir properties. A small change in the pore size distribution, by expansion or compression, will produce large deviations from actual insitu properties (Holditch, 2006). The deviation is greater for tight gas sands because the connecting pore throat sizes are very small; therefore, the relative expansion of the pore throats is considerably greater than it is for high permeability sands when the overburden is released upon extraction of the core. In addition, the measurement of permeability with air must guard against evaporative removal of water by hydration from the minerals. Randolph et al. (1984) conducted air permeability measurements on tight-sand cores using a relative humidity equal to 45% with 1.4 MPa pore pressure and a confining pressure equivalent to the subsurface depth of the core. When the permeability was measured without humidity control, an increase in both porosity and permeability was observed for the same cores. The overburden stress of tight gas sands causes no more than about 5% change of porosity, but the change of absolute permeability is quite large and varies with respect to the type of rock. Permeability changes of up to 75% were observed for compression of cores from ambient pressure to 0.5 MPa. The effect on relative permeability, however, was negligible (Thomas and Ward, 1971).

Soeder and Randolph (1987) used a scanning electron microscope to characterize the relationships between porosity and permeability of the Mesa Verde sandstone in the Piceance basin of Colorado. They identified four distinct types of pore structure that theoretically explained the differences of porosity-permeability relationships of tight gas sands formed in four different environments: fluvia (lenticular sandstones from bars and channel fill interbedded with shales and siltstones); coastal (lenticular channel sandstones originating from sand deposited in low velocity river channels and meander belts); paludal (lenticular sandstones interbedded with coal and shale); and marine (shallow marine sandstones intercalated with shales).

1. Intergranular porosity between grain-to-grain contacts (grain-supported pores) compacted by the overburden pressure. This type of pore structure exhibited very low compressibility because the overburden is supported by quartz grain-to-grain contacts that will not yield to further compression unless the grains are crushed. Small amounts of clay deposited within

the pores and around the grains was the largest contributor to the low compressibility (1.5×10^{-5}/kPa) because rearrangement or further compaction of the quartz grains that were already tightly compacted is severely restricted. The rather clean compacted sandstone has porosities and permeabilities in the upper ranges for tight gas sands ($\phi > 10\%$; $K = 10$ to 100 μd). They did not find sandstones having this type of pore geometry with permeabilities less than 10 μd.

2. Intergranular slots connecting secondary solution pores. The slots contain silica that severely restricts the flow of fluids. The silica was postulated to have originated from the dissolution of quartz sand grains and then precipitated within the slotted primary porosity. The solution pores are connected by entry points that are rectangular slots frequently lined with autogenic clays. Although this dual porosity condition exhibited moderate porosity (5–10%), the permeability associated with this type of pore geometry was 0.1–10 μd because of the restricted pore entry sizes.

3. Matrix-supported grains where the porosity consists of fine pores within the supporting clay matrix. This type of structure exhibits high compressibility (4×10^{-4}/kPa) because some slight movement of the sand grains within the matrix was possible, and the matrix had considerably more compressibility than the quartz grains but it had the lowest range of porosity and permeability ($\phi = 5\%$, $K = 1$ μd).

4. Primary porosity (between grains) that has been degraded to linear slots by precipitation of quartz in the pores.

Images of the detailed structure of small samples can be accurately produced by X-ray tomography and nuclear magnetic resonance (NMR). Water and oil can be distinguished; therefore, the location and movement of reservoir liquids can be studied in addition to the convoluted three dimensional structure of the rocks. Dunsmuir et al. (1991) present a detailed discussion of X-ray tomography and the techniques for its use for the study the pore structure of fused glass beads and Berea sandstone. Images of the dry rock samples are subtracted from the images of the samples completely saturated with an X-ray opaque fluid to secure three dimensional maps of the interconnected fluid-filled pores. In addition, motion pictures of single- and two-phase flow (water and hydrocarbons) also were obtained. The results are excellent for defining the interfacial interactions of fluids moving through networks of pores.

The two properties of rocks that are essential for the storage and transport of hydrocarbons and water are porosity and permeability. There are three ranges of pore sizes where wetting phase saturation limits are established that are characteristic of the rock and wettability of the system; it is customary to express the multiphase saturations in terms of the established wetting phase saturation:

1. When a rock is saturated with a wetting phase and is displaced by a nonwetting, immiscible fluid, displacement (drainage) of the wetting phase will continue until a minimum saturation is attained as the capillary pressure approaches infinity; this is labeled the irreducible wetting phase saturation, S_{wi}. At this saturation, the attractive molecular forces between the walls of the pores and the fluid are greater than the hydrostatic force imposed across the porous medium; thus the initial mobile wetting phase saturation is $1.0 - S_{wi}$.
2. If the displacement process is reversed such that the wetting phase is used to displace a nonwetting phase saturation (imbibition displacement), the wetting phase also will attain a minimum saturation (as the displacement capillary pressure approaches negative infinity) and is the wetting phase saturation at which a residual nonwetting phase is attained ($S_{wor} = 1.0 - S_{or}$) where oil is the nonwetting phase. At this point the nonwetting phase becomes discontinuous within the larger pores and is no longer affected by the imposed hydrostatic force across the porous medium.
3. Both fluids are mobile in the region between S_{wi} and S_{wor}. The initial (discovery) saturation of hydrocarbons in a reservoir may be at any value of S_o (from 0.0–1.0), but some water is always present, either as the wetting or nonwetting phase.

4.2.2 Pore Texture

The pore entry shapes of clastic rocks are infinite in variety but have been described in very general terms as spherical, ellipsoidal, triangular, slotted rectangular, cylindrical, cavernous, and honeycombed. The geometric shapes of the pore entries that can be obtained from microscopic studies are an important aspect of porosity. The pore size distribution obtained from capillary pressure curves are a reflection of the mean hydraulic radii of the pore entries.

The actual void spaces are classified in accord with the changing environmental conditions that modulate the size and rates of accumulating sedimentary particles. A few of the classifications divided into primary (developing during sedimentation) and secondary (developing after sedimentation has occurred because of changing tectonic and environmental conditions): (1) intergranular porosity: voids between point contacts of the grains; (2) intercrystalline: voids produced between individual crystals, cleavage planes of crystals, and voids within the crystal lattice of deposited minerals; (3) cavernous or vuggy spaces formed during deposition from packing of variable size grains or by burrowing organisms; (4) spaces arranged horizontally in an alternating sequence between bedding planes of clays which are generally caused by changes of environmental conditions that moderate the size of grains and type of sediments; and (5) spaces produced by cracks of the rock as cooling took place.

Genesis of the principal secondary types of porosity are: (1) faults created by post-deposition tectonic motions, (2) cracks and joints formed by tension relief at the crests of synclines and anticlines, (3) spaces formed when subsidence rates exceed the rate of deposition by parting of the depositional areas, and (4) voids produced during uplift followed by erosion, precipitation of compounds from meteoric water and subsequent burial, and solution voids produced by circulating meteoric water.

4.2.3 Sandstones

The porosity of sandstones is the end result of a wide variety of sedimentary conditions and cannot be described in precise mathematical terms. Statistical analyses based on grain size distribution and sorting present some means for statistical evaluations. The grain size distribution used to describe sandstones is determined by disaggregating the grains and then using a set of Wentworth-scale sieves with different mesh sizes ranging from coarse sand (1 to 2 mm in diameter) to fine (0.25 mm), silt (0.05 to 0.005 mm) and clay minerals (<.005 mm). The procedures and analyses are detailed by Tiab and Donaldson (2004). Unconsolidated sand beds are geologically young or have not undergone deep burial that results in compaction of the grain and solidification into sandstones by adhesion to minerals codeposited with the sands or precipitated from meteoric water (principally anhydrite, pyrite, calcite, oxides of iron, and clays). The volume of intergranular fine-grained materials known as the matrix may exceed the volume of sand grains and suspend the grains, destroying the grain-to-grain contacts. This type of consolidated sandstone will have severely diminished porosity. The original deposition of clean quartz sand (such as beach sand or river and deltaic sand bars) can have porosity in excess of 35%, but the porosity decreases rapidly as burial, compaction, and cementation of the grains occurs. The decreasing porosity with depth has been found to be a linear trend with respect to depth for specific reservoirs (from Chilingarian and Wolf, 1976; and Loucks et al., 1984).

The classification of clastic rocks is complex because there is considerable variability with respect to the major constituents, particle size, cementing materials, etc. Shales are composites of clay minerals and fine quartz and feldspar minerals that are deposited as laminar planes. Consolidated clastic particles of various types are classified as sandstones. Orthoquartzite sandstone is composed of very pure quartz grains cemented by silicates. Arkose sandstone contains quartz and a majority of feldspar, and is found close to the outcrops of igneous rocks (usually granite) from which it was generated because the particles are soft and cannot survive the abrasion of long distance transport. Arkose sandstones form excellent petroleum reservoirs because they generally have high porosity, but they are water sensitive (containing swelling clays that can cause disaggregation of the rock if it is contacted with fresh water). Graywacke (or litharonite) is a sandstone containing quartz, feldspar, clay and metamorphic rock fragments, and fines; the porosity tends to be low (<15%) but it represents a large segment of oil reservoirs.

Permeability is an approximate log-normal function of porosity and therefore a linear regression equation can be used for the approximate description of sandstones classified by grain size (coarse, medium, fine, silty, and clayey sandstones) (Chilingarian and Wolf, 1975). The log-linear relationship is the most common and can represent segments of the relationship; the second order term can be used to include some of the nonlinear trends, Table 4–2, Eq. 4.1:

$$log(K) = a_0 + a_1\phi + a_2\phi^2 \tag{4.1}$$

Table 4–2 Coefficients for estimation of permeability from correlations of porosity (after Chilingarian and Wolf, 1975).

	a_o	a_1	a_2
Coarse to very coarse-grained	0.4026	29.3078	–87.6729
Coarse to medium-grained: porosity < 0.17	0.1646	18.6216	–10.6615
Coarse to medium-grained: porosity > 0.17	–2.6667	45.0000	–83.333
Fine-grained	–5.4158	71.2453	–166.1990
Silty	0.1040	12.1080	–8.1826
Clayey	–1.1926	17.2853	7.3875

4.2.4 Carbonate Rocks

Carbonates are formed in shallow marine environments on the shelf and slope of quiet margins by deposits of algae-type plants and animals occupying the shallow zones and through precipitation from concentrated solutions. The voids in carbonates are consequently very varied and several authors have proposed terms for general classification which are discussed in detail by Chilingarian et al., 1992. The systems of porosity that develop in carbonate reservoirs can be summarized for simplicity as five principal contributing factors:

1. Intergranular (intercrystalline) porosity: Reservoirs composed of porosity between the grain-like skeletal remains of animals and voids caused by phylloid algae have excellent porosity and permeability. Any of these original voids may have been destroyed by secondary deposition of lime and calcite. Porosity and permeability also are high in dolomite carbonate reservoirs composed of granular rhombic crystals.

2. Intragranular porosity consisting of very fine pores that do not contain hydrocarbons but are filled with immobile water. They register high water saturations (even when the large pores in the system are filled with hydrocarbons); hence, the intergranular porosity is insignificant with respect to the storage and transport of hydrocarbons.

3. Solution porosity: Carbonates are soluble in low pH water (Fig. 4–1) and over geologic time can develop large regions of porosity consisting of solution vugs and channels from circulating marine and meteoric water; these have porosities generally less than 10%, but high permeability.

4. Fracture porosity: the brittle nature of carbonates leads to considerable fracturing if they are disturbed by tectonic pressure and tensile stresses after deposition. The fracture systems in compressive environments are tightly closed and hence do not contribute to porosity or permeability, but in environments that are under tension or have experienced tension which has been relaxed, such as anticlinal folds, exhibit high permeability through the system of fractures, but the fracture's contribution to overall porosity is very low, generally less than 2%.

5. Moldic porosity: This occurs where the carbonate deposit is composed principally of lime mud with inclusions of soft shells that have dissolved, leaving a pattern of pores in the shapes of the plant and animal remains that were dissolved. The porosity of the system as a whole may be high, but the pores are generally isolated or connected only by fine pores partially filled with lime mud or micrite. Pores within the matrix are very fine and do not contain hydrocarbons.

Chilingarian et al. (1992) list correlations (for four carbonate rocks) of permeability to porosity that also include irreducible water saturation and surface area obtained by linear regression. The constants for each rock are different, but all of the correlation coefficients are greater than 0.98. The correlation with the highest degree of confidence ($R = 0.997$) is:

$$\begin{aligned} log(K) = {} & 0.9532 + 1.3309(10^{-1})\varphi - 2.7880(10^{-2})S_{wi} \\ & -5.5597(10^{-4})S_{area} + 1.1709(10^{-5})S_{wi} \bullet S_{area} \end{aligned} \tag{4.2}$$

This correlation may be useful for application to a specific carbonate reservoir if the correlation holds for cores from several locations, Eq. 4.2.

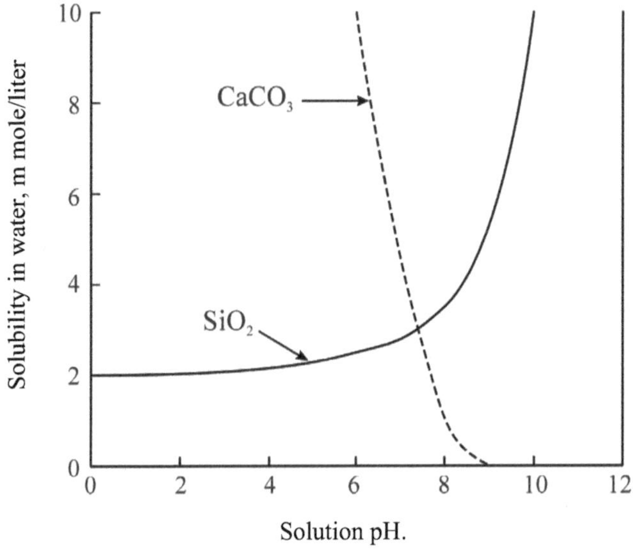

Figure 4–1 *Solubility of silica and calcium carbonate in water at ambient temperature (after Fergusson, 1982).*

Vuggy porosity (>2 mm, irregular in shape) within a solid limestone matrix has variable, irregular shapes several millimeters across. They generally are formed by dissolution of the matrix and many of the vugs are not connected, resulting in an apparent overall porosity which is greater than the interconnected porosity available for fluid-flow.

Intergranular porosity was identified between pellet-shaped grains of calcium carbonate cemented with calcite. The porosity of limestone formed from compacted coral was formed by decay and solution of soft areas of coral particles. Moldic porosity develops from carbonate skeletons that were deposited in the lime mud and later dissolved by circulating water, leaving a mold of its original form.

The most important criteria in the characterization of carbonate reservoirs is the connectivity between the various types of porosity; some may have high overall porosity but very low permeability because the network of connecting tubes have very small diameters. Connectivity has been studied by injection of hardening plastics and Wood's metal (a lead-bismuth alloy) and then dissolving the carbonate matrix with an acid, thus leaving a cast of the interconnected pores (Gleeson et al., 1993). Gleeson et al. describe a method for imaging the three-dimensional pore structure of carbonates using nuclear magnetic resonance. The images identified several categories of carbonate porosity and more importantly allowed the characterization of connectivity in the maze of pores when coupled with observation of the displacement of a fluid from the sample. In one case, only one fluid flow path was found in a sample and the rest of the pores could be clearly observed, but they were not connected with pathways having diameters that allowed the flow of water or oil. This type of analysis coupled with measurement of air permeability and pore size distributions from capillary pressure curves can define the fluid flow properties of rocks and explain the production behavior and residual saturations. These analytical and semiempirical models facilitate rapid estimation and implementation of k_r in simulators.

4.3 Pore Size Distribution

Ritter and Drake (1945) developed a method for determination of a pore size distribution for porous materials based on mercury-air capillary pressure curves (see *Program Listing 6.14 POSIZE.FOR*). They defined the pore size distribution function as:

$$D_{(ri)} = \frac{dV}{dr} \qquad \left[\frac{m^3}{m} = m^2\right]$$
$$D_{(ri)}dr = dV = V_p dS \tag{4.3}$$

from the capillary pressure equation:

$$dr = \left(\frac{r^2}{2\sigma Cos\theta}\right)dP_c = \left(\frac{r}{P_c}\right)dP_c$$
$$D_{(ri)} = \left(\frac{P_c V_p}{r}\right)\frac{dS}{dP_c} \tag{4.4}$$

Air-brine capillary pressure curves were used to obtain pore size distributions of several rock samples, Fig. 4–2 (Donaldson et al., 1991). For air-brine systems, water wets the surface and air is completely nonwetting, therefore the cosine of the contact angle can be assumed to be equal to 1.0. In *POSIZE.FOR* (*Program Listing 6.14*) and *POCEN.FOR (Program Listing 6.9*) a least squares fit of a segment of a hyperbola is used to represent capillary pressure curves and the pore size distribution curves. The curves can then be differentiated or integrated using the hyperbola. By using the least squares fit of the capillary pressure curve, variations because of experimental errors are averaged, producing a smooth curve that can be differentiated without wide deviations resulting from the variance of the experimental data. In addition, the threshold pressure of the capillary pressure curve can be accurately extrapolated. The measured threshold pressure is erroneous because evaporation of fluid from the face of the core cannot be avoided using general laboratory procedures, and this yields a lower estimate of the threshold pressure than the true value obtained from extrapolation of the curve to $S_w = 1.0$. The total pore volume is equal to the total area under the curve, plus the volume occupied by the irreducible saturation ($V_{p(\text{measured})} {}^*S_{wi}$) (Donaldson et al., 1991). The capillary pressure curve is stopped at the practical residual water saturation; therefore, the volume of the residual water must be calculated from the measured pore volume of the sample.

$$P_v = Area_{(size\ dis.)} + P_{v(measured)} * S_{wi} \tag{4.5}$$

The fluid flow characteristics, and practical residual saturation (where $P_c \rightarrow \infty$), are controlled by the pore size distribution and wettability. The irreducible water saturation is located in the small pores, from a size range where the drainage capillary pressure approaches infinity, to the pore diameter equal to the free mean path of the fluid molecules.

4.3.1 Permeability from Pore Size Distributions

Capillary pressure is a function of wettability (through $\sigma Cos\theta$) and the entry size of the tube or pore containing two phases. Permeability is a function of those two parameters; therefore, a pore size distribution obtained from a capillary pressure curve is directly linked to permeability. Purcell (1949) adapted Ritter and Drake's method for calculation of absolute permeability from mercury-air capillary pressure curves. Burdine et al. (1950) incorporated the calculation of the pore size distributions as well as permeability; Burdine (1953) extended the equation to the calculation of relative permeabilities by

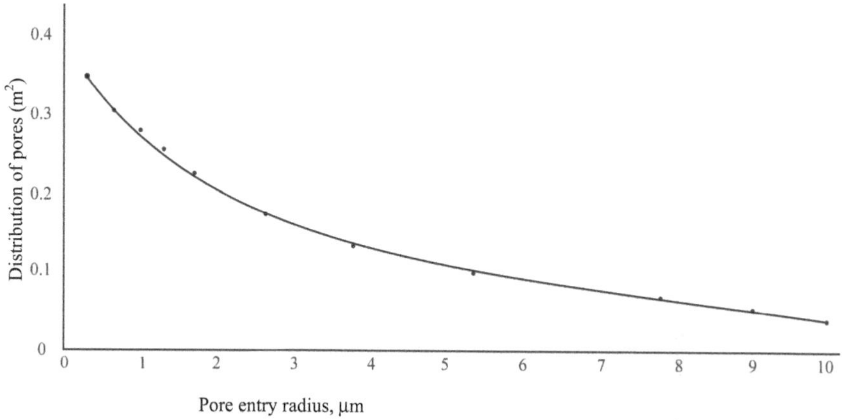

Figure 4–2 *Pore entry size distribution of Berea sandstone.*

Table 4–3 Empirical formulas for the formation resistivity factor for various types of porous media (after Tiab and Donaldson, 2004).

Porous Medium	Formation Resistivity Factor
Carbonate	$1.00/\varphi^{2}$
Calcareous Sands	$1.45/\varphi^{1.7}$
Unconsolidated Sands	$1.00/\varphi^{1.5}$
Unconsolidated Sands (Gulf Coast of USA)	$1.97/\varphi^{1.29}$
Unconsolidated Shaley Sands	$1.65/\varphi^{1.33}$
Sandstones	$0.81/\varphi^{2}$
Sandstones (Humble Formula)	$0.62/\varphi^{2.15}$

interpreting the tortuosity factor in terms of relative fluid saturations. Wyllie and Gardner (1958) showed the development of permeability equations using the Carman-Kozeny equation (Carman, 1937) and interpreted the tortuosity factor as a function of the formation resistivity factor (Table 4–3).

Purcell (1949) began with Poiseuille's equation and validated the equation for mercury-air capillary pressure curves. Poiseuille's equation for straight capillaries may be written as:

$$q = \frac{\pi r^4 \Delta P}{8 \mu L} \tag{4.6}$$

The void volume of the tube, V_p, and the capillary pressure equations may be expressed as:

$$V_p = \pi r^2 L; \qquad r = \frac{2\cos\theta}{P_c} \tag{4.7}$$

Substituting Eq. 4.6 into 4.7 yields an equation for flow of a fluid through individual capillaries that incorporates capillary pressure:

$$q_i = \frac{(\sigma\cos\theta)(\Delta P)(V_i)}{2\mu L^2 P_{ci}^{\ 2}} \tag{4.8}$$

In order to apply Eq. 4.8 to flow through packed beds of sand, Carman (1937) introduced the concept of an equivalent length for the flow path through porous media to account for the longer, actual fluid flow path, L_e, with respect to the physical length of the porous medium, L, where $L_e > L$. Substituting into Eq. 4.8 and using the summation of flow through capillary tubes of various sizes, the total flow rate through the porous medium is:

$$\begin{gathered} q_t = \sum q_i = \frac{(\sigma\cos\theta)^2 \Delta P}{2\mu L_e^2} \sum_0^1 \frac{V_i}{P_{ci}^2} = \frac{KA\Delta P}{\mu L} \\ K = \frac{(\sigma\cos\theta)^2 L}{2AL_e^2} \sum_0^1 \frac{V_i}{P_{ci}^2} \end{gathered} \tag{4.9}$$

Defining V_i and AL of the capillary tubes as follows and substituting into Eq. 4.9 yields an equation for the absolute permeability in terms of capillary pressure and the tortuosity of the system:

$$V_i = S_i V_b \qquad \text{and} \qquad \phi AL = V_b$$

$$K = \frac{(\sigma \cos\theta)^2 \phi}{2\left(\frac{L_e}{L}\right)^2} \sum \frac{S_i}{P_{ci}^2} \tag{4.10}$$

Introducing the definition of tortuosity (Carman, 1937; Wyllie and Spangler, 1953):

$$\tau = \left(\frac{L_e}{L}\right)^2 = F_r^2 \phi^2 S^2 \tag{4.11}$$

where F_r is the formation resistivity factor (Table 4–3) and S is fluid saturation. Substituting Eq. 4.11 into 4.10 allows the calculation of absolute permeability from capillary pressure curves and the formation resistivity factor:

$$K = \frac{(\sigma \cos\theta)^2 \phi}{2(F_r \phi)^2} \int_0^1 \frac{dS_i}{dP_{ci}^2} \tag{4.12}$$

Burdine et al. (1950) used the summation of flow through capillaries of different sizes to obtain the total flow rate from Poiseuille's equation:

$$q_t = \sum q_i = \frac{\pi \Delta P}{8 \mu L_e} \sum_0^n n_i \bar{r}_i^4 = \frac{K A \Delta P}{\mu L}$$

$$K = \frac{\pi}{8 A x_i} \sum_0^n n_i \bar{r}_i^4 \qquad \text{where}: x_i = \frac{L_e}{L} \tag{4.13}$$

Defining the porosity and substituting for π/A yields Eq. 4.14 as an expression of permeability:

$$\phi = \frac{\pi}{A}\sum n_i \bar{r}_i^2$$

$$K = \left(\frac{\phi}{8}\right)\frac{n_i \sum \bar{r}_i^4}{n_i x_i \sum \bar{r}_i^2} \quad (4.14)$$

Defining the volume of an individual capillary (V_i) and the number of pores (n_i) and substituting into Eq. 4.14:

$$V_i = D_{(ri)}\Delta r = \pi x_i L \sum n_i \bar{r}_i^2$$

$$n_i = \frac{D_{(ri)}\Delta r}{\pi x_i L \sum \bar{r}_i^2}$$

and

$$K = \left(\frac{\phi}{8}\right)\frac{\left(\dfrac{1}{\pi x_i L \sum \bar{r}_i^2}\right)\sum_0^n \dfrac{V_i \bar{r}_i^4}{x_i^2 \bar{r}_i^2}}{\dfrac{\sum_0^n D_{(ri)}}{\pi x_i L \sum_0^n \bar{r}_i^2}} \quad (4.15)$$

Eq. 4.15 may be simplified because the summation of the distribution function in the denominator is equal to 1.0; therefore,

$$K = 0.125\sum_0^n \frac{V_i \bar{r}_i^4}{x_i^2 \bar{r}_i^2} \quad [\mu m^2]$$

$$K = 0.127\sum_0^n \frac{V_i \bar{r}_i^4}{x_i^2 \bar{r}_i^2} \quad [md] \quad (4.16)$$

The tortuosity term for a saturated capillary (S = 1.0) (L_e/L) was defined in Eq. 4.11 as $(F\phi)^2$; therefore, Eq. 4.16 can be used to estimate the absolute permeability from the pore size distribution. Burdine (1953) extended Eq. 4.16 for calculation of relative permeabilities. Assuming application of Eq. 4.16 to a two phase system, the effective permeability of the wetting phase is:

$$k_{ew} = 0.127 S_w \phi \sum_0^n \frac{V_{iw} \bar{r}_i^4}{x_i^2 \bar{r}_i^2} \tag{4.17}$$

The tortuosity factor (as defined by Burdine) of the wetting phase, $x_{iw} = L_{iw}/L$, where L_{iw} is the effective length of the wetting phase flow path in the porous media and the relative factor is defined as the ratio:

$$x_{riw} = \frac{x_i}{x_{iw}} \tag{4.18}$$

In addition, Burdine made the assumption that:

$$V_{iw} = \frac{V_i}{S_w} \tag{4.19}$$

Substituting Eqs. 4.16 and 4.16 into 4.15 and dividing by the absolute permeability, Eq. 4.14, yields an expression for the relative permeability of the wetting phase:

$$k_{rw} = \frac{0.127 \phi x_{ri}^2}{K} \sum_0^n \frac{V_i \bar{r}_i^4}{x_i^2 \bar{r}_i^2} \tag{4.20}$$

The term x_{rw} cannot be evaluated for intermediate values of saturation, but was found to be a linear function of saturation from numerous experimental observations where:

$$S_{ew} = aS_w + B \tag{4.21}$$

Eq. 4.21 is solved as follows to yield S_{ew} and S_{en} using the following boundary conditions (where *w* indicates the wetting phase and *n* the nonwetting phase):

$$\begin{aligned} S_{ew} &= 1.0; \quad S_w = 1.0 \\ S_{en} &= 0; \quad\;\; S_w = S_{wi} \end{aligned} \tag{4.22}$$

Thus,

$$\begin{aligned} S_{rw} &= \frac{S_w - S_{wi}}{1.0 - S_{wi}} \\ S_{rn} &= \frac{S_o - S_{or}}{1.0 - S_{or}} \end{aligned} \tag{4.23}$$

Therefore,

$$k_{rw} = \left(\frac{S_w - S_{wi}}{1.0 - S_{wi}}\right) \frac{\sum \frac{V_i \bar{r}_i^4}{\bar{r}_i^2}}{\sum \frac{V_i \bar{r}_i^4}{\bar{r}_i^2}} \equiv \left(\frac{S_w - S_{wi}}{1.0 - S_{wi}}\right) \frac{\int_0^{S_w} \frac{dS_w}{dP_{ci}^2}}{\int_0^1 \frac{dS_w}{dP_{ci}^2}} \tag{4.24}$$

$$k_{rn} = \left(\frac{S_o - S_{or}}{1.0 - S_{or}}\right) \frac{\int_{S_w}^{1} \frac{dS_w}{dP_{ci}^2}}{\int_0^1 \frac{dS_w}{dP_{ci}^2}} \tag{4.25}$$

Corey (1954) equated the saturation terms to the effective saturation defined as:

$$\begin{aligned} S_{ew} &= \frac{S_w - S_{wi}}{1.0 - S_{wi}} \quad \textit{effective saturation of the wetting phase} \\ S_{en} &= 1.0 - S_{ew} \quad \textit{effective saturation of the non-wetting phase} \end{aligned} \tag{4.26}$$

Numerous experiments indicated that (as a good approximation) $1/P_c^2$ is a linear function of the effective saturation, and plots of ($S_w = 0$ and S_w) and ($S_w = 0$ and $S_w = 1.0$) are similar right triangles. From the geometry of the triangles, the ratio of the wetting phase integral becomes $S_{ew}^{\ 2}$ and that of the nonwetting phase is $(1.0 - S_{ew})^2$. Thus Burdine's equations for relative permeability were simplified to:

$$k_{rw} = S_{ew}^4; \quad and \ \ k_{rn} = (1 - S_{ew})^2 (1.0 - S_{ew}^2) \tag{4.27}$$

Corey's equations allow the estimation of relative permeability from the end-points of capillary pressure curves (see *Program Listing 6.6 FRA-FLOW.FOR*). Wettability is intrinsically included because the end-points are controlled by wettability and the pore size distribution, but one must distinguish between the wetting and nonwetting phases (which are not easily recognized) for cores with wettability near neutral. Additional experimental work by Corey, however, showed that the equations produced acceptable results with homogeneous sandstones, but did not represent the relative permeability curves of samples containing random vugs, channels, or stratification by inclusions of thin shale layers. Therefore, the equations were modified by addition of a parameter, α, related to the pore size distribution using the threshold pressure of the capillary pressure curves, expressed as a function of the effective wetting phase saturation. When the parameter, α, is equal to 2.0 the relative permeability equations reduce to the originals, Eq. 4.27 (Brooks and Corey, 1964).

$$\begin{aligned} S_{ew} &= \left(\frac{P_T}{P_c}\right)^{\alpha} \quad \textit{where } P_T \textit{ is the threshold pressure} \\ k_{rw} &= (S_{ew})^{\frac{2+3\alpha}{\alpha}} \\ k_{rn} &= (1.0 - S_{ew})^2 \left(1.0 - S_{ew}^{\frac{2+\alpha}{\alpha}}\right) \end{aligned} \tag{4.28}$$

4.3.2 Carman-Kozeny Equation

Using Poiseuille's equation in a form that defines the approach velocity of a fluid through a group of straight capillaries of complex shapes but uniform size, the mean hydraulic radius of the system is the ratio of: (a) the internal volume per unit bulk volume of the system to (b) the internal surface area per unit bulk volume:

$$r_m = \frac{(Internal\ V)/V_B}{(Internal\ surface,\ S_{area}/V_B} = \frac{\phi}{S_{area}} \tag{4.29}$$

Therefore the area available for fluid flow is the porosity per unit surface area. Because the capillaries have complex shapes, Carman introduced a shape factor (developed from experiments) with a range of values where $k_s = 2.0 - 3.0$. Using the equivalent flow velocity through an equivalent length, Poiseuille's equation can be expressed as:

$$\begin{aligned} u_e &= \frac{m^2 \Delta P}{k\mu L_e} = \frac{uL_e}{\phi L} \\ u &= \frac{m^2 \phi \Delta P}{k\mu \left(\frac{L_e}{L}\right)} = \frac{K\Delta P}{\mu L} \\ K &= \frac{m^2 \phi}{k\left(\frac{L_e}{L}\right)^2} \end{aligned} \tag{4.30}$$

where the mean hydraulic radius is $\frac{d}{4} = \frac{r_m}{2}$. Substituting into Eq. 4.30:

$$K = \frac{\phi^3}{4k_s \left(\frac{L_e}{L}\right)^2 S^2} \tag{4.31}$$

4.3.3 Flow Properties of Porous Media

Donaldson et al. (1991) found that pore size distributions can be used to obtain a realistic representation of the fluid flow characteristics of rocks. The term *relative volumetric flow rate* (at constant pressure) was defined as:

$$q_{re} = \frac{\textit{Rate of flow through a range of pore sizes}}{\textit{Total flow rate of the core}} \quad (4.32)$$

The relative volumetric flow was derived from the theoretical arguments for calculation of absolute permeability from pore size distributions. Beginning with Eqs. 4.13 and 4.15:

$$q_t = \frac{\pi \Delta P}{8 \mu L_e} \sum_1^n n_i \bar{r}_i^4$$
$$n_i = \frac{V_i}{\sum_1^n \pi L_e \bar{r}_i^2} \quad (4.33)$$

Substituting and simplifying yields the relative volumetric flow:

$$q_{rel} = \frac{\sum_1^y V_i \bar{r}_i^2}{\sum_1^n V_i \bar{r}_i^2} \quad (4.34)$$

A relative pore volume (to accompany the relative volumetric flow) was defined as the ratio of the calculated volume for a range of pore divided by the total calculated pore volume of the sample. The area under the pore size distribution represents the volume of the pores ranging from the irreducible saturation to 100%; thus,

$$V_{p(cal)} = Area_{(S_{wi}-->1.0)} + S_{wi} \bullet V_{p(measured)} \quad and$$
$$V_{p(rel)} = \frac{V_i^y}{V_{p(cal)}} \quad (4.35)$$

A Berea core was used to illustrate the analysis (see *Program Listing 6.14 POSIZE.FOR* and *Program Listing 6.9 POCEN.FOR*). The range of pore sizes for the saturation is between 0.0 and S_{wi} and contains the immobile, irreducible saturation whose volume is the total pore volume multiplied by S_{wi}; this range of sizes ($y_I = 0 - 0.27$ µm) is the first range of pore sizes and has a flow rate equal to zero. The remaining groups of pore sizes (represented by saturations from 0.028 to 1.0) were divided into three arbitrary groups, y_2, y_3, and y_4 (Table 4–4). The volume of each group of pore radii was obtained by integration of the distribution curve (Fig. 4–2) between the limits of the radii for each group ($i \rightarrow y$). The sum of the calculated groups of pore volumes is equal to the measured pore volume (1.73 cm^3). The relative pore volumes of each group of pore radii (Fig. 4–3a) are listed in column 3 of Table 4–4 as percentages of the total pore volume and the relative volumetric flow of each group is listed in column 4.

A histogram was used to compare the relative pore volumes to the relative flow rates for the selected ranges of pore sizes allowing an analysis to be made of the fluid flow characteristics of the Berea core sample, Figs. 4–3a and 4–3b. The range of pore sizes from 0 to 0.28 represents the volume of the irreducible saturation (S_{wi}= 0.278) that does not conduct fluid. The range of sizes from 0.28–3.71 μm represent 42.2% of the pore volume, but only conducts 10.5% of the total fluid flow. The range from 3.7–7.15 μm represents 19.3% of the pore volume and conducts 35.9% of the fluid. Finally, the range from 7.15–10.59 μm represents only 10.8% of the pore volume, but because of the larger radii of the pores, it conducts 53% of the fluid. Thus anything that results in a decrease of the larger range of pore sizes will result in an increase of pressure required to maintain a given flow rate by forcing the fluid to flow through pores of smaller sizes.

Table 4–4 Pore volumes and relative flow rates for size ranges of pores (Berea rock; V_p = 1.73 ml).

Group	Pore Size Range, μm	Volume of Pores, cm^3	Relative Pore Vol., %	Relative Flow Rate, %
y_1	0.00–0.28	0.48	27.8	0.0
y_2	0.28–3.71	0.73	42.2	10.5
y_3	3.71–7.15	0.33	19.3	35.9
y_4	7.15–10.59	0.19	10.8	53.9

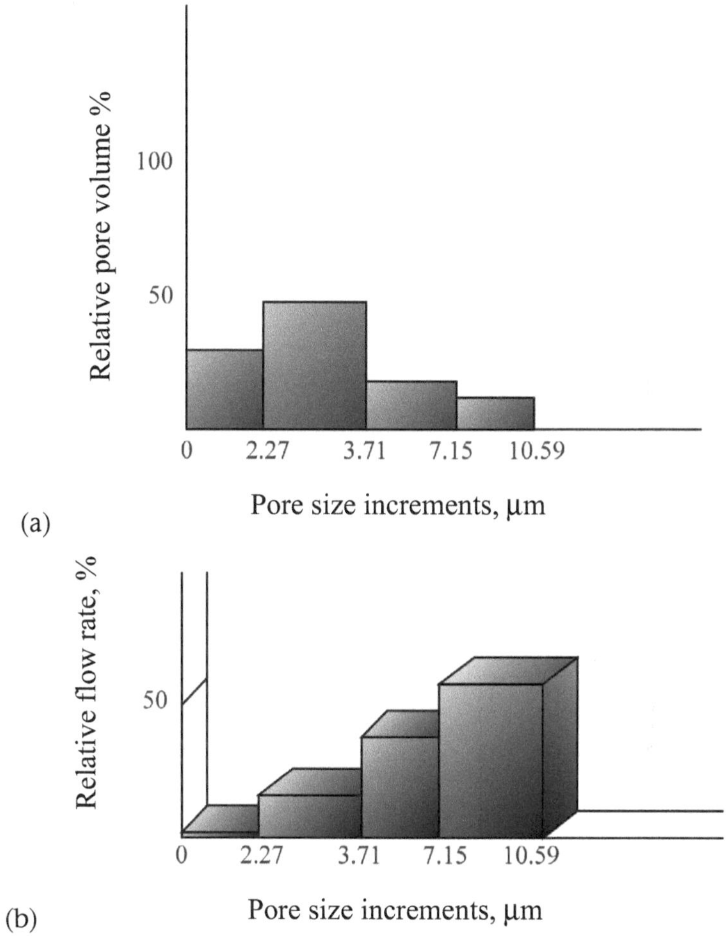

Figure 4–3 *(a) Histogram of the relative pore volumes for groups of pore sizes of a Berea sandstone core, Table 4–4. (b) Histogram of the relative flow rates through groups of pore sizes for a Berea sandstone core, Table 4–4.*

Torbati et al. (1986) used pore size distributions to examine the effects of pore blockage caused by the growth of bacteria in the cores. In one case the largest pore size was reduced from 69 to 28 μm. The reduction of pore sizes by bacteria caused a permeability decrease from 282 to 24 md.

4.4 Wettability Alteration

Wettability alteration of a reservoir can be implemented during the initial waterflood after primary production has diminished, or as a means to produce the residual oil remaining at the end of a waterflood. Many studies of wettability have established three distinct categories of wettability that are readily apparent from capillary pressure curves, fluid flow behavior, and microscopic studies: water-wet, neutral-wet (fractional or mixed), and oil-wet. Consider a water-wet system with water at its minimum irreducible saturation and the balance of the system filled with oil. The water saturation is confined to the small pores and as a film of water lining the pores throughout most of the system; there will be small patches of rock where oil is in direct contact with the rock surfaces. If the oil is displaced by water (brine), water and oil will be observed flowing through individual networks of pores. As the oil saturation diminishes, the network of pores through which oil is flowing will become more and more discontinuous, leaving part of the oil trapped as pockets surrounded by flowing water. Finally, oil will cease to flow except for a trickle (film drainage) that is the result of viscous displacement of molecules of oil at the interface between the flowing water and stationary oil. The residual oil is present within the pores (generally resting on a film of water) as spherical droplets and as continuous ganglia occupying numerous pores, forming three-dimensional pockets of oil surrounded by water.

A change of wettability to a neutral system might join some of the isolated ganglia of oil by filaments and result in some additional production. Complete production requires reduction of the interfacial tension to the point of emulsification where the process changes to miscible displacement.

In neutral-wet systems, when observed by photomicrographs, water and oil appear to be almost evenly distributed in the small pores and crevices between the grains and wetting the rock grain surfaces. The waterflood front has a considerable number of water fingers extending ahead of it and water breakthrough occurs at a lower displacement efficiency than it does for a water-wet system, with a diminishing flow of oil after breakthrough (in contrast to almost complete cessation of oil production after breakthrough from a strongly water-wet system). After passage of the displacing water front, some pockets (or ganglia) of oil behind the water front are connected to oil in pores ahead of the front by a thin filament of oil; oil drains from these pockets through the filament until rupture of the filament occurs, completely isolating the pocket of oil by surrounding water. Thus in this type of neutral system at residual oil saturation, there appears to be equal random areas of the rock that are water-wet and oil-wet; water-wet areas (with no oil present over a distribution of many pores) are interspersed with almost equal areas of the rock that are wet by pockets of residual oil.

Salathiel (1973) described a mixed wettability system where the large pores of the system were oil-wet and the small pores (connected to the large pores) were water-wet. Oil was either in direct contact with the surface of the rock or resting on a film of water that could range from a monolayer to several molecular layers of water in thickness. The neutral systems imbibed small amounts of water or oil, depending on whether they were slightly water-wet or oil-wet. The capillary and viscous forces and the displacement energies, are approximately equal, and the areas under the displacement capillary pressure curves also are approximately equal for neutral systems.

In oil-wet systems, oil occupies the small pores and wets the rock surfaces; water is present in the larger pores and rests on a film of oil in a reversal of the water-wet system. A waterflood is accompanied by rapid formation of water fingers with water breakthrough occurring very soon after initiation of the waterflood. Fingers branch out from the point of injection and rapidly progress to the exit. Production of oil after water breakthrough is apparently one of viscous energy drag on the oil by the moving water phase, with oil production decreasing immediately after the water fingers reach the outlet.

4.4.1 Polar Compounds in Crude Oils

Eight crude oils were separated into saturates (nonpolar hydrocarbons), aromatic compounds, and polar compounds. The acid numbers of the crude oils and their fractions were determined, Table 4–5 and Table 4–6 (Donaldson and Crocker, 1980). The crude oil acid numbers were low, ranging from 0.04 to 0.50, but the polar extract of these crude oils exhibited acid numbers that were many times greater than the acid numbers of the crude oils. Johansen and Dunning (1961) reported a range of acid numbers from 0.10 to 7.0 for a wide variety of oils.

The polar compounds have been identified as asphaltenes by many researchers. Asphaltenes are a group of compounds that are precipitated from many crude oils when the oil is diluted with a low molecular weight alkane such as pentane or hexane. The asphaltenes are held in solution by micelles composed of resin molecules that have a polar group connected to nonpolar alkane and alkyl chains.

Wettability tests on Berea cores were made to examine wettability alteration between a mineral oil and a 5% solution of the polar extract in the mineral oil. The cores were cleaned and tested with the mineral oil, and in every case the USBM wettability index was approximately 0.8 (strongly water-wet). The cores were once more cleaned with toluene followed by steam and then tested with the 5% polar extract from each oil, Table 4–7. In all cases the wettability of the 5% polar extraction solution exhibited a decrease of the USBM index toward a more oil-wet system from 42% to 111% showing that even brief contact with a solution containing polar compounds dissolved in mineral oil causes a significant change toward a more oil-wet system.

The polar compounds in crude oils are principally composed of a characteristic group labeled asphaltenes that have been classified as polyaromatic compounds containing varying amounts of nitrogen, sulfur, and oxygen (NSO compounds) within the ring structures that are linked with saturated short chain alkanes, napthenes (saturated ring compounds), and many other chain and ring types of organic compounds. Thus the asphaltenes are very complex and can only be described as an average structure of compounds with similar physical and chemical properties. They are soluble in benzene but insoluble in pentane. Another general classification of compounds that are part of the polar fraction of crude oils are *resins*. Resins are pentane-soluble, cross-linked compounds with polar groups attached to aliphatic chains. The resins form micelles with the asphaltenes at the center with nonpolar aliphatic groups exposed, thus enabling solubility of the micelle in the bulk phase of the crude oil. When a crude oil is diluted with pentane, the resins are solubilized leaving the insoluble asphaltenes that can be secured by filtration (Yen, 1974; Chiligarian and Yen, 1978).

The rate of adsorption of asphaltenes on rock surfaces displays Langmuir adsorption isotherm behavior as well as some reverse adsorption (Dubey and Waxman, 1991). This is reflected in the aging time that is required for wettability equilibration of cores with crude oils. When in contact with water, the resin/asphaltene micelles and partially separated asphaltenes migrate within the bulk oil to the water-oil interface where they may react with metal ions (iron, copper, nickel, etc.) and oxygen dissolved in the water forming rigid films at the interface. Bartell and Niederhauser (1949) first discovered the rigid films while

Table 4–5 Acid numbers (mg KOH required to neutralize 1.0 g of crude oil to a pH of 7.0, ASTM method No. 664–89) of ractions of several crude oils (McGhee et al., 1979).

Oil	Oil API° & Acid No.	% Saturates & Acid No.	% Aromatics & Acid No.	% Polars & Acid No.	% Asphaltenes
A	48°; 0.04	79.3; 0	17.5; 0.06	3.1; 0.47	0.1
B	30°; 0.08	44.0; 0	35.1; 0.07	20.3; 0.35	0.6
C	38°; 0.18	63.3; 0	26.3; 0.05	10.3; 0.85	0.1
D	32°; 0.41	66.4; 0	24.3; 0.18	9.2; 1.84	0.1
E	33°; 0.43	59.1; 0	27.6; 0.08	13.2; 1.65	0.1
F	39°; 0.31	66.6; 0	24.7; 0.16	8.7; 1.75	0.0
G	40°; 0.14	49.4; 0	35.8; 0.05	14.9; 0.47	0.0
H	36°; 0.05	55.5; 0	32.4; 0.15	12.0; 1.80	0.2

Table 4–6 Acid numbers of various crude oils and their fractions.

Crude Oil No.	Crude Oil	Saturates	Aromatics	Polars
1-76	0.04	0	0.06	0.47
30-76	0.05	0	0.00	0.84
7-76	0.08	0	0.07	0.35
8-76	0.14	0	0.05	0.47
16-76	0.15	0	0.00	2.49
6-76	0.18	0	0.05	0.85
42-76	0.31	0	0.16	1.75
5-76	0.41	0	0.18	1.84
43-76	0.43	0	0.08	1.65
9-76	0.50	0	0.15	1.80

Table 4–7 Percent change of wettability when Berea cores were first examined with mineral oil ($I_w \approx 0.8$) and then using the mineral oil with 5% of the polar extract from each crude oil (after Donaldson and Crocker, 1980).

Oil	I_w of Berea with Mineral Oil	I_w of core with 5% Polars in Min. Oil	Percent Change of I_w
A	0.773	0.450	–41.8
B	0.821	0.454	–44.7
C	0.817	0.410	–49.8
D	0.799	0.286	–64.4
E	0.773	0.150	–80.6
F	0.799	0.145	–81.2
G	0.817	0.118	–85.6
H	0.821	–0.094	–111.4

studying interfacial tension with a pendant-drop interfacial tensiometer. Hasiba and Jenssen (1968) developed a method for isolation of the film material and identified the components as aliphatic esters of aromatic acids, organic nitrogen compounds, and anthracene-type derivatives making up colloidal-size asphaltene particles.

If water and crude oil are present in a water-wet porous system where oil is generally resting on films of water between the oil phase and the mineral surface, groups of resins and asphaltene separate from the micelles and diffuse through the water film to the mineral surfaces. The polar compounds then adsorb physically and/or chemically on the mineral surface until equilibrium is attained. The rate and type of adsorption is controlled by the amount of dissociated micells, the rate of diffusion through the water films, and temperature. Adsorption of the polar compounds on the mineral surfaces drives the wettability toward a more oil-wet condition.

4.4.2 Alkaline Compounds

Basic compounds theoretically have a high probability for the alteration of wettability (from oil-wet to water-wet) and for the mobilization of residual oil because of the large number of potential chemical reactions that are possible. Alkaline compounds can react with dissolved salts, crude oils, and the rock surfaces. If an alkaline reagent is used for wettability alteration, many of the reactions can lead to rapid loss of the reactant in reservoir. Therefore, a considerable amount of analysis is necessary before field application is initiated. Alkali compounds that have been used in laboratory studies and in field applications are some of the most inexpensive reagents that are readily available in bulk quantities: sodium hydroxide, ammonium hydroxide, potassium hydroxide, sodium carbonate, sodium tri-polyphosphate, sodium silicate, and organic amines. Calcium and magnesium hydroxide are not recommended because they do not lower the interfacial tension very much and are not as readily soluble in formation waters.

Calcium and magnesium chloride in the brine react with injected alkalis forming insoluble hydroxides that later re-dissolve in accord with their solubility products, but calcium and magnesium hydroxide do not contribute to production of additional oil:

$$\begin{aligned} Ca^{++} + 2OH^- &\rightleftharpoons \underline{Ca(OH)_2} \\ Mg^{++} + 2OH^- &\rightleftharpoons \underline{Mg(OH)_2} \end{aligned} \tag{4.36}$$

The solubilities of $CaCl_2$ and $MgCl_2$ are 159 g/100 ml and 727 g/100 ml at 100°C, compared to the solubilities of $Ca(OH)_2$ and $Mg(OH)_2$ which are 0.077 g/100 ml and 0.004 g/100 ml, respectively (Weast, 1970).

The common ion effects of specific concentrations of sodium chloride contribute to the decrease of the water-oil interfacial tension (IFT). When the acids in a crude oil are ionized, they establish an ionic double layer with a negative potential on the surface of the oil that is satisfied by the positive sodium ions. The water-oil interfacial tension is reduced when the OH^- concentration at the interface is equal to the pK_a value of the acids. Excess sodium ions at the interface lower the bulk OH^- concentration required for the hydroxyl to equal the pK_a value of the acids (Ex. 4.3). Beyond this point, however, increased addition of excess sodium ions can result in precipitation of the crude oil acid-sodium soaps, resulting in deactivation of the soaps at the interface and an increase of IFT. Donaldson et al. (1989) present a graph showing some of the IFT decreases that can be attained by addition of sodium chloride to sodium hydroxide solutions, Fig. 4–4.

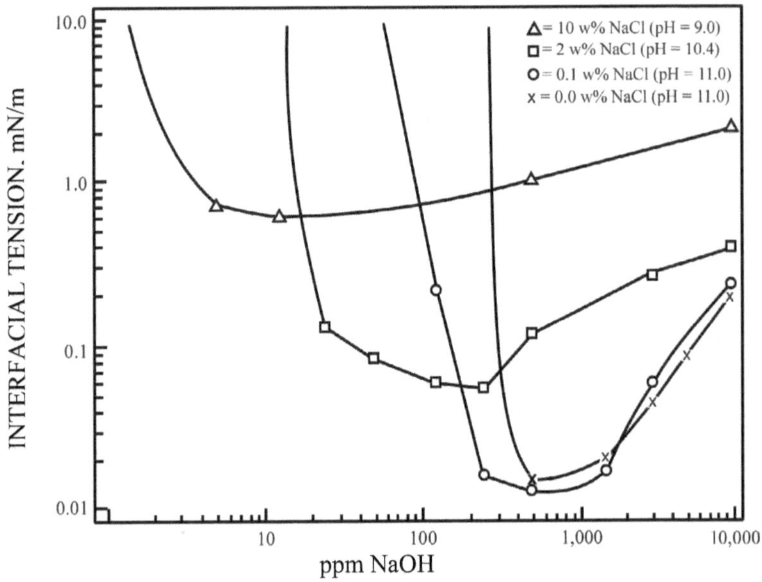

Figure 4–4 *Effect of sodium chloride on the interfacial tension of sodium hydroxide solutions (after Chan and Leu, 1981; in Donldson et al., 1989).*

The addition of sodium chloride moves the curves toward a decrease of sodium hydroxide required to attain a given IFT: 200 ppm of sodium hydroxide is required for an IFT of approximately 0.0, whereas only 40 ppm of sodium hydroxide is required to lower the IFT to 0.1 when 2 w% of sodium chloride is added to the solution.

In addition to reactions with dissolved divalent salts in the brine, alkali solutions react with the silica surface of sands, undergo ion exchange with the clay minerals and react with other minerals present as cementing agents. These are limiting factors that decrease the economic applicability of hydroxides for enhanced oil production.

Ion exchange between sodium and monovalent compounds produce hydrogen ions if the pH of the system is decreased. The exchange with divalent ions in a basic medium can lead to precipitation of calcium and magnesium hydroxide; thus, the reversal of the exchange will require addition of a higher concentration of divalent salts which will lower the pH and induce exchange of the sodium ions:

$$
\begin{aligned}
&MH + Na^{+} + OH^{-} \rightleftharpoons MNa + H_2O \\
&M_2Ca + 2Na^{+} + 2OH^{-} \rightleftharpoons 2MNa + \underline{Ca(OH)_2}
\end{aligned}
\tag{4.37}
$$

where M represents a monovalent compound with an exchangeable hydrogen.

Reactions with the silica in sand are more complex. The silica surface is in equilibrium with water and undergoes reactions with hydroxyl ions as follows:

$$
\begin{aligned}
&SiO_2 + 2H_2O \rightleftharpoons Si(OH)_4 \;\; or\,[H_4SiO_4] \\
&H_4SiO_4 + OH^{-} \rightleftharpoons H_3SiO_4^{-} + H_2O \\
&H_4SiO_4 + 2OH^{-} \rightleftharpoons H_2SiO_4^{=} + 2H_2O
\end{aligned}
\tag{4.38}
$$

The reaction with cementing agents, such as gypsum, results in loss of sodium hydroxide according to the following general reaction:

$$
CaSO_4 + 2Na^{+} + 2OH^{-} \rightleftharpoons \underline{Ca(OH)_2} + Na_2SO_4
\tag{4.39}
$$

Alkaline reactions with the rock can promote particle displacement that will produce permeability damage. Because the reactions can quickly reduce the economic feasibility of using alkali solutions for enhancement of oil recovery, the injection of caustic solutions must be preceded by a preflush of the reservoir with low salinity brine ahead of the caustic solution. The salt concentration of the low salinity preflush must be tailored to the reactive properties of the specific reservoir to avoid loss of permeability from clay swelling and particle movement.

The principal reason for using a caustic solution is the reduction of the capillary pressure that is holding isolated drops and ganglia of oil in single pores and groups of pores. If the system is oil-wet, the alkali reactions with the rock surface at the oil-rock contact (and reduction of IFT aided by saponification of acid compounds) should cause considerable change of local wettability from oil-wet to neutral- or water-wet. This wettability change will be accompanied by a reduction of the capillary forces that will release trapped oil for displacement by the flowing water. Laboratory experiments showed that some reduction of residual oil would take place if caustic solutions were added to the injected water and the result (Fig. 4–5a) shows the trace of capillary pressure curves of a strongly water-wet core that imbibed water at the irreducible water saturation displacing oil to a point close to the residual oil saturation. Forced displacement at the end of the period of spontaneous imbibition produced an additional small amount of oil, ending at S_{or} where capillary pressure assumed greater negative values with production of very minor amounts of oil by film drainage. Introduction of a caustic at that point (1000 ppm NaOH) caused the decrease of IFT and breakup of some of the oil ganglia that are moved into the flowing aqueous stream and produced. This beneficial effect was observed to quickly stop and revert to an established residual oil in pores with smaller entries that were bypassed by the flowing caustic solution. Resting in the hele-shaw models for 24 hours made no difference when the flow of caustic was commenced again (Donaldson and Thomas, 1971; Donaldson and Crocker, 1980).

Figure 4–5b is the trace of the capillary pressure curve for an oil-wet system. In this case when the system is contacted with water, it did not spontaneously imbibe and displace oil; instead, a finite pressure (threshold pressure) was required for entry of water before displacement of oil occurred to a residual oil saturation much greater than the one left behind by waterflood of the water-wet system (S_{or1}). Similarly, when the residual oil saturation was established, the capillary pressure immediately assumed large negative values as water continued to be injected. When a caustic was introduced, the trend of the capillary pressure curve was reversed and some oil production occurred as oil was displaced from larger pores to S_{or2}, but like the water-wet system, production of oil soon returned to film drainage while the capillary pressure assumed large negative values.

After waterflood of an oil-wet system, when a mixture of a caustic and surfactant solution (1000 ppm *NaOH* and 1000 ppm of a nonionic detergent) was introduced at residual oil saturation, the capillary pressure was completely reversed and production of all of the oil took place, Fig. 4–5c. Apparently some emulsification of oil at the interface took place, thus allowing entry of the mixed reactant into the smaller pores and fostering complete displacement of the oil. The oil was dispersed into the flowing aqueous stream as small spherical drops. The use of sodium hydroxide or the nonionic detergent alone only yielded 1–3% additional oil, as indicated in Figs. 4–5a and 4–5b.

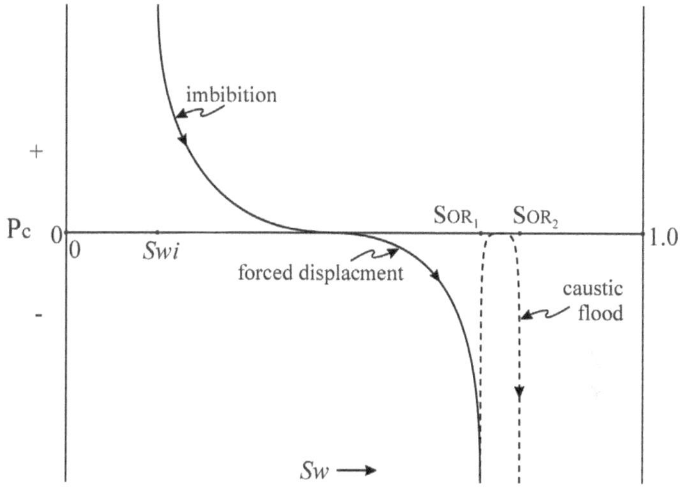

Figure 4–5a *Capillary pressure curves for a water-wet system. Introduction of caustic at SOR_1 causes a reduction of negative capillary forces trapping the oil and production of a small amount of additional oil followed by assumption of large negative values once more.*

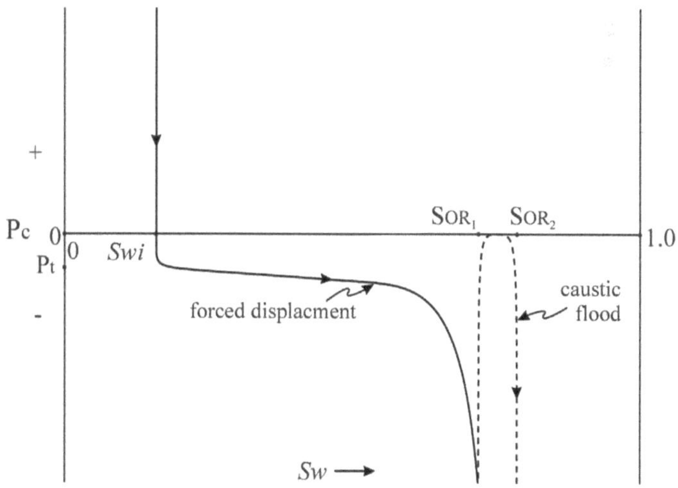

Figure 4–5b *Capillary pressure curve of an oil-wet system. Pt is the threshold pressure required for entry of water. A caustic flood after SOR_1 produces a small amount of oil as the negative capillary pressure trapping oil is diminished but returns to large negative values as a second residual oil is established.*

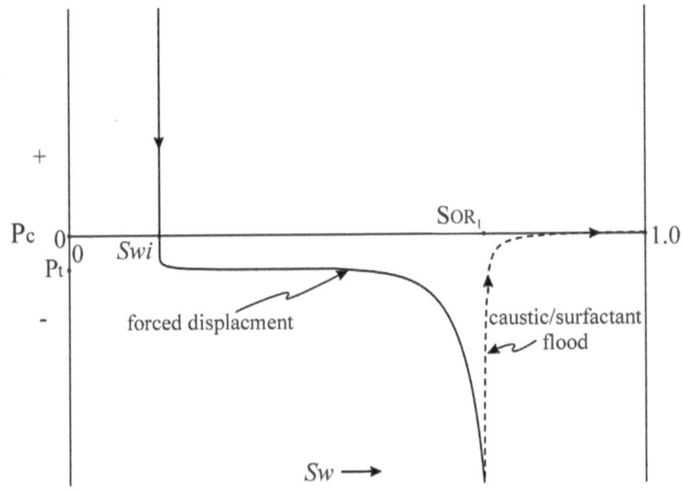

Figure 4–5c *Capillary pressure curve of an oil-wet system showing complete displacement of oil by a mixture of caustic and a non-ionic detergent as the negative capillary pressure trapping the oil is diminished.*

The displacement of oil by a caustic depends on the ability of the caustic to contact the acidic polar compounds of the crude oil that are in contact with the rock surfaces and isolated in pores with small entry sizes. The surfactant probably furnished access to trapped oil allowing reaction of caustic at the rock-oil interface, thus causing oil to disperse into small drops and spread out into the aqueous flowing stream.

Reactions with crude oil acids present at the water-oil interface produces surfactants that decrease the surface tension and promote emulsification of the oil (producing an oil-in-water emulsion) at the interface. The reactive polar compounds are generally carboxylic acids, phenols, porphyrins, and asphaltenes with acid functional groups. Because of the diverse mixture of reactive compounds, ionization of the active compounds occurs at different concentrations (or pH) of the caustic solution which is the reason why a direct correlation of the acid number to lowering of IFT or increase of recovery has not been found. Crude oils exhibit different optimum concentrations or pH for attainment of the lowest IFT. Mungun (1981) reported that the minimum IFT for a series of crude oils occurs at sodium hydroxide concentrations between 0.05 to 0.20 w%.

4.4.3 Capillary Number

The most obvious characteristic of caustic solutions is the decrease of interfacial tension and its effect on recovery efficiency. Wagner and Leach (1966) reported that increased oil recovery of waterfloods (at the point of water breakthrough for both water-wet and oil-wet systems) occurred after the IFT was reduced to 0.07 mN/m. Oil displacement efficiency, however, is a function of the interaction of viscous and capillary forces which is expressed as the capillary number that is related to Darcy's law through the fluid velocity (u_w) as follows:

$$N_c = \frac{u_w \mu_w}{\varphi \sigma_w} \qquad \left[\frac{(m/s)(N \bullet s/m^{2)}}{N/m}\right] = 1.16(10^{-4}\left[\frac{(m/D)(cP)}{mM/m}\right] \tag{4.40}$$

In order to correlate oil displacement to the capillary number, Melrose and Brander (1974) defined displacement efficiency as the ratio of the mobile oil saturation to the initial oil saturation of the porous medium:

$$E_d = \frac{1 - S_{or} - S_{wi}}{1 - S_{wi}} = 1 - \frac{S_{or}}{1 - S_{wi}} \tag{4.41}$$

The capillary number for waterfloods is in the range of 10^{-6} to 10^{-5} for various pore size distributions of cores; however, a critical value of N_c is observed at approximately 10^{-4} for water-wet systems, Fig. 4–6 (Tiab and Donaldson, 2000). Capillary numbers less than the critical value have no observable effect on the displacement efficiency of waterfloods (the residual oil saturation remains almost constant, except for a very slow decrease by a mechanism described as film drainage). When the capillary number is increased above the critical value, the displacement efficiency begins to increase, attaining complete displacement of the oil at values near 10^{-2}. The shape and sizes of pores (expressed as pore size distributions) have an effect on the displacement efficiency of waterfloods which is more pronounced for values of N_c less than 10^{-4} showing a difference in displacement efficiency of 0.4 (for a rock with a relatively large pore size distribution) to 0.6 displacement efficiency (for rocks with small pore size distributions).

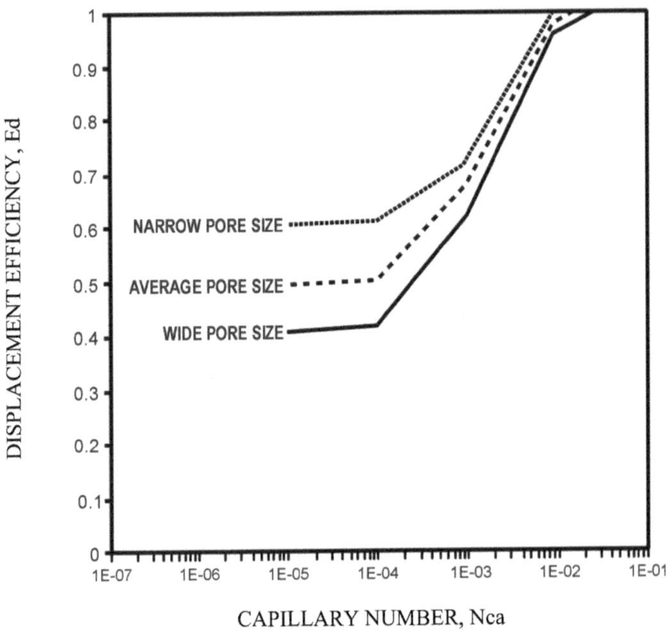

Figure 4–6 *The effect of capillary number on displacement efficiency (Tiab and Donaldson, 2000).*

The velocity of the advancing displacement front cannot be changed enough to substantially affect the value of the capillary number. Viscosity increase of the displacing fluid can be made using water soluble synthetic polymers and biopolymers; but used alone they have limited success for enhancement of oil recovery. The simplest way to increase the capillary number is by reduction of the IFT using caustics and/or surfactants. When combined with viscosity increase, it should be an effective means for increasing recovery. The low IFT that can be obtained from formulations of surfactants is difficult to maintain at the advancing front of a tertiary flood because of adsorption and dispersion. In addition, the cost of the additives, equipment, and labor associated with these processes can become the limiting factor. Therefore, this area of wettability alteration is receiving a considerable amount of research activity in the search for methods to overcome these difficulties.

The capillary number can be related to fluid flow properties by substitution of Darcy's law for the fluid velocity term (u_w) in Eq. 4.40 as follows:

$$N_c = \frac{Kk_{rw}\Delta P}{\phi\sigma_{ow}\Delta L} \tag{4.42}$$

This term refers to the mobilization of residual oil trapped in single pores and groups of pores (oil ganglia or oil blobs) after the waterflood of a water-wet system. In this case the pressure difference in the water phase is across the length of the ganglia. Melrose and Brander (1974) related this to the difference in pressure between the capillary pressure hysteresis curves for drainage and imbibition ($P_{dr} - P_{im}$) at a specific water saturation. The pressure difference because of hysteresis diminishes and completely disappears as the capillary number is increased to approximately 10^{-2} where 100% production of oil is indicated (Batychy and McCaffery, 1978). The problem for quantitative evaluation, however, is determination of the length of the trapped oil ganglia. This length can vary considerably with respect to the pore size distribution and wettability, as shown in color micrographs from microscopic observations of Hele-Shaw type cells by Donaldson and Crocker, 1977. Furthermore, residual oil patterns of intermediate- and oil-wet systems exhibit continuous filaments and ganglia from entry to exit of the packed beds, and these systems have not been studied in relation to the capillary number.

4.4.4 Surfactants

There are three types of surfactant molecules: anionic, nonionic, and cationic, Fig. 4–7. The first two have been used in enhanced oil recovery (EOR) process because they are more soluble in brines and exhibit lower loss by precipitation on the rock surfaces. They consist of two distinctively different parts: a hydrophobic portion composed of hydrogen saturated carbon atoms that thus are readily soluble in oils, and a hydrophilic group (or groups) that are ionic and are expelled by oils and tend to accumulate at the interface depending on the hydrophilic/hydrophobic balance of the molecule. In general, surfactants with less than 12 straight chain methyl groups will be soluble in water and those with 12 or more methyl groups will be soluble in oils; thus, the surface active properties (accumulation at the interface, lowering of interfacial tension, etc.) are controlled by the hydrophilic/hydrophobic properties. The molecules bridge the water-oil interface with the hydrophobic portion of the molecule imbedded in the oil phase and the hydrophilic portion in the water, Fig. 4–8.

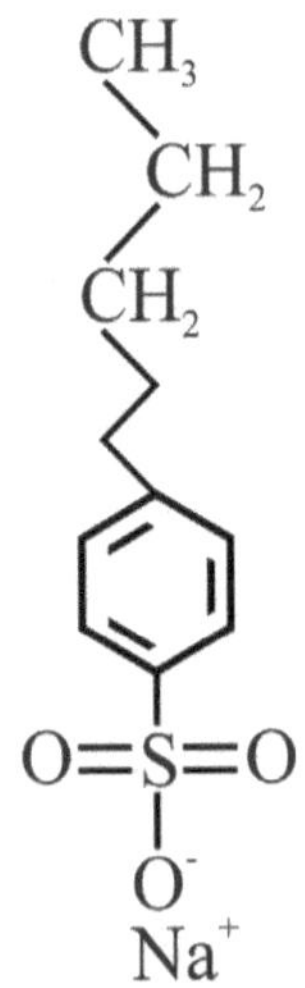

alkyl benzene sulfonate
(anionic)

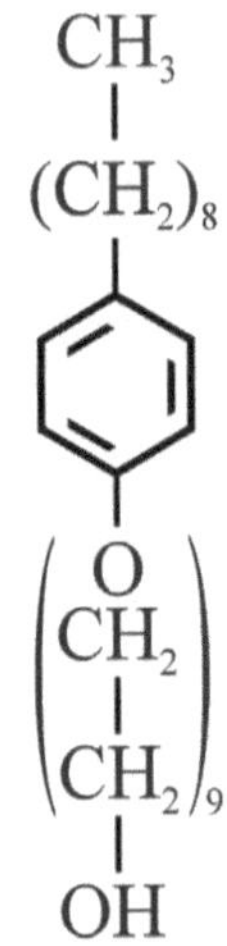

nonoxynol
(non-ionic)

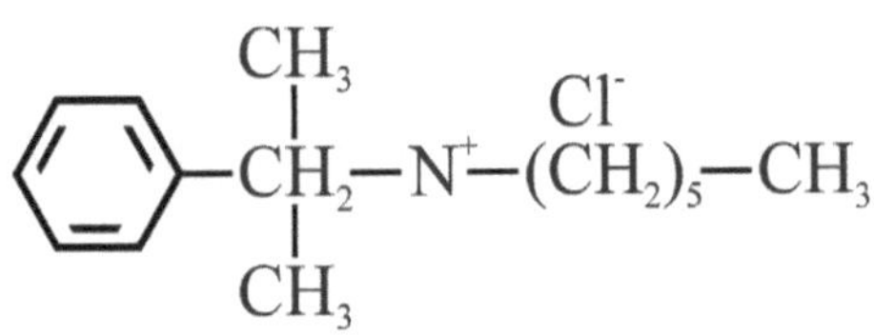

Cationic chloride, benzylcety-dimethyl ammonium

Figure 4–7 *Chemical formulas of the three types of surfactants.*

When the concentration of a surfactant in water reaches a specific value, it will form spherical aggregates (micelles) of 50–100 molecules with the ionic ends exposed to the water on the outside of the sphere. The concentration at which this occurs is called the critical micelle concentration (*CMC*), Ex. 4.4. The micelles are actually a stable second phase that can incorporate oil molecules on the inside (non-polar portion), thus holding the oil molecules in an aqueous phase. An oil-in-water emulsion forms when oil is dispersed within water as micelles of oil within the aqueous phase, Fig. 4–9. The micelle is suspended in water because its exterior is covered by the hydrophilic parts of the surfactant.

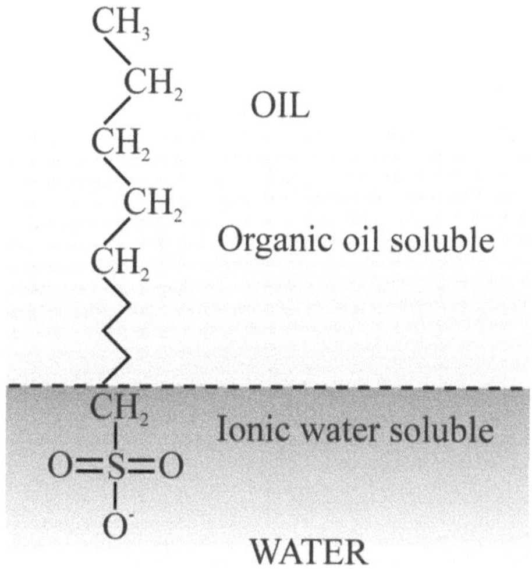

Figure 4–8 *Structure of an alkyl sulfonate surfactant at an oil-water interface.*

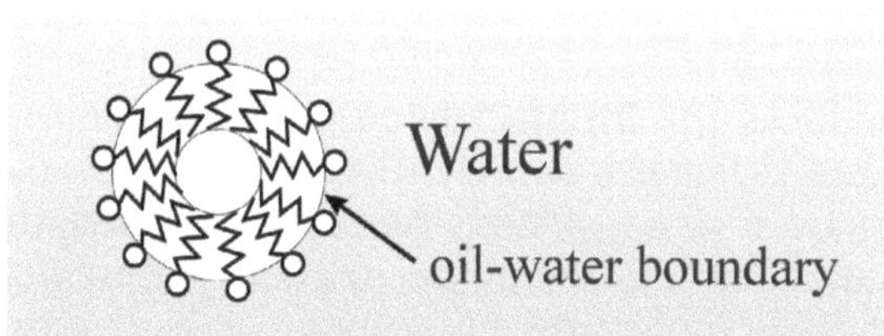

Figure 4–9 *Structure of a surfactant micelle in water showing the lipophobic heads in contact with water and the lipophilic tails clustered within a drop of oil.*

Sodium and potassium salts of carboxylic acids (soaps) are easily protonated in acidic aqueous solutions causing the acids to precipitate and float to the top of the water interface. In addition calcium, magnesium, and iron salts of the carboxylic acid are not soluble in water, Fig. 4–10. Synthetic sulfonate detergents are used for EOR because they are stronger acids than carboxylic acids, hence their acids are not protonated in acidic solutions and remain in solution; in addition, the calcium, magnesium, and iron salts of the sulfonate surfactants are water soluble.

Residual oil, whether in water-wet or oil-wet reservoirs, is left in the reservoir as isolated drops, globules, or ganglia, surrounded by water

$$CH_3(CH_2)_n{-}\overset{\overset{\Large O}{||}}{C}{-}O^-Na^+ + H^+ \rightarrow \underline{CH_3(CH_2)_n{-}\overset{\overset{\Large O}{||}}{C}{-}OH} + Na^+$$

$$2CH_3(CH_2)_n\overset{\overset{\Large O}{||}}{C}{-}O^-Na^+ + Ca^{++} \rightarrow \underline{\left[CH_3(CH_2)_n{-}\overset{\overset{\Large O}{||}}{C}{-}OH\right]_2Ca} + 2Na^+$$

Figure 4–10 *Protonation of a linear carboxylic acid causing it to precipitate and reaction with a divalent cation.*

in adjacent pores; therefore, for efficient recovery the isolated pockets of oil must be coalesced and moved into a mobile oil-bank by injected fluids for production, Fig. 4–11a. The fundamental problem is to reduce the capillary pressure and interfacial viscosity of the oil and furnish a moving mass of oil within the reservoir. The negative charges of synthetic surfactants when used in acidic solutions are repelled by negative charges on silica-based reservoir. In addition, the adsorption of the hydrophobic portion of the surfactant molecules into the oil droplet interface displaces the natural crude oil polar compounds (asphaltenes) that have both positive and negative charges and reduces the interfacial tension and viscosity; thus, the capillary pressure is reduced sufficiently to allow movement through the pore restrictions and coalescence of drops when they come in contact with each other. On a macro-scale, a mobile oil-bank is formed that gathers increasing amounts of oil as it moves toward a production well, Fig. 4–11b.

Donaldson and Crocker (1977, 1980) found that use of a surfactant alone produced less than 5% of the oil in place under ideal laboratory conditions; although complete displacement of the oil was theoretically feasible, the surfactant solution only produced oil that was not wetting the grain surfaces. The surfactant could not promote the release of oil droplets in contact with or chemically bound to the sand grain surfaces. Apparently the surfactant alone did not alter the wettability; it only mobilized oil that was resting on films of water by decreasing the interfacial tension and thus facilitating passage through the pore throats. Changing the conditions of the displacement process by introducing sodium hydroxide and running the flood with a slightly basic solution resulted in alteration of wettability and production of 90 to 98% of the residual oil in place.

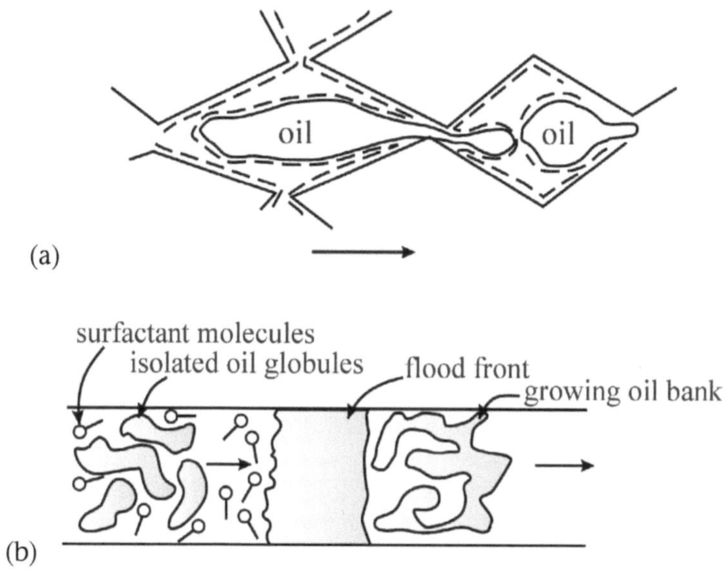

Figure 4–11 *(a) Residual oil droplets, in negatively charged pores that have sequential negatively charged interfaces from a surfactant. The interfacial tension and viscosity of the oil is reduced and electrical repulsion between oil and the pore surface has been increased. The reduction of capillary pressure enables displacement of adjacent pores where coalescence and formation of an oil-bank takes place. (b) A surfactant in injected water implementing coalescence of oil droplets followed by formation of a mobile oil-bank that grows by miscibly incorporating oil droplets during advancement.*

Rocks with net positive surface charges are oil-wet, leave higher residual oil saturations, and cannot efficiently be displaced with surfactants alone. Early work on wettability has long established that it is necessary to alter the wettability of the system by addition of an alkaline component that can promote the release of adsorbed oil droplets (Leach et al., 1962; Mungan, 1970).

It also is necessary to control the salinity of surfactant solutions used for waterflooding. Because of the dual hydrophilic/hydrophobic character of surfactants, they can dissolve in both oil and water. Salt concentration controls the partitioning (concentration in oil/concentration in water) of the surfactant between the two phases. At low concentrations of salt (<1%), most of the surfactant is dissolved in the water phase, but as the concentration of salt increases the surfactant is excluded from water, increasing the concentration of surfactant at the interface and in the oil. Near the critical micelle concentration the effect produces a sharp minimum value of the interfacial tension as a function of salt concentration; this minimum IFT is the optimum condition for displacement of oil but is difficult to maintain because of mixing with interstitial water already present in the rock, Fig. 4–12 (see *Program Listing 6.1 DIS2.FOR*).

The hydrocarbon chain length of the surfactant controls the absolute value of the minimum interfacial tension, Fig. 4–13. As the chain length of any specific type of surfactant is increased, the minimum value of the interfacial tension decreases to an absolute minimum and then begins to increase again. Hence, surfactants can be selected for various salt concentrations to yield the lowest interfacial tension.

Each crude oil must be examined with specific brine solutions because the aromatic/saturate ratio of the crude oil also moderates the interfacial minimum with respect to salt type and concentration. Crude oils with higher amounts of aromatic compounds produce lower interfacial tensions.

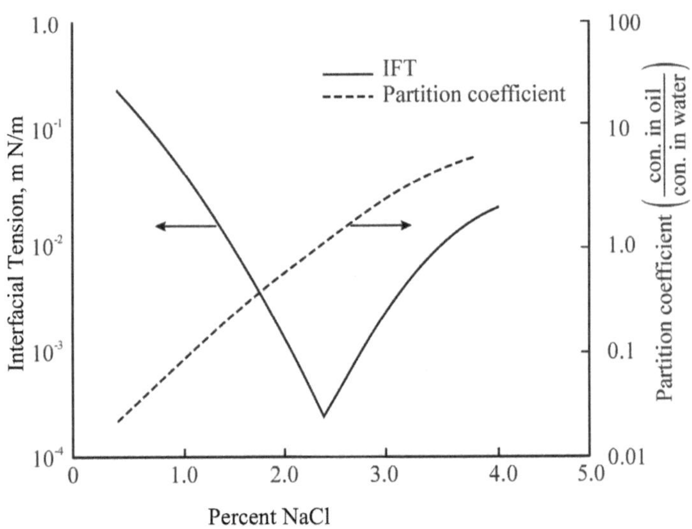

Figure 4–12 *Variation of interfacial tensions and the surfactant partition coefficient as a function of sodium chloride concentration for an anionic surfactant.*

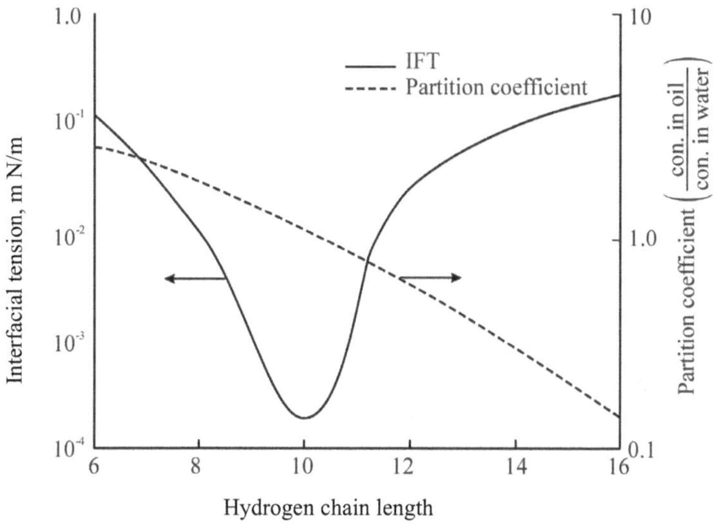

Figure 4–13 *Variation of interfacial tension and partition coefficient (for a surfactant in 1.5% NaCl) as a function of hydrocarbon chain length (after Sharms and Sheh, 1989. In Donaldson et al. Enhanced Oil Recovery, V.II. Elsevier Science Publications).*

The wettability of calcium and calcium/magnesium carbonate rocks can be reversed from oil-wet to water-wet with cationic surfactants to promote countercurrent imbibition. Standnes and Austat (2000) found that the amine surfactant dodecyltrimethylammonium bromide (DTAB) at concentrations slightly greater than the CMC (~0.4 w%) reversed the wettability of chalk cores saturated 100% with crude oil, as well as those that were saturated first with brine and displaced to S_{wi}. As the reversal of wettability takes place, the negative capillary pressure existing in an oil-wet system changes to a positive value and imbibition of the surfactant-brine solution initiates spontaneously. The rate of imbibition approaches zero at production efficiencies of 50% to 70% of original oil in place (OOIP).

An anionic surfactant solution (such as an alkyl aryl sulfonate) did not imbibe after a week in contact with the oil saturated cores; however, nonionic ethoxylated sulfonates produced up to 5% OOIP after a month in contact with the cores. The ethoxyl groups produce effects that are characteristic of the nonionic surfactant promoting temporary, weak adsorption with advancing contact angles less than 90° along with passive emulsification of polar compounds at the interface.

The oil-wetting properties are because of adsorption and reaction of the negatively charged polar compounds of the crude oil (predominantly carboxylic and phenol groups) with the excess positive charges on carbonate rocks. Anionic surfactant compounds lower the oil-water interfacial tension, react with positively charged polar compounds and promote some emulsification of the crude oil at the oil-water interface. This accounts for the minor amount of countercurrent imbibition. The anionic compounds, however, cannot desorb or react with the chemically adsorbed polar compounds on the carbonate rock.

The cationic surfactants have a greater reaction potential toward the anions of the crude oil than the surface of the rock. Consequently, they can desorb and react with the polar compounds forming anionic bonds between the positive cationic surfactant and the negative anions. The compounds that result are large, neutral hydrocarbons that can dissolve into the oil phase and accumulate in micelles of the surfactant in the water phase. The desorption of the polar compounds allows penetration of water along the pore walls accompanied by displacement of oil as the water film advances. The capillary pressure changes from negative to positive as the system wettability is altered from oil-wet to water-wet and spontaneous imbibition displaces oil from the rock. The process is both sequential and simultaneous: (1) desorption of negative polar compounds, (2) reaction to form stable bonds between the cationic surfactant and crude oil anions, (3) alteration of wettability from oil-wet to water-wet, (4) change of capillary pressure from negative to positive, and (5) spontaneous imbibition of the aqueous solution driven by capillary forces accompanied by displacement of oil (Standnes and Austat, 2000, 2003; Freer et al., 2003; Spinder et al., 2000).

Seetheqailli et al. (2004) found that solutions of 0.3 M sodium carbonate and 0.05 w% alkylarylsolfonates promoted imbibition into limestone cores with up to 50% production of OOIP. The solutions had low IFT (~0.001 mN/m) and a pH equal to approximately 10. The most probable mechanism of wettability reversal is reaction of sodium carbonate with the rock surface causing release of the calcium-polar compound pond followed by saponification of the polar molecule:

$$\succ \underline{\overset{+}{Ca}} \bullet \overset{-}{O} - polar + Na_2CO_3 + H_2O \rightarrow \succ \overset{+}{Ca} \bullet \overset{+}{H}CO_3 + \overset{+}{Na} \bullet \overset{-}{O} - polar + NaOH$$
$$\succ \overset{+}{Ca} \bullet \overset{-}{O} - polar + NaOH \rightarrow \succ \overset{+}{Ca} \bullet \overset{-}{O}H + \overset{+}{Na} \bullet \overset{-}{O} - polar \tag{4.43}$$

Table 4–8 Change of USBM wettability index sproduced by pure bacteria cultures in Hele-Shaw cells packed with water-wet and oil-wet sand and saturated with brine and crude oil (from Donaldson and Kianipey, 1991).

Sand Wettability	Bacteria	Initial I_w	I_w after treatment with Bacteria	Percent change of Wettability
Water-wet	Bacillus	0.247	0.285	15
Oil-wet	Bacillus	–0.269	0.100	63
Water-wet	Pseudomonas	0.530	0.481	–9
Oil-wet	Pseudomonas	–0.160	–0.160	0
Water-wet	Clostridium	0.281	0.332	18
Oil-wet	Clostridium	–0.970	–0.378	61

The saponified polar compounds will be slightly soluble in water but will play a larger part at the oil-water interface by lowering the IFT and interfacial oil viscosity. The role of the alkylarylsulfonate is reduction of IFT to a very low value that in turn lowers the threshold pressure for entry of the aqueous solution into the pores and contributes to some emulsification of the oil, thus facilitating its displacement. The alteration of wettability from oil-wet to water-wet changes the capillary pressure from negative to positive which initiates spontaneous imbibition; once wettability alteration takes place, the reversal of capillary forces enables spontaneous imbibition.

4.5 Wettability Alteration Caused by Microbes

The insitu growth of microbes in petroleum reservoirs can enhance the recovery of residual oil from both water-wet and oil-wet reservoirs. The microbes alter wettability from oil-wet to water-wet, produce gas (H_2, N_2, CH_4, and CO_2) at pressures up to 23 MPa in a closed system, and they produce biosurfactants, biopolymers, and co-surfactants. Reservoirs that have been under waterflood contain an indigenous population of bacteria that produce various metabolic products (depending on the species) when their growth is promoted by injection of microbial nutrients. The generation of high pressure has been used to clean wellbores removing paraffins and asphaltenes (Donaldson et al., 1989; Donaldson, 1991).

Kianipey and Donaldson (1991) observed wettability changes to more water-wet conditions caused by microbes at ambient conditions. Capillary pressure curves were measured before and after treatment with pure cultures using Hele-Shaw cells. The cells were packed with either water-wet or oil-wet sand, saturated first with brine, and then the brine was displaced with crude oil to S_{wi}. After a period of aging, the oil was displaced by brine until residual oil saturation was attained. Capillary pressure was measured as a function of saturation for all of the saturation changes and used to calculate the USBM wettability index (Table 4–8). *Bacillus* and *Clostridium* bacteria changed the systems toward more water-wet conditions; complete reversal from oil-wet to water-wet was not observed, however, possibly because the sand was treated with a silicone solution to produce the oil-wet conditions.

Cottage Grove cores 25 cm long and 3.8 cm in diameter were saturated with brine (including a microbial nutrient). Crude oil was injected and then reduced to S_{or} by waterflood with a brine containing either *Bacillus* or *Clostridium* cultures; the cores were sealed and incubated until the pore pressure leveled off (at pressures as high as 23 MPa). Oil was then produced by tertiary waterflood, resulting in consistent production of 23 to 61% of the residual oil. The incremental production was the result of the combination of microbial metabolic activity that occurred during their growth in the core:

1. wettability change
2. production of gas that served as a solution gas drive during the production phase
3. production of biosurfactants that reduced the interfacial tension

Alteration of wettability has been shown to consistently enhance the recovery of oil. A change of wettability from strongly water-wet ($I_w > 0.4$) or strongly oil-wet ($I_w < -0.4$) to neutral wettability results in greater production (lower S_{or}). A change of wettability from oil-wet to water-wet will change the capillary forces from negative to positive and initiate spontaneous imbibition accompanied by countercurrent displacement of oil.

4.6 Chemistry of Silica Sand

Silicon ($^{28}_{14}Si$) is the second most abundant element in earth's crust (28%, compared to oxygen at 46%) occurring as silicon dioxide (sand, flint, quartz) and various silicates. It dissolves in alkali solutions from which it may be precipitated by neutralizing with an acid.

Silicon has a valence of four, forming a small ion (Si^{+4}, 4.5 nm). It is similar to carbon ($^{12}_{6}C$) which is just above it in Group IV of the periodic table and most of its chemical properties are nonmetallic. It combines readily with oxygen and halogens forming very strong covalent bonds that require strong reducing agents and heat for reduction to elemental silicon. Small amounts can be reduced by heating with magnesium:

$$SiO_2 + 2Mg \rightarrow Si + 2MgO \tag{4.44}$$

Silicon forms a covalent bonded crystal, similar to carbon in diamonds, but its shared electrons are not as strongly bonded as those of carbon. The bonding to oxygen has both ionic and covalent characteristics; the bonding appears to involve both ionization and covalent electron sharing, forming macro-molecules linked by strong bonding, Fig. 4–14a. The basic building structure of silicate is the small silicon atom at the center of four large oxygens forming a tetrahedron. Each of the oxygens shares one of its electrons with the silicon atoms leaving an excess negative charge on the surface of each oxygen atom. Indeed, the surface charge on Ottawa sand is negative for all values of pH ranging from 2 to 12, as shown from measurements of the zeta potential (double electrical layer surface potential at the shear plane), Fig. 4–14b and Fig. 4–15. The negative charges of the tetrahedron can be shared with other tetrahedral groups forming SiO_2 arrays of polymer-chains, sheets or three-dimensional frameworks with net negative surface charges, Fig. 4–14c.

In clays, aluminum occupies some of the tetrahedral central positions in place of silicon. Sodium, potassium, and calcium may enter the structural framework between the multiple tetrahedral groups to form the feldspars: $Na(AlSi_3)O_3$, $K(AlSi_3)$, and $Ca(Al_2Si_2)O_8$. With the exceptions of carbonate and basic igneous and metamorphic groups, rocks generally contain quartz and silicates.

Clay minerals are regarded as composed of sheets of ($Si_2O_5^{2-}$) anions of magnesium and aluminum hydroxide, Fig. 4–14d.

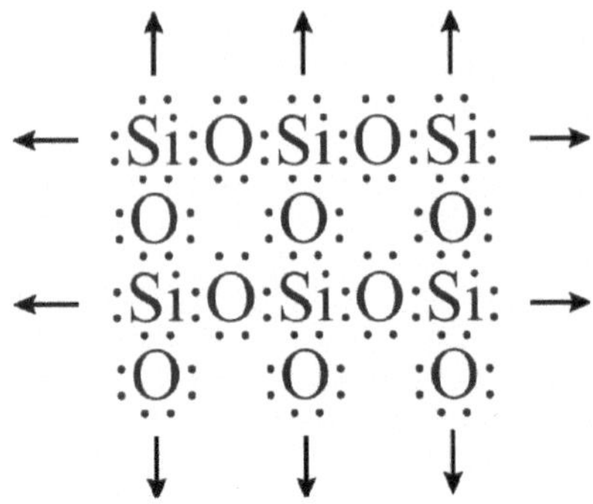

Figure 4–14a *Double tetrahedral linked through an oxygen.*

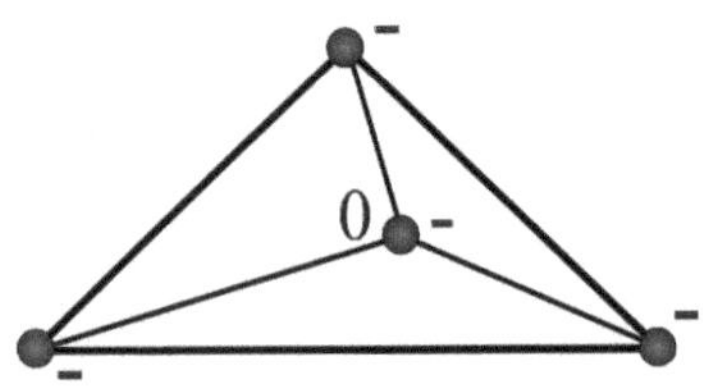

Figure 4–14b *Orthosilicate. Oxygen (filled circles) molecules occupy each corner of the tetrahedral structure, each with an exposed negative charge. The silicon atom (open circle) is centered within the structure.*

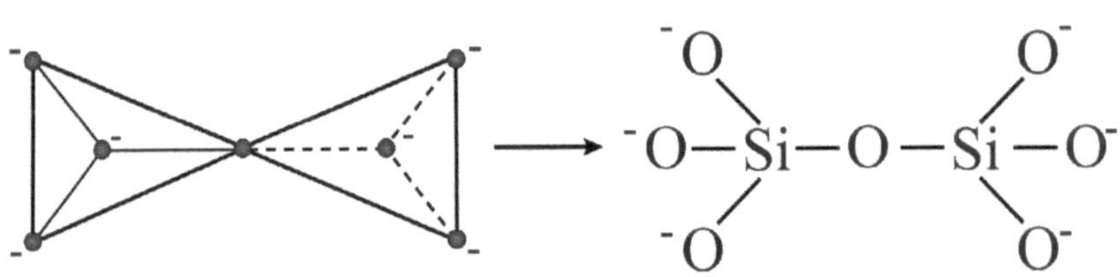

Figure 4–14c *Pyrosilicate ($S_2O_7^{6-}$). Schematic macro-molecular structure of silicon dioxide.*

$$
\begin{array}{ll}
Al_2O_3 \bullet 2SiO_2 \bullet 2H_2O & \text{kaolinite} \\
Al_2O_3 \bullet 4SiO_2 \bullet 9H_2O & \text{montmorillonite} \\
5Mg \bullet 8SiO_2 9H_2O & \text{attapulgite} \\
2K \bullet 3Al_2O_3 \bullet 6SiO_2 \bullet 2H_2O & \text{muscovite (mica)} \\
3Mg \bullet 4SiO_2 \bullet 4H_2O & \text{talc}
\end{array}
\tag{4.45}
$$

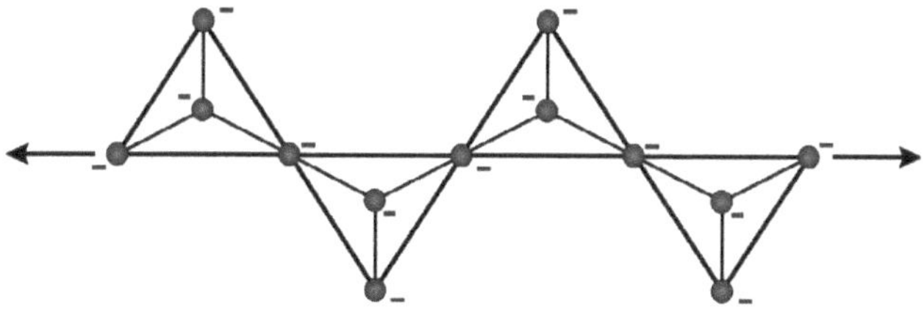

Figure 4–14d *Pyroxene* $(SiO_3^{2-})_n$. *Chain Structure.*

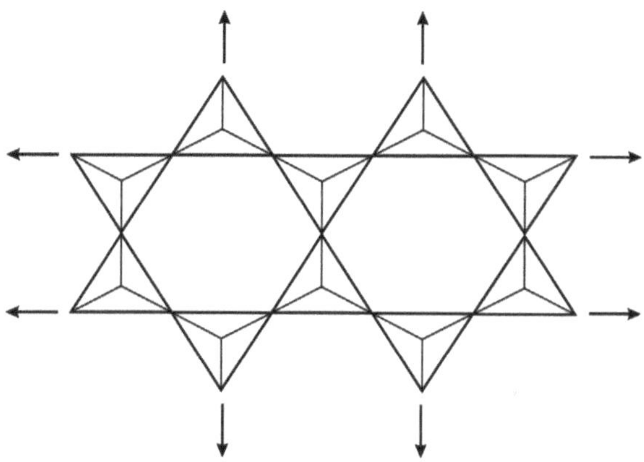

Figure 4–14e *Dual ring structure that can form into sheets* $(Si_2O_5^{2-})_n$.

When they are electrically neutral, the sheets of clay are soft and easily cleaved. If they are combined with monovalent ions (Na^+, K^+) they are harder (such as mica), and when combined with calcium, the sheets are brittle. The layers in montmorillonite are hydrophilic and can take water into the lattice structure, swelling as it acquires water molecules.

The interactions of silica with water, dissolved salts, acidic and alkaline solutions, and the polar compounds of crude oils that moderate wettability are all functions of the net negative surface charges. Each of the oxygens on the surface of silica has one negative charge that can be neutralized by a positive cation (in acidic solutions the negative charges are satisfied by the hydronium ion):

$$-\overset{|}{\underset{|}{Si}}-\overset{-}{O}:\overset{+}{H_3O}$$

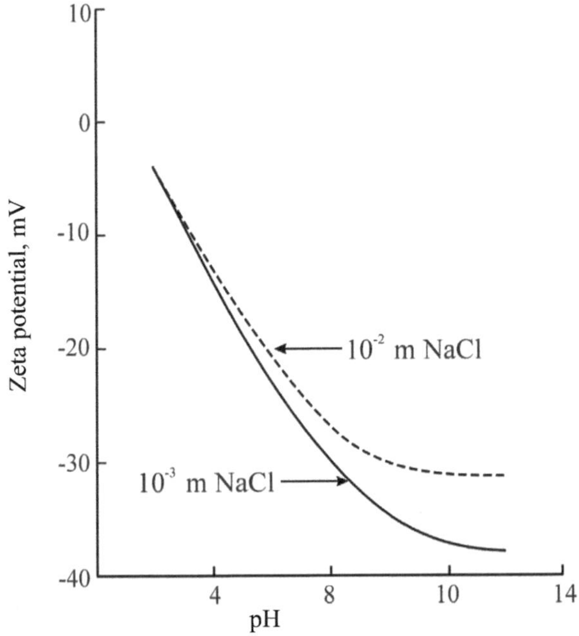

Figure 4–15 *Zeta potential of Ottawa sand as a function of pH and sodium chloride concentration. The higher concentration of sodium surface cations suppresses the negative surface charge.*

Silicon is slightly soluble in water at pH between 2 and 7 (~2 mmol/L) forming silicic acid with a $pK_a = 9.8$. When sodium hydroxide is used in a waterflood at pH < 8, ion exchange with clays and other minerals consumes a large portion of the caustic. When the pH is increased above 8, the loss of caustic is attributed to reactions with the silica sand grains. The reaction rates increase with respect to temperature and decrease if the dissolved salts in the brine are increased:

$$\underline{Si\overset{-}{O}_2} \bullet \overset{+}{H}_3O + Na^+ + OH^- \rightleftharpoons \underline{SiO_2} \bullet Na + 2H_2O \text{ ion exchange}$$

$$\underline{SiO_2} + 2H_2O \rightleftharpoons H_4SiO_4 \qquad [Si(OH)_4]$$
$$H_4SiO_4 + OH^- \rightleftharpoons H_3SiO_4^- + H_2O$$
$$H_3SiO_4^- + OH^- \rightleftharpoons H_2SiO_4^{-2}$$

(4.46)

4.7 Chemistry of Limestone, Chalk and Dolomite

Approximately 50% of petroleum reservoirs are carbonates (limestone, calcite, chalk and dolomite). The pore structure of the various types of carbonate reservoirs is more heterogeneous than that of sands and sandstone. They contain considerable fractures, vugs, and dead-end pores. The production characteristics are thus much different from those of sands. In naturally fractured reservoirs water must imbibe, countercurrent to oil, into the matrix of fracture-blocks to displace oil into the fractures. Thus, the matrix permeability and wettability are governing parameters for oil production from carbonate reservoirs. Chalk can have high porosity but very low permeability because of the small pore size range and distribution. The net surface charge on carbonate rocks is positive which results in the overall wettability ranging from neutral to strongly oil-wet (in contrast to the wettability of sands that ranges from neutral to strongly water-wet). The principal elements involved are magnesium, calcium, and oxygen. Magnesium $\left({}^{24}_{12}Mg\right)$ and calcium $\left({}^{40}_{20}Ca\right)$ occur in Group II of the periodic table. They each have two outer electrons that must be lost in order to form a stable outer octet of electrons resulting in small, doubly charged positive cations (Mg^{+2}, 6.5 nm, and Ca^{+2}, 9.7 nm).

Calcite crystals are composed of complex ions of calcium and carbonate ions held in a cubic structure with each calcium ion centered between two carbonate ions, Fig. 4–16. Calcium derived from continental rocks and carbon dioxide from the atmosphere are dissolved in surface waters and the oceans where invertebrates and algae use them to make shells and skeletons which are later deposited, along with precipitates of calcium carbonate, to form limestone beds. Magnesium carbonate precipitating along with calcium carbonate, or dissolved in water percolating through limestone beds, can replace calcium carbonate within the lime beds, leaving magnesium alternating with calcium ions in the structure to form dolomite.

The calcite surface has alternating calcium and carbonate exposed. The oxygens are bonded to the carbons through strong covalent bonds with completed electron octets, leaving a very weak negative electron cloud, compared to the exposed calcium atoms with net strong positive charges. Thus calcite, dolomite, and other carbonate deposits have an overall electrically positive surface. Zeta potential measurements of calcite in water exhibit a constant value (about +32 mV) as the solution pH is changed from 6.5 to 11. In the presence of 0.1 NaCl, the zeta potential remains linear with respect to increasing pH, but at a value equal to about +15 (Legens et al., 1999).

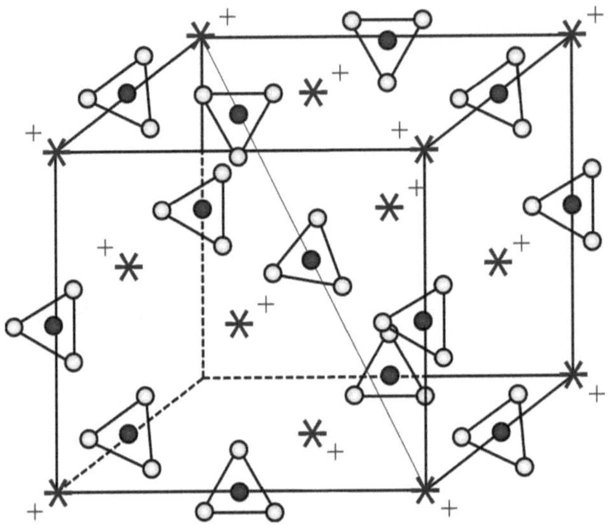

Figure 4–16 *Cubic structure of calcium carbonate: calcium *, carbon •, oxygen °.*

The equilibrium relationships of carbon dioxide and water are:

$$
\begin{aligned}
&CO_2 + H_2O \rightleftharpoons HCO_3^- + H_3O^+ && K_1 = 4.2 * 10^{-7} && pK_1 = 6.5 \\
&HCO_3^- + H_2O \rightleftharpoons CO_3^{-2} + H_3O^+ && K_2 = 4.8 * 10^{-11} && pK_2 = 10.3 \\
&H_2CO_3 + 2H_2O \rightleftharpoons CO_3^{-2} + 2H_3O^+ && && pK_3 = -16.7
\end{aligned}
\tag{4.47}
$$

Calcium, dolomite, and magnesium carbonates, in equilibrium with water, have the following constants:

$$
\begin{aligned}
&\underline{CaCO_3} \rightleftharpoons Ca^{+2} + CO_3^{-2} && pK = 8.3 \\
&\underline{MgCO_3} \rightleftharpoons Mg^{+2} + CO_3^{-2} && pK = 8.2 \\
&\underline{CaMg(CO_3)_2} \rightleftharpoons Ca^{+2} + Mg^{+2} + 2CO_3^{-2} && pK = 17.0
\end{aligned}
\tag{4.48}
$$

Aqueous solutions containing calcium and magnesium salts will decrease the rate of dissolution and precipitation by driving the reactions to the left (Fergusson, 1982; Somasundaran and Agar, 1997; Zou et al., 1998).

4.8 Properties of Crude Oils that Affect Wettability

Crude oils can be separated into four fractions that, for analysis of their interactions with rock surfaces, have a direct influence on wettability, as discussed by McGhee (1979) and Skauge et al. (1999). Although the crude oils originate from very different locations, the amounts of each fraction in the crude oil are quite similar. For example, Skauge et al. report ranges of saturates from 47.3 to 88.6% (McGhee et al., 44.0–79.3%); aromatics 9.7–40.9% (McGhee et al., 24.3–35.8%); polar NSO compounds 1.8 to 10.0% (McGhee et al., 3.1–20.3%); and asphaltenes 0.0–6.2 (McGhee et al., 0.0–0.6). Asphaltenes have an overriding influence on the alteration of wettability; therefore, the quantities reported above may be low. The procedures that were used could have been incomplete, resulting in the inclusion of asphaltenes with the polar fractions.

McGhee et al. only determined the acid numbers of the crude oils and fraction to evaluate wettability alteration caused by interactions of the basic components with the electro-negative Berea sandstone surface. Buckley et al. (1998) and Skauge et al. (1999) measured the acid and base numbers of crude oils, revealing the duality of both negative and positive compounds as well as zwitterions, Fig. 4–17. It is therefore evident that polar type reactions form strong bonds with polar resins and asphaltenes, either with the negative silica-based charges of rocks and minerals or the positive charges exposed on calcium and magnesium based rocks. Buckley et al. (1998) identify three other types of interactions between crude oil components, water, and the rock surfaces: surface precipitation of asphaltenes, acid/base reactions at interfaces and ion binding, Fig. 4–18a–c.

Brine interfacial behavior is more complex because stable films of water can maintain a water-wet system. Once established, the balance of van der Waals and electrostatic forces at the water-oil and water-solid interfaces acting through intervening films of water (DLVO theory) can maintain a stable water film. Many oil reservoirs are strongly water-wet and even the Athabasca tar sands in Canada (which are intuitively expected to be oil-wet) are strongly water-wet. The stabilizing DLVO forces can be maintained with relatively dilute brines and with predominantly monovalent dissolved salts. At high brine concentrations or when the brine contains divalent cations, other forces (hydration and structural) become important and DLVO theory alone is insufficient to explain the interfacial activity.

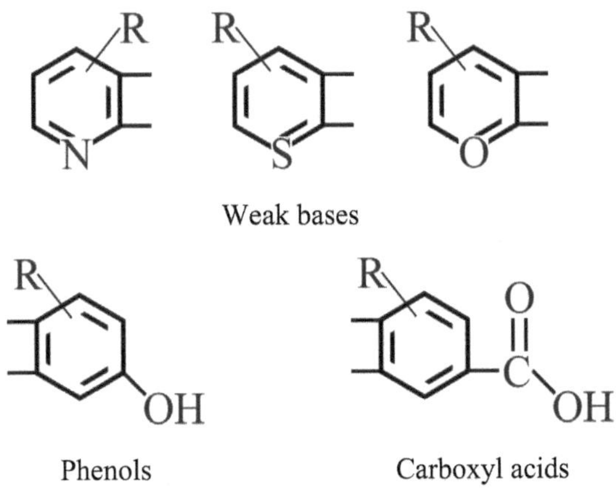

Figure 4–17 *Major functional groups of interfacially active components of crude oils.*

Changes of salinity and the type of salt (and/or the salt concentration) can disturb the balance of forces causing rupture of a water film separating oil from contact with the surface. This type of interaction will result in electrostatic acid-base reactions and physical contact can promote chemical interactions of precipitated or interfacially present polar compounds in the oil, Fig. 4–18b.

A change of salts (caused by leaching of minerals, reactions, osmotic migration, or solution injection) can result in counter-ion concentrations at the water-oil and water-rock interfaces that are identical, and hence repulsive, which will stabilize the water film and maintain the water-wet condition. If the invading salts establish oppositely charged counter ions in the electrical double layers, however, the forces between the oil and solid become attractive and collapse of water films will take place. Removal of the water film will cause direct contact between the oil and electrically charged solid surfaces causing alteration of the wettability to a more oil-wet condition. At a pH greater than 2, silica-based surfaces are negatively charged (Fig. 4–15) and carbonate surfaces have a net positive charge; thus these charge conditions prevail for almost all of the petroleum reservoirs.

Divalent ions in the water film can dominate the interactions through electrostatic reactions with the polar oil interface and/or the rock surface between the polar compounds at the oil interface and the surface charges of the rocks, establishing a bridge between the oil and surface (Buckley et al.,1998), Fig. 4–18c.

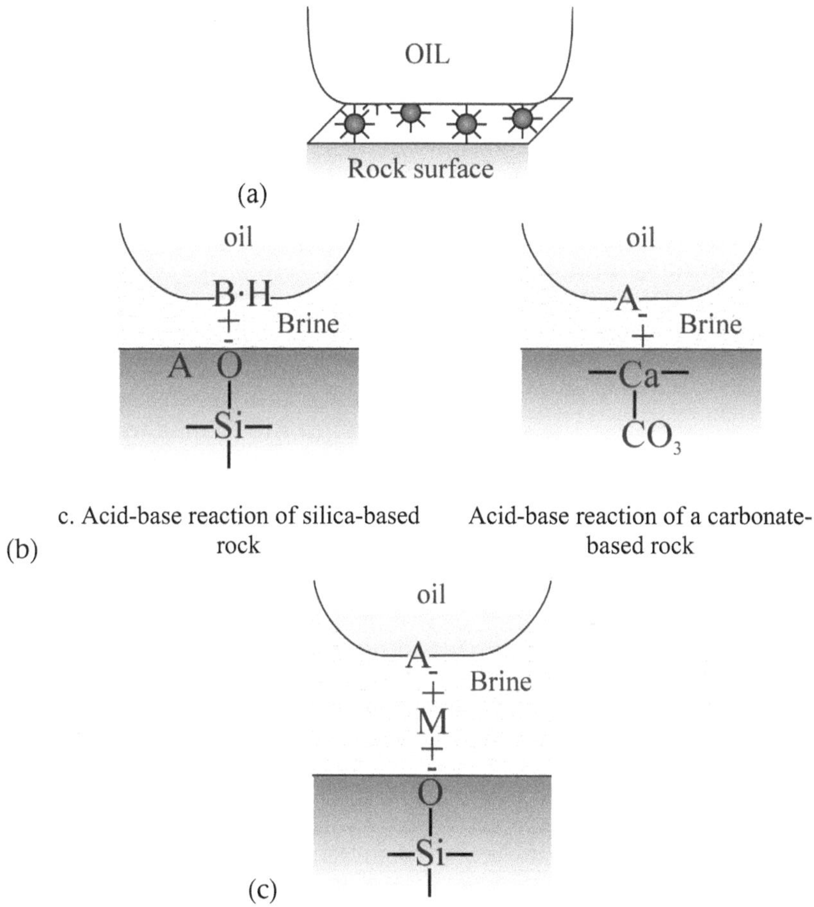

Figure 4–18 *(a) Oil in direct contact with the rock surface showing physical absorption of polar compounds with like charges and chemical adsorption of polars with opposite charges. (b) Interfacial acid-base reactions of ionized oil polar compounds, and negative and positive rock surfaces. (c) Divalent calcium or magnesium (M^{2+}) ion binding oil to the negative silica surface through a water film.*

4.9 Asphaltene Solvency of Crude Oils

Asphaltenes are highly polar heterocyclic compounds containing nitrogen, sulfur, and oxygen within the polynuclear ring structures; hence they are insoluble in saturated alkanes. In fact the asphaltenes form micelle aggregates with resins that are soluble in aliphatic solutions; they have polar components attached to nonpolar alkane and alkene chains making them amphiphilic. Thus the polar ends can associate with the polar asphaltenes surrounding the asphaltenes with extended nonpolar groups that are soluble in crude oils. The solubility of the micelles is a function of the ratio of aromatic compound to saturated compounds. The addition of low molecular weight alkanes (pentane, hexane) to a crude oil containing micelles decreases the aromatic to saturate ratio and results in disaggregation of the micelles followed by precipitation of the insoluble asphaltenes. The nonpolar groups of the resins are sufficient to maintain them in solution (Kaminsky and Radke, 1997).

Buckley et al. (1998) found that the tendency of crude oils to alter wettability toward a more oil-wet condition can be attributed to the solvent properties of the oils with respect to the asphaltene content. They conducted measurements of spontaneous imbibition, adsorption of polar compounds and contact angles, and found that alteration of wettability to more oil-wet conditions was more pronounced as the crude oil solvency for asphaltenes decreased. In addition, when glass surfaces were treated with deasphalted crude oils, wettability alteration did not take place, but when the glass surfaces were treated with the crude oils, more oil-wet conditions developed immediately (Buckley and Morrow, 1990).

4.9.1 Adsorption on Reservoir Rocks

Assuming that Langmuir isosteric adsorption (where the adsorbed atoms are arranged equally in all directions in a homogenous monolayer on the surface) takes place for molecules in a liquid that is in contact with a solid surface, the adsorption rate constant and heat of adsorption can be obtained. The "aging time" for attainment of equilibrium wettability with respect to adsorption of polar components on the rock is a function of the rate constant for adsorption. The heats of adsorption produce an indication of the type of adsorption (physical van der Waals or chemical reaction); if the heat of adsorption is less than 10 kCal/g-mole, only physical adsorption has occurred and at heats of adsorption greater than 10 kCal/g-mole, chemical adsorption has taken place.

Adsorption isotherms for aqueous solutions of various organic compounds on Cottage Grove sandstone were obtained by determining the amount (mg/kg of rock) that was adsorbed as a function of time for specific concentrations, Fig. 4–19 (Donaldson et al., 1975). The isotherms show a rapid initial rate of adsorption that declines to zero

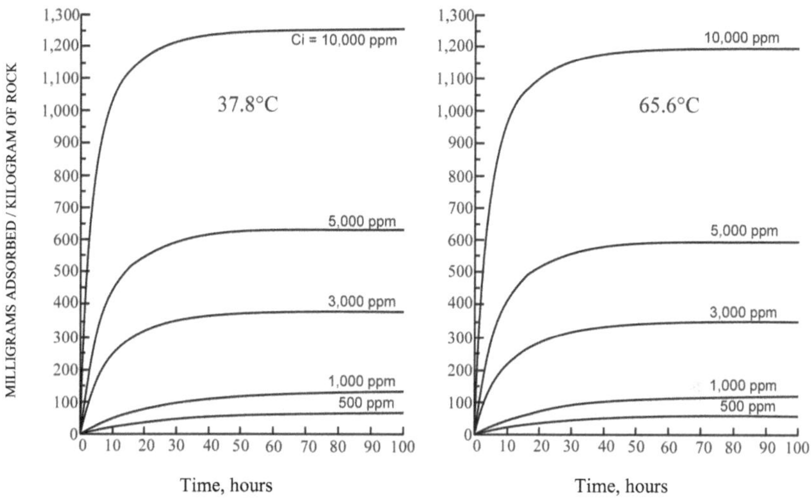

Figure 4–19 *Adsorption isotherms for n-hexylemine from aqueous solutions onto Cottage Grove sandstone at 37.8 and 65.6°C (Donaldson et al., 1975).*

when the final equilibrium adsorption is attained. The rate at which adsorption takes place is used to calculate the rate constant for a given concentration of reactant in the solution as follows:

$$\frac{dC}{dt} = K(C_\infty - C)$$
$$K = \frac{1}{t}\ln\left(\frac{C_\infty}{C_\infty - C}\right) \tag{4.49}$$

Four general mechanisms control the rate of adsorption: (1) diffusion of the compounds from the bulk solution to the interface, (2) transfer to the surface, (3) physical adsorption on the surface, and (4) chemical interaction, if bonding criteria are present. The heats of adsorption for the compounds and rock that were used were less than 10 kCal/g-mol; therefore, the conclusion was that only physical adsorption took place. Physical adsorption occurs rapidly when the molecules contact the surface. Therefore, the rates of adsorption were controlled by the first two mechanisms: diffusion to the surface and transfer across the interface to the solid (Smith, 1968).

4.10 Summary and General Conclusions

The fluid flow characteristics of all reservoirs are governed by a complex relationship between: (1) pore size range and distribution, (2) matrix and fracture permeability, (3) gravity segregation, (4) wettability, (5) pressure and temperature, and (6) gravity. Gravity and capillary forces govern the initial invasion of hydrocarbons into water saturated geologic traps, and over the long range of time local and vertical saturation distributions are established in regions within the reservoir that depend on the six parameters.

Density differences, pressure, and temperature distribute the light components at the top of the hydrocarbon trap and dissolve within the liquid phases. Therefore when oil recovery decreases reservoir temperature and pressure, the saturation of the various phases changes with a higher proportional loss of gas and lighter hydrocarbons. Capillary pressure changes (which are saturation dependent) take place and together with a decrease of temperature distribute the crude oil solubility properties, resulting in flocculation and precipitation of heavy molecular weight polar compounds and an overall change of wettability toward a more oil-wet reservoir. If the altered wettability causes a change from strongly water-wet to neutral ($I_w > 5$ to $I_w \sim 0$), secondary production will achieve a higher S_{wor} (lower residual oil saturation) and an improved waterflood (addition of a polymer, polymer-surfactant flood, etc.) should be economically feasible. If the reservoir is initially oil-wet ($I_w < 0$) alteration of overall wettability toward a more oil-wet system by precipitation of polar components would cause a decrease in the rate of production, an increase in water/oil production rate and a decrease in the final practical S_{wor}. Improved secondary oil recovery would not be effective.

Microbial EOR, by introduction of nutrients followed by a 3 to 6 month shut-in period, can affect a change of wettability because of more water-wet conditions and development of solution gas brought about by the bacteria. This can induce an additional period of production using controlled pressure maintenance.

A small range of pore sizes dictates low permeability and a low overall volume of hydrocarbons because when S_{wi} is high, S_{wor} $(1-S_{or})$ is low. Water was the initial fluid saturating the reservoir, and therefore water is occupying the smaller pores that could not be displaced by invading hydrocarbons because the water drainage threshold capillary pressure is too high. The reservoir will tend to be strongly water-wet with oil in the larger pores isolated from the surface of the pores by relatively thick water films. Thus the probability is great that a low permeability sandstone reservoir will be water-wet, have a low volume of oil, exhibit low production efficiency, and after water breakthrough there will be negligible production of oil.

Reservoirs with relatively larger pores (higher permeability) have a broader range of mobile oil saturation ($S_{wor} - S_{wi}$). Sandstone reservoirs will tend to range from moderately water-wet to neutral, and a decrease of reservoir pressure and temperature during production will tend to drive the reservoir toward a more oil-wet condition. Secondary production will have longer periods of production after water breakthrough that can be enhanced by improved waterfloods. Carbonate reservoirs trend from oil-wet to neutral in wettability. Water breakthrough will occur early after initiation of secondary recovery and the production rate will be lower than it is for a comparable sandstone reservoir. Fractured carbonate reservoirs depend on countercurrent imbibition into the fracture blocks, as well as a decrease of interfacial tension and wettability alteration to a more water-wet condition.

The relentless force of gravity establishes a vertical gradient of the saturation distribution and vertical change of wettability and capillary pressure. These variations with respect to the heights above the water-oil contact and many of the observations noted previously were confirmed by laboratory tests and by Hamon (2000) and also were presented by Archer and Wall (1987). Rocks close to the water-oil contact (WOC) are generally water-wet, but become more oil-wet with respect to the distance above the WOC. Near the WOC, the water drainage capillary pressure is low; consequently if the range of pore sizes is relatively large, the rock permeability is high and spontaneous imbibition will be low. Oil initially migrating into the reservoir will invade the larger pores and leave the small pores filled with water, establishing a mixed-wettability system with water films isolating oil from the solid surface in the pores. The drainage capillary pressure increases with the height above the WOC, and therefore oil invading the reservoir displaces water from ever smaller pores. The film of water between oil and the pore walls is thin, leading to conditions where interactive molecular forces result in development of a more oil-wet system. Forced displacement will leave residual oil isolated in globules throughout the network of pores.

Water and oil saturated rocks near the WOC tend to leave a lower residual oil saturation after primary and secondary production. Thus oil displacement efficiency depends on: (1) rock permeability (large range of pore sizes), (2) height above the WOC (for a homogeneous pore network, displacement efficiency decreases as the height above the WOC increases), and (3) greater displacement efficiency. Lower residual oil is obtained from the portions of the reservoir where the wettability is close to neutral.

The factors that control fluid-flow properties and displacement efficiency are a complex combination of wettability, interaction of molecular force, pore size range and distribution, permeability, gravity, capillary pressure, location in the reservoir, temperature and pressure. All must be considered for correct production analyses of reservoir performance. The dominant parameters extant in a specific reservoir can be determined, but there is no simple controlling property.

4.11 Example Calculations

Example 4–1

Using the coefficients of Eq. 4.1 presented in Table 4–3, estimate the permeability of a fine-grained rock with a measured porosity equal to 18% (note that porosity is expressed as percent).

$$log(K) = 0.4074 + 0.0808(18) + 0.0024(324)$$
$$log(K) = 2.6394$$
$$K = 436\ Darcies$$

Example 4–2

Use Eq. 4.2 to estimate the permeability of a carbonate rock with the following properties: $\phi = 0.20$; $S_{wi} = 0.18$; $S_{area} = 1.5$.

$$Log(K) = 0.95320 + 0.13309(0.20) - 0.02788(0.18) - 5.5597(10^{-4})(1.5) + 1.1709(10^{-5})(0/18(1.5)$$
$$Log(K) = 0.9665$$
$$K = 9.3\ Darcies$$

Example 4–3

Determine the strength of an acid dissolved in water.

The strength of an acid is defined by the extent of its ionization:

$$HA + H_2O \rightleftharpoons H_3O^+ + A^- \qquad K_a = \frac{[H_3O^+][A^-]}{[HA]}$$

where K_a is the acid-dissociation constant that indicates the strength of the acid. The larger the value of $K_{a,}$ the stronger the acid. The range of the dissociation constants for acids is very large; therefore it is expressed as a logarithmic scale using the negative logarithm of the constant which is labeled the pK_a constant. The range of the pK_a-values is the opposite of the dissociation constants: strong acids (such as mineral acids) have small values near zero and weak acids exhibit large pK_a values. For example the K_a of hydrochloric acid is 1.6×10^2 ($pK_a = -2.2$), and the weaker acetic acid has a $K_a = 1.8 \times 10^{-5}$ ($pK_a = 4.7$).

Example 4-4

What is the solubility product of calcium hydroxide in water? The solubility product is the concentration of the ions of a compound in a saturated solution. In a saturated aqueous solution, the dissociation of calcium hydroxide is as follows:

$$Ca(OH)_{2(solid)} \rightleftharpoons Ca^{+2} + 2OH^- \qquad K = \frac{[Ca^{+2}][OH^{-1}]^2}{Ca(OH)_{2(solid)}}$$

The denominator is the molar concentration of the solid calcium hydroxide in contact with the solultion.The amount of solid is constant regardless of the quantity present; therefore, the solubility product (K) and a new constant labled the solubility product constant (pK_{sp}) are defined as:

$$K\left[Ca(OH)_{2(solid)}\right] = K_{sp} = [Ca^{+2}][OH^-]^2 = 6.5x10^{-6}$$
$$pK_{sp} = -\log(6.5x10^{-6}) = 5.2$$

The solubility product constants of a few other compounds are: $Al(OH)_3$ 3×10^{-34}; $CaCO_3$ 4.5×10^{-9}; $Mg(OH)_2$ 7.1×10^{-12}; and $MgCO_3$ 3.5×10^{-8}.

Example 4–5

Determine the critical micelle concentration (CMC) of an oil-in-water surfactant emulsion.

Refer to Figure Ex.4–5. The interfacial tensions of increasing concentrations of surfactant are measured and plotted versus the log of the surfactant concentration. The IFT-log(C) plot goes through an abrupt change of slope at the CMC.

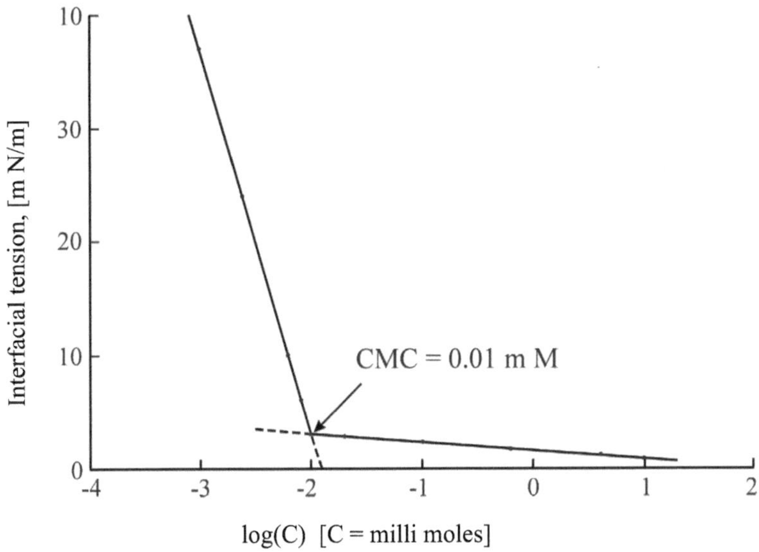

Figure Ex.4–5 *Determination of the critical micelle concentration of a surfactant emulsion.*

CHAPTER 5

Practical Applications of Wettability

5.1 Introduction

It is well known that wettability plays an important role in the recovery efficiency of a petroleum reservoir. This chapter presents some practical applications of the wettability concept. As discussed in the preceding chapters, wettability is a major factor in controlling the location, flow, and distribution of fluids in the reservoir because the partitioning of water and oil between the smaller and larger pores is controlled by the local conditions of wettability. Generally, water-wet reservoirs have brine in the smaller pores with oil occupying the larger pores of the rock, while in oil-wet reservoirs the opposite condition exists. Numerous experiments and the extensive experience with petroleum reservoir have shown that the efficiencies of primary, secondary, and tertiary oil production are governed by the overall wettability of the rock matrix. Reservoirs often consist of mixed wettability systems because of the heterogeneity in the rock matrix and fluid composition; however, reservoir wettability is best defined as the overall average characteristic of the reservoir. In some reservoirs, wettability may be a function of height above the oil-water contact with a water-wet layer near the OWC ranging to oil-wet near the top of the reservoir. In other words, reservoirs may have a complex wettability characteristic, but simplistic reservoir models can be used to represent the average wettability of a reservoir. For example, sandstone reservoirs are generally characterized as water-wet reservoirs. Sandstones behave like weak acids in contact with water and repel acidic polar compounds in crude oil and thus develop a preferentially water-wet surface, whereas carbonate reservoirs have a basic surface and attract acidic polar compounds in the crude oil, making the surface preferentially oil-wet.

5.2 Wettability and Primary and Secondary Recovery Efficiency

Recovery efficiency of oil is dependent upon the wettability of the rock matrix. In a water-wet system, the oil (which is the nonwetting phase) resides predominantly in the larger pores of the rock matrix and is relatively more mobile than if it resided in the smaller pores. Therefore, in a primary recovery with a pressure drop at the wellbore, the oil phase moves towards the production well with relative ease compared to oil in the oil-wet system that is trapped in smaller pores of the rock matrix. The efficiency of drive mechanisms (solution gas, gas cap, or natural water drive) in primary recovery is also dependent on the wettability of the rock. Under most circumstances, the recovery efficiency is higher in a water-wet reservoir because of higher relative permeability. In a study conducted on relative permeabilities at simulated reservoir conditions (Donaldson, Civan, and Alam, 1988) it was observed that with the increase of temperature the relative permeability increased and residual oil saturation decreased as the wettability shifted towards more water-wet conditions at higher temperatures.

Secondary recovery involves the injection of water (sometimes with dissolved additives) to displace reservoir oil to nearby production wells. In a water-wet reservoir water imbibes into the matrix pores, including the small pores, and displaces the resident fluids. The water resides in the smaller pores and is displaced by the injected water, which in turn displaces the oil from the larger pores towards the production well. With the continued injection of water, the water phase saturation increases and the capillary pressure decreases. Water saturation continues to increase until the differential pressure between water (wetting) and oil (nonwetting) becomes zero. To increase the water saturation beyond this point (meaning to produce more oil) the pressure of the water phase has to be greater than the oil phase or, in other words, there has to be negative capillary pressure. This is observed in mature waterflood cases in a water-wet reservoir. It is also observed in the initial phases of a waterflood in an oil-wet reservoir.

In an oil-wet reservoir, the pressure in the nonwetting phase (water phase) is increased with the injection of water. It displaces the oil phase from pore spaces (oil resides in the smaller pores) that are increasingly more difficult for the water (nonwetting) phase to enter because of pore throat restrictions or adhesion force of the oil phase to the matrix. Subsequently, a point is reached when the water phase cannot enter the remaining pore spaces where the oil phase resides (even at very high injection pressure), resulting in a relatively smaller recovery efficiency compared to a water-wet system.

5.3 Wettability and Tertiary Recovery Efficiency

Recovery efficiencies can be enhanced by wettability reversal and some tertiary recovery processes can achieve these reversals. Research has shown that, in general, the recovery efficiency is higher in a water-wet system; however, reservoirs are seldom completely water-wet. In general, reservoirs have a combination of water- and oil-wet regions with overall preferential wettabilities that are either water- or oil-wet. Therefore injection of water (even in a water-wet reservoir) may not be as efficient as expected. In order to make the recovery process more efficient, application of wettability modifiers (chemicals or heat) are used. These modifiers can reverse the wettability of the portions of the reservoir that are not water-wet. Although heat applications generally change reservoir wettability towards more water-wet conditions (Belvins, 1984), heat has also been shown to change the wettability toward more oil-wetting conditions in some cases (Sharma, 1985).

Tertiary recovery processes such as surfactant flooding can alter the wettability of the reservoirs in some cases which can improve oil recovery. Though the primary additional recovery mechanism in the surfactant flooding process is attributed to the lowering of the interfacial tension between the injected fluid and the oil, wettability alteration can also improve recovery significantly. Selection of the type of surfactant is essential in ensuring improved recovery because of wettability alteration. Some surfactants could even adversely affect the wettability and lower ultimate recovery. Surfactants also are used extensively in drilling mud and completion fluids that can change the wettability at the wellbore and affect ultimate recovery (Menezes, Yan, and Sharma, 1989). They, however, observed that any attempt at predicting changes of wettability because of surfactants must account for structural forces. These forces rely on the solvent structure and are referred to as salvation, or hydration (if the solvent medium is water), forces.

5.4 Thermally-Induced Wettability Alteration for Improved Recovery

Wettability of a reservoir can, in general, be altered from oil-wet to water-wet by application of heat. In carbonate reservoirs where natural fractures are common, secondary recovery by waterflood is almost ineffective because carbonate reservoirs have been found to range in wettability from neutral to oil-wet; thus, injected water does not imbibe into the rock matrix. It takes the path of least resistance through the fractures without coming in contact with the bulk of the oil. In water-wet reservoirs, however, the wettability index shifts from weakly water-wet to strongly water-wet at higher temperatures, resulting in a substantial increase in oil recovery because of spontaneous imbibition of water as observed in a laboratory simulated study (Schembre, Tang, and Kovscek, 2006). They found that, with the increase of temperature (temperature varied between 45°C to 230°C) there was (1) a substantial increase of the rate of imbibition rate and extent of oil recovery, (2) some reduction in the residual oil saturation, and (3) significant alteration of wettability from weakly water-wet to strongly water-wet, possibly resulting from the detachment of fines coated with oil from pore surfaces at high temperatures.

There are different viewpoints of the understanding of wettability alteration with temperature. Tang and Morrow (1997) demonstrated that the Amott wettability index always changed toward a more water-wet system for Berea sandstone at higher temperatures resulting in higher oil recovery. Similar observations were made using North Sea carbonate rocks (Dangerfield and Brown, 1985). They found that the rock surface in its original form was water-wet; however, the oil that deposited on the rock surface changed its wettability towards oil-wet because of adsorption of polar asphaltene compounds from the crude oil. At higher temperatures, the polar compounds detached from the rock surface and the rock wettability changed to water-wet again.

5.5 Ghaba North Field

The Ghaba North field in Oman is an extensive, naturally fractured, carbonate formation that is oil-wet (Al-Hadhrami and Blunt, 2001); Fig. 5–1 shows the location of the Ghaba North. The oil reservoir is a fractured, chalky limestone and produces by strong natural water drive mechanism. The oil-saturated column is approximately 64 m, and original oil in place was about 119 million cubic meter of 27°API gravity oil. The viscosity of the oil is 7 mPa-s. Average porosity of the oil-bearing formation is about 30% and the average reservoir matrix permeability varies between 10 md (unfractured) to 100 md (fractured). Only 2% oil recovery efficiency was achieved from this field after more than 20 years of production. The low efficiency is attributed to the nonwetting nature of the driving fluid (natural water drive) that does not imbibe into the rock matrix to displace oil. In order to improve recovery efficiency, the operators attempted to use a gas drive mechanism instead of water drive. Although gas is still nonwetting to oil, the gas-oil interfacial tension is lower than water-oil interfacial tension and would lower the capillary pressure required for oil displacement. The production mechanism involves drainage of oil by gravity from an oil-filled reservoir matrix that is surrounded by a gas-filled fracture system. The production rate was increased to reduce reservoir pressure that would create the secondary gas cap that could be used for driving the oil out of the reservoir. Because of the strong natural water drive, however, the reservoir pressure could not be reduced. The secondary gas cap that was expected to shift the drive mechanism from water to gas drive to take advantage of the wettability of the reservoir did not form. Other gas injection attempts also failed.

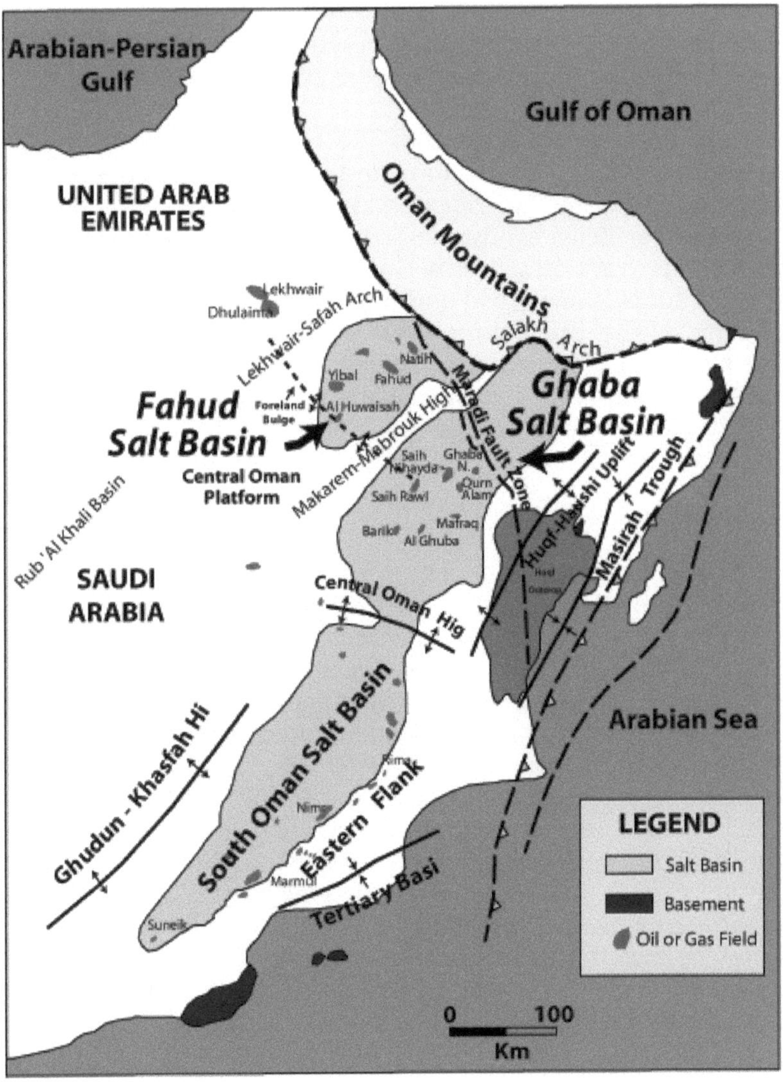

Figure 5–1 *Map showing Ghaba salt basins, other structural elements, and major oil and gas fields of Oman. Ref: USGS Open-File report 99-50D.*

Because the field could not be produced with a wetting fluid, resulting in a very low recovery efficiency, it was recommended that the wettability of the rock be altered from oil-wet to water-wet. Though the Ghaba North field seems to be a poor candidate for thermal recovery (steam or hot water injection) because of the characteristics of the oil (medium gravity and low viscosity) and rock matrix (wide fractures) as shown in Table 5–1, it was recognized that application of heat in the thermal recovery process could change the wettability from oil-wet toward a more water-wet system. If the wettability of the Ghaba field could be altered toward water-wet, imbibition of the condensed water into the rock matrix could perhaps be used favorably for improving recovery efficiency. An analytical solution was developed for the Ghaba field using a thermally-induced wettability reversal technique for oil recovery. In general, the recovery rate in media with permeabilities lower than 0.1 md are limited by capillary forces, while diffusion of heat limits imbibition for permeabilities higher than 1 md. In the 0.1 to 1 md range, both capillary and thermal effects are important. For the Ghaba field, in particular, it was determined that the recovery efficiency could be increased to 30% with steam, or hot water flooding, for a period of 700 days in contrast to the limited 2% recovery that had occurred in more than 20 years.

The amount of heat required to reach the critical temperature for the wettability change depends on several factors including the size, porosity, thermal capacity, thermal conductivity, and fracture widths of the reservoir system. Cost effectiveness of heat application will depend on these parameters. Table 5–1 shows that although the heat capacities of the rock and the reservoir fluids are moderate, the wettability transition temperature is quite high (180°C), which would require a substantial amount of heat to be applied. Figure 5–2 shows the conceptual model of water imbibition because of wettability alteration in a typical oil-wet reservoir.

Figure 5–2 shows the imbibition of water into the rock matrix following a wettability change. Steam flows through the fractures and condenses into hot water in the lower parts of the fractures. Heat diffuses into the rock matrix from the fractures and changes the wettability to water-wet, allowing water to imbibe into the matrix. As a result, the oil is displaced into the fractures and can be produced.

5.6 Qarn Alam Field

Qarn Alam, another fractured oil field in Oman (refer to Fig. 5–1) is also oil-wet. Similar to Ghaba North, this is a carbonate field that lies atop a salt diaper. The field was discovered in 1972 and started producing in 1975. Water breakthrough occurred in early 1978, and oil production declined substantially. Oil recovered by conventional

Table 5–1 Physical properties of the Ghaba North field (after Al-Hadhrami and Blunt, 2001).

Parameter	Value
Original oil in place	119×10^6 m^3
Oil recovered by conventional methods	2%
Porosity	0.30
Fracture permeability	~100 md
Matrix permeability	10 md
Initial water saturation	0.14
Residual oil saturation	0.56
Liquid column height	64 m
Initial reservoir temperature	50°C
Hot water temperature	240°C
Wettability transition temperature	180°C
Oil API gravity	27
Oil viscosity	7 mPa.s
Oil density	890 kg m^{-3}
Water (brine) density	1122 kg m^{-3}
Rock density	2710 kg m^{-3}
Heat capacity of oil	2.08 kJ $kg^{-1}K^{-1}$
Heat capacity of water	4.18 kJ $kg^{-1}K^{-1}$
Heat capacity of rock	0.8 kJ $kg^{-1}K^{-1}$
Oil thermal conductivity	0.15 J $m^{-1}s^{-1}K^{-1}$
Water thermal conductivity	0.6 J $m^{-1}s^{-1}K^{-1}$
Rock thermal conductivity	1.7 J $m^{-1}s^{-1}K^{-1}$

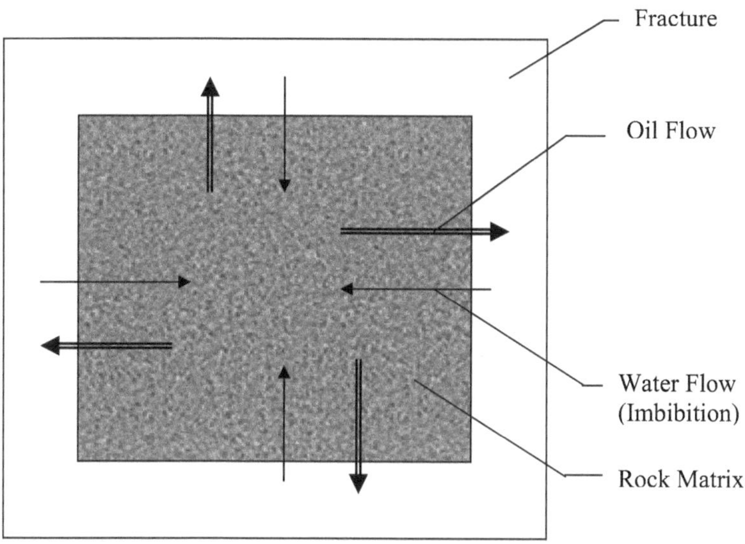

Figure 5–2 *Schematic of fluid flow in a fractured reservoir such as Ghaba North.*

methods was only 2% of the original oil in place. Table 5–2 characterizes the Qarn Alam field.

Qarn Alam is considered a better candidate for improved thermal recovery than Ghaba North because of its reservoir and fluid characteristics. The permeability range of this field is smaller, indicating smaller average fracture widths from which steam and heat cannot escape easily. In a pilot study conducted at this field, the oil production rate increased by more than four times, from 300 m^3/D to 1400 m^3/D, after steam injection. Core studies from this field indicated that at higher temperatures (240°C and 50 atm) the core would spontaneously imbibe water, a phenomenon not observed at lower temperatures (50°C). This indicated that at these higher temperatures, the core changed its wettability from oil-wet to water-wet. The recovery efficiency at the higher temperature increased to about 35% (a significant improvement over primary recovery of about 2% observed for both Qarn Alam and Ghaba North oilfields). The wettability reversal occurred only in the portions of the rock matrix where the temperature reached a critical point required for the reversal to occur. Table 5–2

Table 5–2 Physical characteristics of the Qarn Alam field (after Al-Hadhrami and Blunt, 2001).

Parameter	Value
Field size	6 km × 3 km
Oil recovered by conventional methods	2%
Porosity	0.30
Permeability	5–20 md
Initial oil saturation	0.95
Liquid column height	165 m
Initial reservoir temperature	50°C
Oil API gravity	16
Viscosity	200 mPa.s
Wettability transition temperature	240°C

shows that the wettability transition temperature at this field was 240°C. Though this temperature was higher than the temperature required at Ghaba North (180°C), the thermal efficiency at this field is better because of the lower permeability range that is indicative of narrower fracture width and hence smaller heat loss.

To understand wettability reversal phenomena in a rock matrix, it is important to know the thermodynamic properties of the system. The critical temperature depends upon composition of the crude oil components that are deposited on the rock matrix and the matrix grain composition making the rock oil-wet. Figure 5–3 schematically shows the effect of temperature on wettability of the rock system. The steam or the hot water enters the fractures of the reservoir at temperature T_i and the heat transfers to the rock matrix. The temperature increase in the rock matrix surrounding the fractures is proportional to the distance from the fracture conduits (assuming that the composition of the grain and crude components are nonvariables). The wettability of the rock matrix changes where the temperature reaches the critical temperature T_c and the water imbibes into the matrix, displacing the oil from the matrix into the fracture for production. At greater distances from the fracture conduits where the temperature does not

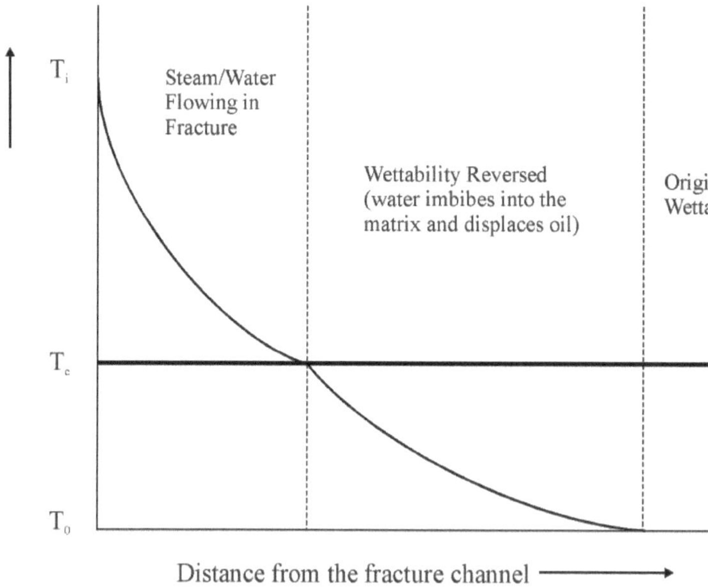

Figure 5–3 *Effect of temperature on the wettability of rock matrix.*

reach T_c, the wettability does not change to water-wet; therefore, water does not imbibe into that portion of the matrix and oil is not displaced by this mechanism. There is no effect on the wettability where the temperature remains at the reservoir's original temperature, T_0.

The reservoir characteristics of the Qarn Alam field can be considered better than the Ghaba North because of relatively narrower fractures in the Qarn field that allow for efficient heat transfer throughout the rock matrix, inducing favorable wettability alteration to water-wet rock. Other factors at Qarn Alam also may play a significant role in the increased recovery over the Ghaba North as was noted in the pilot study conducted at Qarn Alam. This indicates higher viscosity of oil (200 mPa-s at Qarn Alam compared with 7 mPa-s at Ghaba North), resulting in a higher mobility ratio at normal reservoir temperatures at Qarn Alam, indicating poor sweep efficiency. With the application of heat, the viscosity of oil decreased to about 2 mPa-s and the mobility ratio improved, resulting in relatively higher recovery of oil (because of low viscosity of the oil in Ghaba North, the effect of temperature is not substantial). The thickness of the oil saturated column at Qarn Alam was much greater than the Ghaba North (165 m compared with 64 m in Ghaba North). This increased relief allowed the creation and expansion of a gas cap in Qarn Alam which was not possible at Ghaba North.

Based on favorable results from the Qarn Alam pilot and core studies, a full-scale steam injection project is to be initiated at this field. This will be a pioneering project where the purpose of steam injection will not only be to reduce the viscosity of oil and use steam condensate as the drive mechanism to produce the oil, but also to change the wettability of the rock matrix to water-wet by the application of heat so that the water may imbibe and displace the oil from the rock matrix into the fractures for production. The injection of steam and production of oil will be accomplished by about 150 wells and installing facilities to treat water and generate around 18,000 tons per day of steam (Petroleum Development Oman, announcement, May 14, 2007). Additional facilities will be built to process the incremental oil and gas produced at the field as well as disposing of excess produced water in deep reservoirs. Close to 220 km of pipelines and flow lines will be installed to connect these facilities with the wells that supply water, oil production, steam injection and water disposal.

5.7 Preventing Adverse Wettability Shift During Enhanced Recovery Process

Rao (1999) evaluated the effect of temperature that can shift the wettability, which may or may not be advantageous for greater ultimate recovery of oil. He observed that higher temperatures could shift the wettability towards water-wet (calcite surface) or oil-wet (quartz surface). As a result, when the wettability shifts adversely towards oil-wet in a thermal recovery process, the relative permeability to the oil phase decreases. For example, wettability reversal from water-wet to oil-wet was observed because of steaming in a heavy oil sandstone reservoir in Northern Alberta that resulted in less recovery efficiency than projected (Karyampudi, 1995). The question is whether the wettability reversal occurred because of the effect of elevated temperature on sandstone or because of deposition of surface active compounds in the heavy oil that are stripped from the crude at elevated temperatures and deposited on the rock surface, making it oil-wet. Whatever the reason, under certain conditions of higher temperatures, the rock surface would change wettability which can be detrimental to production of oil.

In the laboratory, Rao (1999) also observed that precipitation of $CaCO_3$ on quartz surface protects the surface from becoming oil-wet even at higher temperatures. The $CaCO_3$ deposit, even at relatively low concentrations (in the range of 30 to 120 ppm), changed the rock surface to water-wet. As a result, oil relative permeability improved significantly.

5.8 Field Test of the $CaCO_3$ Deposition Process at Elk Point Thermal Pilot

The process of preserving and improving wettability to improve oil recovery was successfully applied at PanCanadian's Elk Point operations by depositing a $CaCO_3$ layer at the wellbore in 1990. PanCanadian Petroleum Limited of Calgary conducted a field test of this wettability control process at Well 4A-03. The well had gone through four cycles of steam flood that would alter the wettability significantly. The objectives of this test were to see if the process of wettability control works under field conditions and if the wettability alteration towards oil-wet can be stopped.

$CaCO_3$ at low concentrations of 90 ppm was applied to the wellbore region of approximately 400 m^3 of pore volume in a zone of 6.5 m radius from the wellbore. Therefore, the quantities of sodium bicarbonate and calcium chloride salts were 60 kg and 40 kg respectively, as shown below.

$$2NaHCO_3 + CaCl_2 = CaCO_3 + 2NaCl + CO_2 + H_2O \quad (5.1)$$

2 gm-mole of $NaHCO_3$ + 1 gm-mole of $CaCl_2$ is required for 1 gm-mole of $CaCO_3$

2 gm-mole of $NaHCO_3 = 2 \times (23 + 1 + 12 + 48) = 2 \times 84 = 168$ gms

1 gm-mole of $CaCl_2 = 40 + 2 \times 35.5 = 111$ gms

1 gm-mole of $CaCO_3 = 40 + 12 + 48 = 100$ gms

Volume of $CaCO_3$ needed = 12,720 cc = 0.01272 m^3

Density of $CaCO_3$ = 2.83 gm/cc

Weight of $CaCO_3$ needed = 36 kg

Therefore, weight of sodium bicarbonate = $168/100 \times 36$ kg = 60 kg

Weight of $CaCl_2 = 111/100 \times 36 = 40$ kg

To ensure that the $CaCO_3$ would adhere to the rock matrix in the wellbore, the wellbore was cleaned of crude oil by injection of individual salt solutions near the end of the steam injection phase, just before the soak period. The sequence of operations included: (1) injecting approximately 80% of the steam volume designed that would clean the wellbore region of any crude oil; (2) injecting half the desired amount of sodium bicarbonate (30 kg) dissolved in softened water, directly into the steam pipe at the steam generator discharge; (3) injecting the desired quantity of $CaCl_2$ (40 kg) dissolved in softened water directly into the formation, and flushing the steam line with hot water before and after calcium chloride injection to avoid scale formation; (4) resuming steam injection and injecting the remaining quantity of sodium bicarbonate (30 kg); and, (5) completing the rest of the steam injection phase. Results of the field test are shown in Table 5–3. The data shows substantial improvement of the oil production rate because of the application of $CaCO_3$.

Table 5–3 Cyclic steam production data with and without the application of a wettability modifier ($CaCO_3$) at the wellbore (after Rao, 1999).

Parameter	Without $CaCO_3$ at the wellbore	With $CaCO_3$ at the wellbore
Production cycle length (days)	34	33
Total fluids production rate (m^3/d)	21.2	35.5
Total fluids (m^3 per m^3 of steam)	0.26	0.51
Water cut (volume %)	83	77
Oil cut (volume %)	17	23
Oil rate (m^3/d)	3.6	8.2
Steam injected (m^3)	2799	2300
Steam-oil ratio (m^3/m^3)	22.9	8.5

5.9 Improved Oil Recovery by Altering Wettability by Chemical Injection

Improved oil recovery is often achieved by altering the interfacial forces between the fluids. Improved oil recovery also can be achieved by modifying the interfacial forces between the fluids and the rock matrix. Laboratory and field studies have been conducted that successfully improve oil recovery by modifying the wettability using chemical treatment during waterfloods (Leach, Wagner, Wood, and Harpke, 1962). A study was conducted using caustic solution (NaOH) in the Muddy "J" sand of the West Harrisburg field, Banner County, Nebraska. The field was determined to be preferentially oil-wet and susceptible to wettability reversal through chemical injection (susceptibility may be determined by injecting chemicals under simulated field conditions in the laboratory and measuring the wettability before and after the injection).

The "J" sand is located at a depth of approximately 5,900 ft, and the average pay thickness is 10 ft; average porosity and permeability are 15% and 119 md, respectively. The field had gone through waterflooding before the caustic solution was injected. A 40,000-bbl slug of 2% NaOH (0.5 molar) was injected into the field followed by injection of floodwater. Chemical analysis of the produced water indicated that the injected NaOH had reacted with the rock to produce a highly caustic silicate. Incremental oil production was observed very soon after wettability modification because of injection of NaOH. As expected, the incremental amount of oil produced from each well was proportional to the residual saturation of oil in that portion of the reservoir after waterflooding and before injection of NaOH. Greater oil recoveries were obtained from wells that had not been so completely flooded out. The study indicated that wettability reversal oil production started early at a low rate which gradually increased with time. The total oil production that was attributed to wettability reversal was 8,700 bbl.

Based on the field results, an empirical relationship was developed between the cumulative volume of oil Q_o released and moved by wettability reversal in the reservoir to the cumulative volume of water W_t injected after caustic flooding.

$$Q_o = cW_t^{n/2+1} \tag{5.2}$$

The constant, *c*, is a function of porosity, reservoir thickness, oil saturation, and conformance. If the amount of fluids produced by wells were some constant fraction of the fluids being moved in the reservoir, then according to the Eq. 5.2, a log-log plot of cumulative oil produced by wettability-reversal versus cumulative water injection would be a straight line with a slope of $(n/2 + 1)$ and an intercept of *c*. For the Muddy "J" sand, the values for the slope and the intercept were: 2 and 1×10^{-4}, respectively.

Wettability reversal should be conducted early in the life of the field to take advantage of the favorable production mechanism. In addition, the extent of wettability reversal will depend on reservoir-fluid properties. The reaction of a caustic solution with high molecular weight acids in the crude oil is required for wettability alteration in order to soponify the acidic polar compounds in contact with the rock surface. In order for the caustic process to be effective, it is necessary that the reservoir conditions be favorable for its applications. Acid gases (H_2S and CO_2) in a reservoir may be tolerated at low concentrations because their reaction products with excess NaOH may still be sufficiently alkaline. Very high concentrations of acidic crude oil components in the reservoir, however, will not be favorable for wettability reversal (Ehrlich, Hassibi, and Raimondi, 1974). In addition, the presence of large quantities of divalent ions in connate water can deplete a caustic slug by precipitation of insoluble hydroxides. The adverse effect of divalent ions can be minimized by injecting fresh water or NaCl-brine buffer ahead of the caustic slug. Other factors that can be detrimental to caustic flooding for wettability reversal include the presence of large quantities of gypsum ($CaSO_4 \cdot 2H_2O$) that can neutralize the effect of caustic by forming calcium hydroxide, and the presence of clays with high ion-exchange capacity that can deplete the caustic slug by exchanging hydrogen for sodium. In other words, there are several factors that can affect the efficiency of wettability reversal in a caustic flood, and it is recommended that a laboratory study followed by a pilot test be conducted to determine sight-specific efficiency of the technology before a full-scale application is undertaken.

5.10 Determination of Applicability of Wettability Reversal

Chemicals that react with the rock matrix to achieve wettability reversal are generally injected into reservoirs. Chemicals that induce a wetting change undergo either an adsorption reaction or a chemical reaction. These reactions can range from simple to complex depending upon the composition and environment (temperature, pressure) of the rock matrix. The rock matrix consists of more than silica (in sandstone reservoirs) and calcite (in carbonate reservoirs). Many mineral species are present that can react with the injected chemicals. These reactions can change the characteristics of the chemical bank responsible for altering the wettability and render it ineffective. The amounts of chemicals required depend on the pore volume and the reaction(s) that take place in the reservoir, and in many cases can be cost prohibitive (Froning and Leach, 1967). Some of the chemical agents injected are lost in the reservoir system. Before embarking on a wettability reversal project using chemicals, it is therefore imperative that a reservoir characterization study and field application pilot study be conducted.

Applicability of this process to specific reservoirs is determined by a series of tests to ascertain susceptibility of the reservoir to alteration of its wettability in order to indicate the degree of recovery improvement and to estimate the amount of chemical required. In general, laboratory simulation of the reservoir to ascertain wettability reversal potential is not accurate and leads to erroneous conclusions. Pilot tests on the other hand, although they provide good field data, are often too expensive and time consuming because of wide well spacing that creates a very long time-period for acquisition of data. A low-cost, time-efficient procedure for pilot study has been developed, as shown in Fig. 5–4, that utilizes the process of injecting and producing from the same well. The procedure has been applied at various fields to assess the wettability reversibility potential of a reservoir. The process involves injecting the chemicals into a producing well, and therefore selection of the well is based on the criteria that the reservoir matrix near the well will not be altered by injected wettability reversal chemical compounds. A leading bank of water is initially injected to reduce the oil saturation and to act as a buffer between the formation water and the injected water, which helps reduce chemical loss because of reaction with the formation water. After that, chemicals and tracers (nonadsorbable) of known volume and concentrations are injected. This is followed by fresh water (chaser water) to move the chemical bank deeper into the reservoir from the wellbore. The well is then placed on production and the volume and chemical composition of the produced liquid is measured and monitored. It is recommended that the selected well in which the test is conducted should have good productivity so that

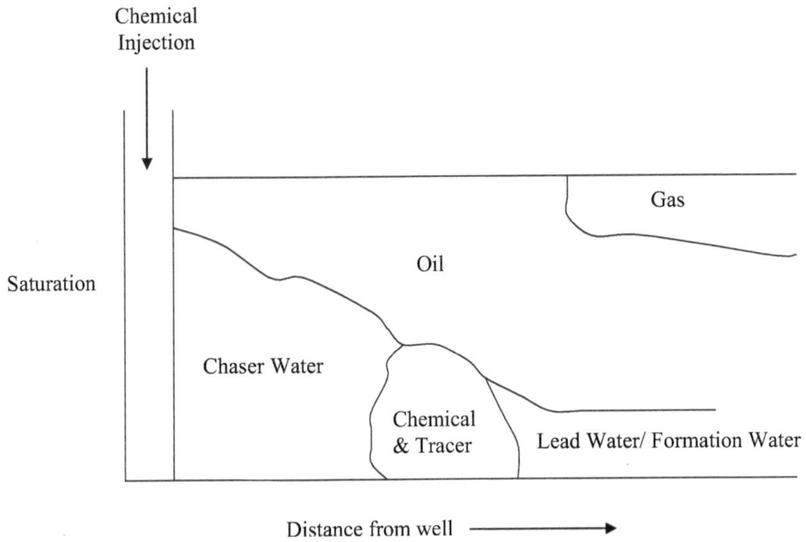

Figure 5–4 *Single well pilot test for determining applicability of chemical flooding for wettability reversal.*

the pilot test can be completed within a reasonable time frame (one week of injection and 12 weeks of production is generally recommended).

Reservoirs that can utilize wettability reversal favorably for enhanced recovery of oil are generally selected based on contact angle tests using reservoir crude oil and simple mineral surfaces such as quartz, calcite, and dolomite. The test provides an indication of the degree and rate of wetting change and conditions under which the treatment is effective. Several chemical agents are used for altering the wettability, and they vary depending upon the rock matrix and the conditions of the surrounding environment such as reservoir fluid characteristics, temperature, and pH. Several low-cost, water-soluble agents such as hydroxides, silicates, sulfides, sulfites, carbonates, and phosphates are used to alter reservoir wettability. An alkaline carbonate solution or a combination of carbonates and phosphates often changes the wettability of a carbonate reservoir from oil-wet to water-wet; however, each reservoir should be tested separately to identify the best combination of chemicals that work under the specific environment.

A single-well pilot test was conducted at the Clearfork limestone formation in a West Texas field to determine reservoir sensitivity to wettability reversal and loss of chemical agent (ammonium carbonate-sodium tripolyphosphate). A 16 ft section of a producing zone with negligible water cut was selected for the test. A total of 10,000 bbl of fresh water was injected as the lead water bank, followed by a slug of chemicals (4,500 lb of ammonium carbonate, 1,289 lb of sodium triphosphate and 100 lb of sodium iodide) and 1,280 bbl of fresh chaser water. Though no sulfate was injected in the well, relatively high concentrations of sulfate were found in the produced water, resulting from reactions between carbonates and phosphates with the reservoir rock that contained sulfates of calcium in the form of gypsum and anhydrite. Chemical loss because of this type of reaction in the reservoir can be substantial and can make a project cost prohibitive. Chemical losses in the Clearfork test were 1 lb of ammonium bicarbonate and 0.2 lb of sodium tri-polyphosphate per barrel of pore volume contacted.

5.11 Effect of Drilling and Completion Fluids on Wellbore Wettability and Its Stability

Wettability at the wellbore can be altered by the composition of drilling fluids. A water-wet formation can be altered to oil-wet during drilling activities and change the reservoir fluid flow characteristics at the production well. Culec (1989) observed that oil-based drilling fluids can alter the wettability to oil-wet near the wellbore. How far beyond the wellbore this alteration takes place depends upon the permeability of the rock formation. Effects of wettability alterations are especially remarkable on core samples taken from the wellbore when oil-based drilling fluids are used. These cores are generally used to study the reservoir behavior. If proper measures are not taken, these cores can give erroneous test results because of an altered wettability state. The wettability of the wellbore and the core can, however, be restored in most cases by cleaning procedures discussed by Tiab and Donaldson (2004).

Another study was conducted to evaluate the effect of cationic and anionic surfactants present in the drilling and completion fluids on wettability (Menezes, Yan, and Sharma, 1989). These surfactants could, in some cases, drastically alter the wettability of the rock formation. Most oil-based mud components greatly changed the wettability. The cationic surfactant caused a decrease in the zeta potential of the silica-water interface. The anionic surfactant caused an increase in the zeta potential.

Note that ionic surfactants (especially cationic) are extremely sensitive to divalent ions in the water and are generally not useful for EOR.

Stability of the wetting phase can be determined by evaluating the zeta potential. In a study conducted on the stability of aqueous wetting films (water-wet system) in the Athabasca Tar Sands (Hall, Collins, and Melrose, 1983) the stability of these thin wetting films was critically dependent on whether the zeta potential for the rock-brine interface and brine-oil interface were of like signs. The zeta potential for the sand-brine interface will be negative in most cases. Therefore, the requirement for the stability of the wetting phase at the Athabasca field was that the zeta potential at the brine-oil interface also be negative. They observed that pH had a significant effect on the zeta potential. At very low pH (~1), the zeta potential had significant positive values and therefore the aqueous film would not be stable in that environment. As the pH increased, the zeta potential became increasingly more negative, a condition that would support stable aqueous films around the sand grains. The same wettability stability concept can be applied to oil reservoirs.

5.12 Minimizing Formation Damage by Controlling Wettability

Reversal of wettability often occurs near the wellbore during drilling, completion, and workover activities. These wettability reversals often result in formation damage that can reduce production. The use of oil-based muds and synthetic-based muds, though very efficient in drilling operations, can alter the wettability of the wellbore and its vicinity. This can cause formation damage. Also, additives of drilling fluids such as amido-amine and crude, oil-fatty acid surfactants, can cause alteration of wettability at the wellbore that promotes formation damage (Patel and Growcock, 1999).

In order to protect the formation from such damage, reversible invert emulsion drilling fluids have been developed whose characteristics can be changed from oil-wet (water-in-oil emulsion) to water-wet (oil-in-water emulsion) (Patel, 1998). This reversible emulsion maintains oil-wet characteristics during drilling for higher drilling efficiency and can be reversed to water-wet characteristics after drilling for making the formation water-wet for higher production rates. The formulated drilling fluid system is similar to conventional OBMs in terms of both composition and performance except for the emulsifier that plays a critical role in enhancing the performance characteristics of the fluid. The surfactants used in this reverse system form a stable invert emulsion (water in oil) in a basic (alkaline) environment, while under acidic environment they form direct emulsions (oil in water). Therefore with acid/base treatment, this type of surfactant can be converted between direct and invert emulsifier to enhance both drilling and production of a well. Figure 5–5 shows the conversion between direct and invert emulsion under acid/base environment.

The reversible invert emulsion drilling fluid formulations use emulsifiers, wetting agents, viscofiers, and fluid-loss control agents. The emulsifier should allow the conversion between the invert and direct emulsion without destroying the surfactant.

The reversible invert emulsion drilling fluid has been successfully applied in the Central Graben area of the North Sea. Two intervals (3,346 ft of 16 in. hole and 8,016 ft of 12 ¼ in. hole) were drilled using this type of fluid. The fluid was easy to handle and provided high drilling efficiency. The drilled cuttings were oil-wet and firm. Disposal of drill cuttings that are oil-wet in a traditional oil-based mud system or in an invert emulsion system creates an environmental hazard, especially in the offshore environment. It does not dissolve in water because of its oil-wet surface that inhibits water interaction. However, reversible invert emulsion fluids can overcome this problem. Oil-wet cuttings are converted to water-wet by treating them with acid. The resulting water-wet cutting readily hydrates and/or biodegrades. The

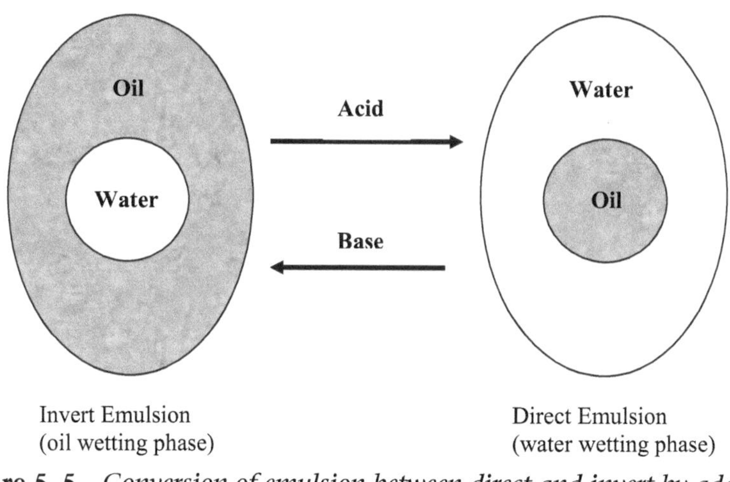

Figure 5–5 *Conversion of emulsion between direct and invert by adding acid/base.*

process also was tested in a North Sea field using a reversible invert emulsion drilling fluid; the oil-wet cuttings were converted to water-wet by soaking them in citric acid. The rate of softening of the drill cuttings was increased by conversion of the oil-wet cuttings to a water-wet state that reduced the time required for grinding the cuttings before injecting them into the ground. The process helped in increasing the rate of drilling and protecting the environment.

Use of reversible invert emulsion drilling fluids in addition to high drilling performance also provides superior completion and cementing of the well. Once the drilling activity is completed, the wettability of the drilling fluid is reversed to water-wet by adding a weak acid such as citric or acetic acid. The water-wet conditions in the wellbore allow for higher production rates because of favorable relative permeability to the oil phase. The water-wet conditions also help by producing stronger cement bonds with the pipe or casing and the formation. To demonstrate the enhanced cement bonding resulting from the switch to water-wet conditions, shear bond tests were conducted on reversible invert emulsion drilling fluids and compared to the oil-based muds. In all tests, 16.4 lb/gal Class H cement slurry was cured at 60°C for 72 hours before the shear bond was measured. Results of the test are shown in Table 5–4.

Test results show that the shear bond strength is much higher in the reversible mud compared to the conventional mud. Also, with the increment of acid concentration and time of treatment, the shear bond strength increases. This is expected because under these conditions the system becomes more water-wet.

Table 5–4 Cement shear bond test (after Patel, 1998).

Conditions	Shear Bond Strength (psi)	
	Conventional Mud	Revesible Mud
Baseline with dry pipes	428	
No wash	0	0
5% acid-water, 2-min wash	0	9
15% acid-water, 2-min wash	0	95
15% acid-water, 10-min wash	0	183

5.13 Example Problems

Example 5–1

It is estimated, based on the pore volume of the wellbore region, that 0.05 m^3 of calcium carbonate is needed to protect the region from changing the wettability from water-wet to oil-wet in an enhanced recovery process. The chemicals to be injected to form calcium carbonate are sodium carbonate and calcium chloride. How much of these chemicals will be required?

Solution to Problem 1:

The chemical reaction is:

$$Na_2CO_3 + CaCl_2 = CaCO_3 + 2NaCl$$

1 gm-mole of Na_2CO_3 + 1 gm-mole of $CaCl_2$ is required for 1 gm-mole of $CaCO_3$

1 gm-mole of Na_2CO_3 = (46+12+48) = 106 gms

1 gm-mole of $CaCl_2$ = 40 + 2 × 35.5 = 111 gms

1 gm-mole of $CaCO_3$ = 40 + 12 + 48 = 100 gms

Volume of $CaCO_3$ needed = 0.05 m^3 = 50,000 cc

Density of $CaCO_3$ = 2.83 gm/cc

Weight of $CaCO_3$ needed = 141.5 kg

Therefore, weight of sodium carbonate = 106/100 × 141.5 kg = 150 kg

Weight of $CaCl_2$ = 111/100 × 141.5 = 157 kg

Example 5–2

An oil-wet reservoir has been converted to a water-wet reservoir by the injection of alkali that was followed by water injection. Data indicated that the water injection, W_t, produced incremental oil, Q_o, as shown in the table below. Estimate the incremental oil produced after injecting 100,000 bbl of water.

W_t (bbl)	Q_o (bbl)
10,000	4,000
18,000	6,500

Solution to Problem 2:

Cumulative volume of oil, Q_o, released and moved by wettability reversal in the reservoir to the cumulative volume of water, W_t, injected after caustic flooding can be estimated with the empirical equation:

$$Q_o = cW_t^{n/2+1}$$

or

$$\log Q_o = \log c + (n/2+1) \log W_t$$

This may be solved graphically or analytically to determine the values of the slope ($n/2+1$) and the intercept ($\log c$). From the solution of this linear equation, it is determined that:

$$n/2 + 1 = 0.83, \text{ and}$$

$$\log c = 0.3$$

Therefore, when $W_t = 100{,}000$ bbl,

$$\log Q_o = 0.3 + 0.83 \log 100{,}000$$

$$Q_o = 28{,}184 \text{ bbl of oil}$$

CHAPTER 6

Fortran Computer Programs

6.1 Introduction

This chapter is composed of computer programs designed for the calculation of the various useful parameters discussed in the preceding chapters. The programs are entered in alphabetical order with a brief discussion of each. Input data to the programs are made through data files that can be copied along with the output files to be retained with the results of each calculation; the program thus remains undisturbed during input operations. Graphical output is not used because there are so many different graphical routines available, and the output files can be used directly by graphical routines found in other publications.

Tiab and Donaldson (2004) present detailed discussions of 20 petrophysical experiments, many of which are addressed by the programs in this book; therefore, the design of testing equipment and experimental procedures are not included. Many of the programs use subroutines and functions that must be incorporated when the program is compiled.

6.2 Dispersion

Miscible phase dispersion. DIS2.FOR (DIS2IN.DAT, DIS2OUT.DAT, MOVE.FOR, TRIDAG.FOR)

The dispersion of two miscible fluids flowing in a porous medium occurs by the simultaneous actions of convective and molecular diffusion. They are largely influenced by the pore size distribution, pore constrictions, and the tortuosity of the rock.

Convective dispersion also is affected by the density and viscosity differences between fluids as well as velocity. There are two types of molecular diffusion: (1) bulk diffusion which is concentration-dependent and will take place in the absence of fluid flow as long as a concentration gradient exists between two miscible phases in a porous medium, and (2) Knudsen diffusion that takes place (according to Wheeler, 1951) when the mean free path is more than 10 times greater than the radius of the pore, and it occurs in pores that are less than 10^{-3} mm in radius (Wheeler, 1951). Pore size distributions of rocks show that most of the rock pores are greater than 10^{-3} mm; therefore, the influence of Knudsen diffusion may be neglected.

If turbulent flow occurs, dispersion is increased by eddy diffusion, but flow in rocks is only turbulent in the immediate vicinity of injection or production wells. The analysis of dispersion represented by this program is under laminar conditions of fluid flow, and the effects of density and viscosity differences are eliminated by using two aqueous phases different only in the concentration of a tracer salt, or with one phase containing a small concentration of a soluble tracer compound such as tritium. Thus, the dispersion is a function of the physical properties of the porous medium, velocity, and the rate of molecular diffusion of the tracer being used.

This program calculates the effluent, S-shaped curve that depicts the dispersion of a tracer compound injected into a saturated core at constant rate. The output S-shaped curve (normalized concentration, C/C_o, versus pore volumes injected) depicts the dispersion of the tracer that will occur for a given value of the coefficient of dispersion, *DL*. The output curve can be compared to an experimental curve, and if it does not match the experimental curve, the value of *DL* can be changed until a match between the curves is obtained. The coefficient of dispersion then describes the dispersion taking place in the specific porous medium at the experimental flow rate that was used.

If the output curve concentration equal to 0.5 does not occur when exactly 1.0 pore volume has been injected, the S-shaped curve is moved (subroutine *MOVE.FOR*) horizontally so that the points will coincide. If the experimental data also is moved, comparison of the curves is facilitated.

Input Variables

CCO	normalized concentration (C_i/C_o)
DIAM	core diameter (cm)
DL	coefficient of dispersion
DT	time increment (second)
N	number of data points
PHI	porosity
QRATE	flow rate (cm/s)
XLEN	core length (cm)

Program Listing 6.1 DIS2.FOR

```
C      DIS2.FOR
C
C      Subroutine TRIDAG is required; data files are DIS2IN.DAT AND DIS2OUT.DAT
C
```

```
C       Solution of the dispersion equation (Crank-Nichlson FD discretization)
C
C       Computes the effluent concentration curve versus pore volumes injected
C         for the Coefficient of Dispersion, DL, flow rate and core porperties.
C
        REAL CON(110),CIN(110),A(110),B(110),C(110),D(110),V(110)
        REAL CC(110), VOL(110)
C
        NX= 110
        OPEN (UNIT= 16, FILE= 'DIS2IN.DAT', STATUS= 'OLD')
        READ(16,*) N, DL, XLEN, DIAM, PHI, DT, QRATE
        CLOSE (16)
C
        VBULK= 0.7854*XLEN*(DIAM**2.0)
        PV= PHI*VBULK
C
        M= N + 1
        H= 1.0/FLOAT(N)
        U= 0.0877*QRATE/PHI
        GAM= XLEN*U/DL
        ALFA= DT/(2.0*H*H)
        BETA= DT*GAM/(4.0*H)
C
        DO 2 I= 2,M-2
          A(I)= -ALFA-BETA
          B(I)= 1.0 + 2.0*ALFA
          C(I)= -ALFA+BETA
2       CONTINUE
        A(M-1)= -2.0*ALFA
        B(M-1)= 1.0 + 2.0*ALFA
C
        DO 3 I=1,M
          CON(I)= 0.0
3       CONTINUE
C
        T= 0.0
C
4       T= T+DT
        CON(1)= 1.0
        TIME= T*XLEN*XLEN*PHI/DL
        PVI= TIME*U/(PHI*XLEN)
C
        DO 5 I= 2,M-2
          D(I)= (ALFA+BETA)*CON(I-1)+(1.0-2.0*ALFA)*CON(I)+
       &(ALFA-BETA)*CON(I+1)
5       CONTINUE
        D(2)= D(2)+(ALFA+BETA)*CON(1)
        D(M-1)= 2.0*ALFA*CON(M-1) + (1.0-2.0*ALFA)*CON(M)
C
        CALL TRIDAG(M,NX,A,B,C,D,CON)
        CON(M)= CON(M-1)
C
        IF(PVI .LT. 0.5) GO TO 4
C
        DO 6 I= 1,M
          CIN(I) = 1.0-CON(I)
          CC(I)= CIN(I)
6       CONTINUE
C
        V(1)= 0.0
        DO 7 I= 2,M
          V(I)= V(I-1) + 0.02
          VOL(I)= V(I)
7       CONTINUE
C
```

```
      DO 8 I= 1,M
        IF(CIN(I) .LE. 0.0005) GO TO 8
        NSTOP= I
        GO TO 10
8     CONTINUE
10    DO 12 I= 1,M
        IF (CIN(I) .GE. .999) GO TO 13
12    CONTINUE
13    LAST= I
C
      CALL MOVE(M, CC, VOL, NS, NT, COR)
      DO 14 I=1,M
        VOL(I)= VOL(I) + COR
14    CONTINUE

      CONTINUE
      OPEN (UNIT= 17, FILE= 'DIS2OUT.DAT', STATUS= 'OLD')
      WRITE (17, 900)
900   FORMAT (3X, 'DISPERSION:  CONCENTRATION VS VOLUME INJECTED CURVE',/)
      WRITE (17, 902)
902   FORMAT (5X, 'Curve computed from Coefficient of Dispersion, Flow
     & Rate, and Core Properties',//)
C
      WRITE (17, 904) DL, QRATE
904   FORMAT ('Coefficient of Liner Dispersion=' F8.5, 3X,'Flow rate,
     &ml/s =' F8.5,//)
C
      WRITE (17, 906) XLEN, DIAM, PHI
906   FORMAT ('Core Length =', F6.2, 3X, 'Core Diameter =',F6.2,3X,
     & 'Porosity =', F6.4,//)
C
      WRITE (17, 907)
907   FORMAT (3X'Curve as Computed' 10X, 'Curve Moved')
      WRITE (17, 908)
908   FORMAT (5X,'VOLIN', 6X, 'CON',10X, 'VOLIN', 6X, 'CON')
      WRITE (17, 910) (V(I), CIN(I), VOL(I), CC(T), I= NSTOP,LAST)
910   FORMAT (2F10.3,4X, 2F10.3)
C
999   STOP
      END
```

Program Listing 6.2 DIS2IN.DAT

```
100,0.003,50.8,3.81,0.1909,0.0001, 0.0155
```

Program Listing 6.3 DIS2OUT.DAT

```
DISPERSION:  CONCENTRATION VS VOLUME INJECTED CURVE

    Curve computed from Coefficient of Dispersion, Flow Rate, and Core Properties


Coefficient of Liner Dispersion= 0.00300   Flow rate, ml/s = 0.01550


Core Length = 50.80   Core Diameter =  3.81   Porosity =0.1909
```

Curve as Computed		Curve Moved	
VOLIN	CON	VOLIN	CON
0.480	0.001	0.454	0.001
0.500	0.001	0.474	0.001
0.520	0.002	0.494	0.002
0.540	0.002	0.514	0.002
0.560	0.004	0.534	0.004
0.580	0.005	0.554	0.005
0.600	0.007	0.574	0.007
0.620	0.010	0.594	0.010
0.640	0.013	0.614	0.013
0.660	0.018	0.634	0.018
0.680	0.024	0.654	0.024
0.700	0.031	0.674	0.031
0.720	0.041	0.694	0.041
0.740	0.052	0.714	0.052
0.760	0.065	0.734	0.065
0.780	0.082	0.754	0.082
0.800	0.101	0.774	0.101
0.820	0.122	0.794	0.122
0.840	0.147	0.814	0.147
0.860	0.175	0.834	0.175
0.880	0.206	0.854	0.206
0.900	0.240	0.874	0.240
0.920	0.277	0.894	0.277
0.940	0.316	0.914	0.316
0.960	0.357	0.934	0.357
0.980	0.400	0.954	0.400
1.000	0.443	0.974	0.443
1.020	0.487	0.994	0.487
1.040	0.531	1.014	0.531
1.060	0.575	1.034	0.575
1.080	0.617	1.054	0.617
1.100	0.658	1.074	0.658
1.120	0.697	1.094	0.697
1.140	0.734	1.114	0.734
1.160	0.768	1.134	0.768
1.180	0.799	1.154	0.799
1.200	0.828	1.174	0.828
1.220	0.853	1.194	0.853
1.240	0.876	1.214	0.876
1.260	0.896	1.234	0.896
1.280	0.914	1.254	0.914
1.300	0.929	1.274	0.929
1.320	0.942	1.294	0.942
1.340	0.953	1.314	0.953
1.360	0.962	1.334	0.962
1.380	0.970	1.354	0.970
1.400	0.976	1.374	0.976
1.420	0.981	1.394	0.981
1.440	0.985	1.414	0.985
1.460	0.989	1.434	0.989
1.480	0.991	1.454	0.991
1.500	0.993	1.474	0.993
1.520	0.995	1.494	0.995
1.540	0.996	1.514	0.996
1.560	0.997	1.534	0.997
1.580	0.998	1.554	0.998
1.600	0.998	1.574	0.998
1.620	0.999	1.594	0.999
1.640	0.999	1.614	0.999

Program Listing 6.4 MOVE.FOR

```
C     Subroutine MOVE
C
c     Moves the effluent concentration profile of dispersion data
c
      SUBROUTINE MOVE (N,CON,V,NS,NT,COR)
      REAL CON(N), V(N)
C
      J= 0
      DO 6 I=1,N
        IF(J .GT. 0) GO TO 6
        IF (CON(I) .GE. 0.5) GO TO 5
        GO TO 6
5       J= I
6     CONTINUE
C
      NS= J
      NT= N - NS
      TERM= (CON(NS-1) - 0.5)/(CON(NS-1) - CON(NS))
      T1= TERM*(V(NS-1) - V(NS))
      VX= V(NS-1) - T1
      COR= 1.0 - VX
C
      RETURN
      END
```

Program Listing 6.5 TRIDAG.FOR

```
C     SUBROUTINE TRIDAG
C
C     M= number of nodes in the grid
C
      SUBROUTINE TRIDAG (M,NX,A,B,C,D,CON)
      REAL A(NX), B(NX), C(NX), D(NX), CON(NX)
      REAL BETA(910), GAMMA(910)
      L= M-1
      IFAC= 2
      BETA(IFAC)= B(IFAC)
      GAMMA(IFAC)= D(IFAC)/BETA(IFAC)
      IFP1= IFAC + 1
C
      DO 20 I= IFP1,L
        BETA(I)= B(I) - (A(I)*C(I-1))/BETA(I-1)
        GAMMA(I)= (D(I) - A(I)*GAMMA(I-1))/BETA(I)
20    CONTINUE
C
      CON(L)= GAMMA(L)
      LAST= L - IFAC
      DO 30, K= 1,LAST
        KK= K
        I= L - KK
        CON(I)= GAMMA(I) - C(I)*CON(I+1)/BETA(I)
30    CONTINUE
C
      RETURN
      END
```

6.3 Two-Phase Fractional Flow

Two-phase fractional flow. FRAFLO.FOR (FRAC.DAT, FRAOUT.DAT)

The fractional flow of water and oil at the terminal end of a core is a function of wettability and saturation gradients within the porous medium. The curves of the fraction of water flowing with respect to the terminal water saturation vary in shape and position in accord with the average wettability of the fluid saturated system.

Welge (1945) discovered the relationship between the average saturation and the saturation at the terminal end of the core:

$$S_{w,av} - S_{w,w} = f_o W_i$$

where f_o is the fraction of oil flowing and is equal to the slope of the average water saturation with respect to the cumulative water injected:

$$f_o = \frac{dS_{w,av}}{dW_i}$$

In addition, the fraction of water flowing at the terminal end is related to the relative permeability ratio (Eq. 3.10) which is a function of the terminal fluid saturation. Analysis of a curve of the fraction of water flowing (at the terminal end of the core) versus the water saturation at the terminal end is shown in Fig. 3–9. The tangent of the fractional flow curve is extrapolated to $f_{w2} = 1.0$ to obtain the average saturation of the core:

$$\frac{df_{w2}}{dS_{w2}} = \frac{1 - f_{w2}}{S_{w,av} - S_{w2}}$$

The oil displacement efficiency for the waterflood is given by:

$$E_d = \frac{1 - S_{or} - S_{wi}}{1 - S_{wi}} = \frac{S_{wor} - S_{wi}}{1 - S_{wi}}$$

Individual relative permeabilities are calculated using Corey's (1954) equations (Eq. 4.26), and are introduced here in order to add the estimated values of the relative permeabilities. Corey's equations have considerable validity because wettability and the pore size distribution of the core govern the values of the end-points (S_{wi} and S_{or}).

The assumed laboratory procedure is: (1) the core is saturated to 100% with brine, (2) oil is injected, displacing water to S_{wi}, and (3) the core is water-flooded to S_{or}.

Input variables

D	diameter of the core (cm)
L	length of the core (cm)
MUOIL	oil viscosity (cP)
MUW	water viscosity (cP)
N	number of data points
OILOUT	cumulative oil produced (ml)
PHI	porosity
SWI	irreducible water saturation

Program Listing 6.6 FRAFLOW.FOR

```
C       FRACTIONAL FLOW
C
C       Calculates: (1) the fractional flow of water flowing at the terminal end as a function
C          of the terminal water saturation; (2)  the relative permeability ratio,
C          (3) the individual relative permeabilities by Corey's method and (4) the displace-
        ment efficiency.
C       N= number of data points; LEN and DIAM are the lenth and diameter of the core;
C          PHI= porosity; MUW and MUOIL= viscosity (cP) of brine and oil
C       OILOUT(I)= ml oil produced (fractions), WOUT(I)= ml water produced (fractions)
C       FO2= fraction of water flowing at the terminal end;  SW2= terminal water saturation,
C          SWAV= average water saturation;KRATIO= relative permeability ratio;
C          KRW AND KRO = individual relative permeabilities, ED= displacement efficiency
C
        REAL MUW, MUOIL, KRATIO(50), KRW(50), KRO(50)
        REAL CUMW(50), CUMOIL(50), SW2(50),FW2(50), FO2(50), SWAV(50)
        REAL WINJ(50), OILOUT(50), WOUT(50),LEN
C
        OPEN (UNIT= 10, FILE= 'FRAC.DAT', STATUS= 'OLD')
        READ (10,*) N, LEN, DIAM, PHI, MUW, MUOIL, SWI
```

```
      READ (10,*) (OILOUT(I), WOUT(I), I= 1,N)
      CLOSE (10)
C
      VBULK= 0.7854*LEN*(DIAM**2)
      VPORE= PHI*VBULK
C
      SUM1= 0.0
      SUM2= 0.0
      DO 20, I=1,N
         WINJ(I)= OILOUT(I)+WOUT(I)
         SUM1= SUM1+OILOUT(I)
         CUMOIL(I)= SUM1
         SUM2=SUM2+WINJ(I)
         CUMW(I)= SUM2
         SWAV(I)=(CUMOIL(I)/VPORE)+SWI
         FO2(I)= OILOUT(I)/WINJ (I)
         SW2(I)= SWAV(I)-(FO2(I)*CUMW(I)/VPORE)
         FW2(I)= 1.0-FO2(I)
         KRATIO(I)= (FW2(I)/FO2(I))*(MUW/MUOIL)
20    CONTINUE
C
      SWOR= SWAV(N)
      SOR= 1.0-SWOR
      ED=(SWOR-SWI)/(1.0-SWI)
C
      DO 22, I= 1,N
         TERM= (SW2(I)-SWI)/(1.0-SWI)
         KRW(I)= TERM**4
         X= (1.0-TERM)**2
         Y= 1.0-TERM**2
         KRO(I)= X*Y
22    CONTINUE
C
      OPEN (UNIT= 12, FILE= 'FRAOUT.DAT', STATUS='OLD')
      WRITE (12, 900)
900   FORMAT (5X,'FRACTIONAL FLOW AND RELATIVE PERMEABILITY', //)
C
      WRITE (12, 902) SWI, SOR, ED
902   FORMAT (1X, 'SWI= ',F6.3, 2X, 'SOR= ',F6.3,3X,
     &'DISPLACEMENT EFF (ED)= ',F6.3, //)
C
      WRITE (12, 904)
904   FORMAT (4X,'SW2',5X,'FO2',4X,'SW(AVE)')
      WRITE (12, 906) (SW2(I), FO2(I), SWAV(I), I=1,N)
906   FORMAT (3F8.3)
C
      WRITE (12, 908)
908   FORMAT (//,4X,'SW2',5X,'FW2',3X,'KRATIO',4X,'KRW',5X,'KRO')
      WRITE (12,910) (SW2(I),FW2(I), KRATIO(I), KRW(I), KRO(I), I=1,N)
910   FORMAT (5F8.3)
      CLOSE (12)
      STOP
      END
```

Program Listing 6.7 FRAC.DAT

```
13, 3.175, 1.905, 0.285, 1.0, 1.89, 0.396

0.5130    0.0000
0.0980    0.0521
0.0191    0.1280
0.0139    0.1339
0.0111    0.1161
0.0080    0.1450
0.0601    1.2539
0.0279    1.0188
0.0410    2.0447
0.0199    1.3429
0.0111    1.3269
0.0480    9.6667
0.0570   18.900
```

Program Listing 6.8 FRAOUT.DAT

```
FRACTIONAL FLOW AND RELATIVE PERMEABILITY


SWI=  0.396  SOR=  0.244   DISPLACEMENT EFF (ED)=  0.596


   SW2     FO2    SW(AVE)
  0.396   1.000   0.595
  0.465   0.653   0.633
  0.600   0.130   0.640
  0.611   0.094   0.646
  0.613   0.087   0.650
  0.628   0.052   0.653
  0.631   0.046   0.676
  0.650   0.027   0.687
  0.660   0.020   0.703
  0.671   0.015   0.711
  0.688   0.008   0.715
  0.699   0.005   0.734
  0.713   0.003   0.756


   SW2     FW2   KRATIO     KRW     KRO
  0.396   0.000   0.000   0.000   1.000
  0.465   0.347   0.281   0.000   0.774
  0.600   0.870   3.546   0.013   0.390
  0.611   0.906   5.097   0.016   0.363
  0.613   0.913   5.534   0.017   0.357
  0.628   0.948   9.590   0.022   0.323
  0.631   0.954  11.039   0.023   0.316
  0.650   0.973  19.321   0.031   0.276
  0.660   0.980  26.387   0.036   0.257
  0.671   0.985  35.705   0.043   0.235
  0.688   0.992  63.249   0.055   0.204
  0.699   0.995 106.555   0.063   0.186
  0.713   0.997 175.439   0.076   0.164
```

6.4 Pore Size Distribution

Centrifuge displacement of air by water. POCEN.FOR (POIN.DAT, POZOUT.DAT, AREA.FOR, HYPER.FOR)

This program is used to calculate: (1) the air displacing brine capillary pressure curve from $S_w = 1$ to $S_w = S_{wi}$, (2) the pore size distribution (pore radius versus distribution function), and (3) the relative flow properties of the core.

Centrifuge displacement of water by air is used to calculate the capillary pressure curve as a function of the inlet saturation using the Hassler-Brunner method (1945; also detailed in Tiab and Donaldson, 2004). Subroutine *HYPER* is called several times to model the capillary pressure versus saturation curve and the size-distribution versus pore radii by least squares solution of the data. Function *AREA* is used to calculate areas under the size-distribution versus pore-radii curve for ranges of pore radii.

The pore entry size distribution is calculated using Burdine's equations (Eqs. 4.3–4.20). The pore size distribution is then used to evaluate the fluid flow characteristics of the core (the relative volume and volumetric flow: Eqs. 4.31–4.35, Table 4.4, and Figs. 4–3 and 4–4). Finally, the absolute permeability of the core is estimated from the pore size distribution.

Input variables

DIAM	core diameter (cm^3)
DRO	density of air at 25° (0.0012)
N	number of data points
NROCK	type of rock that is being tested: 1 = sandstone, 2 = unconsolidated sand, 3 = carbonate, and 4 = calcareous sand
PHI	porosity
RE	distance from the center of the centrifuge rotation to the end of the core
RPM	revolutions per minute (step increases in RPM are used to measure the amount of fluid displaced at each RPM)
SIGMA	interfacial tension (mNewton/meter)
WOUT	amount of water displaced at each increment of RPM
XLEN	length of the core (cm)

Program Listing 6.9 POCEN.FOR

```
C       POCEN.FOR
C
C       PORE-SIZE DISTRIBUTION ** CENTRIFUGE DISPLACEMENT OF BRINE BY AIR **
C
C       FUNCTION AREA and SUBROUTINE HYPER are required
C
C       Calculates:(1) air displacing brine capillary pressure curve and develops
C          a least squares model of the capillary pressure curve using SUBROUTINE
C          HYPER;  (2) computes the pore-entry size distribution (3) the relative
C          fluid flow properties, and (4) the absolute permeability.
C
C       Uses data file POIN.DAT for the input data and POZOUT.DAT for the output.
C
C       N= Number of data points;  DRHO= density difference (density of air at 25C=
C          0.0012);  RE= distance from the center of rotation to the end of the
C          core; SIGMA= interfacial tension (IFT water/air at 25C= 72).
C
C       NROCK= type of rock being tested:  1= sandstone, 2= unconsolidated sand, 3= carbonate,
C          4= calcareous sand
C
C       XLEN= core lenght; DIAM= core diamter, PHI= porosity;  RPM= revolutions per
C          minute; WOUT= brine displaced
C
        REAL RPM(50), WOUT(50),SAV(50),PC(50),PCH(50),PCH1(50),QREL(5)
        REAL RAD(50), DIS(50),V(100), RELV(5), X(100), Y(100), H(100),Z(5)
C
        OPEN (20, FILE= 'POIN.DAT', STATUS= 'OLD')
        READ (20,*) N, DRHO, RE, SIGMA
        READ (20,*) NROCK, XLEN, DIAM, PHI
        READ (20,*) (RPM(I), WOUT(I), I= 1,N)
        CLOSE (20)
C
        VBULK= 0.785*XLEN*DIAM**2
        VPORE= VBULK*PHI
        SWI= 1.0 - WOUT(N)/VPORE
C
        DO 20 I= 1,N
           SAV(I)= 1.0 - WOUT(I)/VPORE
           XX= 1.578E-07*RPM(I)**2
           PC(I)= XX*DRHO*(RE-XLEN/2.0)*XLEN
20      CONTINUE
C
        CALL HYPER (N,SAV,PC,P,Q,R)
        P1= P
        Q1= Q
        R1= R
C
        J= INT((1.0 -SWI)/0.05) + 2
        SAV(1)= 1.0
        DO 24 I= 2,J
           SAV(I)= SAV(I-1) - 0.05
           IF((SAV(I) - SWI) .LE. 0.0) GO TO 26
24      CONTINUE
26      SAV(J)= SWI
C
        DO 28 I= 1,J
        PCH(I)= (P + Q*SAV(I))/(1.0 + R*SAV(I))
28      CONTINUE
C
        DO 30 I= 1,J
           PCH1(I)= 6.895*PCH(I)
```

```
         RAD(I)= 2.0*SIGMA/PCH1(I)
         SLOPE= -((1.0 + R*SAV(I))**2/(Q-P*R))
         DIS(I)= PCH1(I)*(VPORE/RAD(I))*SLOPE
30    CONTINUE
      THP= PCH(1)
C
C     Calculation of the relative volumes, flow rates and absolute permeability
C
      CALL HYPER (J, RAD, DIS, P,Q,R)
      X1= RAD(1)
      XN= RAD(J)
      VT= AREA (X1,XN,P,Q,R)
      VTOT= (VT/10) + VPORE*SWI
C
      DRAD= (RAD(1) - RAD(J))/3
      ZERO= 0.00
      AP= RAD(J)
      BP= AP + DRAD
      CP= BP + DRAD
      DP= RAD(1)
C
      V(1)= VPORE*SWI
      T1= AP
      T2= BP
      XN= T1
      X1= T2
C
      SUM= 0.0
      DO 50 I= 2,4
         XN= T1
         X1= T2
         V(I)= (AREA(X1,XN,P,Q,R))/10
         SUM= SUM + V(I)
         T1= T2
         T2= T2 + DRAD
50    CONTINUE
C
      SUM= SUM + VPORE*SWI
      RATIO = SUM/VTOT
C
      DO 52, I= 2,4
         V(I)= RATIO*V(I)
52    CONTINUE
C
      DO 54 I= 1,4
         RELV(I)= (V(I)/VTOT)*100
54    CONTINUE
C
C     Calculation of the relative flow rates and absolute permeability
C
      SELECT CASE (NROCK)
      CASE (1)
         F= 0.81*(PHI**2)
         GO TO 44
C
      CASE (2)
         F= 1.0/(PHI**1.5)
         GO TO 44
C
      CASE (3)
         F= 1.0/(PHI**2)
         GO TO 44
C
      CASE (4)
         F= 1.45/(PHI**1.7)
```

```
C
      END SELECT
C
44    TAU= SQRT(F*PHI)
      T= SQRT(TAU)
C
      FAC= 0.2
8     NR= ABS((RAD(1)-RAD(J))/FAC)
      IF(NR .GE. 100) GO TO 10
      GO TO 12
10    FAC= FAC + 0.1
      GO TO 8
12    CONTINUE
C
      SUM1= 0
      X(1)= AP
      Y(1)= (P + Q*X(1))/(1.0 + R*X(1))
C
      DO 70 I=2,NR+1
         X(I)= (X(I-1) + X(I))/2
         Y(I)= (P + Q*X(I))/(1.0 + R*X(I))
70    CONTINUE
         X(NR+1)= RAD(1)
         Y(NR+1)= (P + Q*X(NR+1))/(1.0 + R*X(NR+1))
C
      DO 72 I= 1,NR+1
         XAV= (X(I) + X(I+1))/2
         H(I)= (Y(I) + Y(I+1))/2
         V(I)= H(I)*0.2
         SUM1= SUM1 + V(I)*(XAV**2)
72    CONTINUE
C
      TEMP= 7.854*(DIAM**2)*XLEN*T
      PERM= TEMP*SUM1
C
      T1= AP
      T2= BP
      QREL(1)=0.0
C
      FAC= 0.2
35    ND= ABS((T1 - T2)/FAC)
      IF(ND .GE. 100) GO TO 36
      GO TO 37
36    FAC= FAC + 0.1
      GO TO 35
37    CONTINUE
C
C     ND= ABS((T1-T2)/0.2)
C
      TOTAL= 0.0
      DO 78 I=2,4
C
         X(1)= T1
         Y(1)= (P + Q*X1)/(1.0 + R*X1)
         DO 74 k=2,ND+1
            X(k)= (X(k-1) + X(k))/2
            Y(k)= (P + Q*X(k))/(1.0 + R*X(k))
74    CONTINUE
         X(ND+1)= T2
         Y(ND+1)= (P + Q*X(ND+1))/(1.0 + R*X(ND+1))
         SUM2= 0.0
         DO 76 k= 1,ND+1
            XAV= (X(k) + X(k+1))/2
            H(k)= (Y(k) + Y(k+1))/2
            VR= H(k)*0.2
```

```
         SUM2= SUM2 + VR*(XAV**2)
76     CONTINUE
       Z(I)= SUM2
       TOTAL= TOTAL + SUM2
       T1= T2
       T2= T2 + DRAD
       ND= ABS((T1-T2)/0.2)
78    CONTINUE
C
      DO 80 I= 2,4
         QREL(I)= (Z(I)/TOTAL)*100
80    CONTINUE
C
      OPEN (UNIT= 11, FILE= 'POZOUT.DAT', STATUS= 'OLD')
      WRITE (11,900)
900   FORMAT(10X,'PORE ENTRY SIZE DISTRIBUTION',/)
      WRITE (11, 902)
902   FORMAT(7X, 'CENTRIFUGE----AIR DISPLACING BRINE',//)
C
      WRITE(11,914), THP
914   FORMAT('Threshold Pressure (kPa) =',F5.2,/)
      WRITE(11,916), VPORE
916   FORMAT('Measured Pore Volume =',F5.2,/)
      WRITE(11,918),VTOT
918   FORMAT('Calculated Pore Volume =',F5.1,/)
      WRITE (11,920), PERM
920   FORMAT('Calculated Permeability (md) =',F5.0,//)
C
      WRITE(11,906)
906   FORMAT(9X,'SW',9X,'PC(psi)',3X, 'RADIUS(mm)',4X,
     &'DIS(m**2)',/)
      WRITE (11, 908), (SAV(I), PCH(I), RAD(I), DIS(I), I= 1,J)
908   FORMAT (4F12.2)
      WRITE(11,930)
930   FORMAT(//)
C
      WRITE(11,910)
910   FORMAT(3X,'RANGE(1) - RANGE(2)',3X,'REL VOLUME',3X,'REL
     & FLOW RATE',/)
      WRITE(11,912) ZERO,AP,RELV(1),QREL(1)
      WRITE(11,912) AP,BP,RELV(2),QREL(2)
      WRITE(11,912) BP,CP,RELV(3),QREL(3)
      WRITE(11,912) CP,DP,RELV(4),QREL(4)
912   FORMAT(F9.2,F11.2,F12.1,F15.1)
      STOP
      END
```

Program Listing 6.10 POIN.DAT

```
14, 1.0156, 8.60, 72.0
1, 3.23, 1.89, 0.202
   0  0.000
 510  0.05
 552  0.18
 606  0.31
 654  0.44
 732  0.62
 900  1.03
1134  1.18
1404  1.30
1644  1.37
```

```
2052  1.43
2298  1.48
2592  1.50
2844  1.52
```

Program Listing 6.11 POZOUT.DAT

```
 PORE ENTRY SIZE DISTRIBUTION

       CENTRIFUGE----AIR DISPLACING BRINE


Threshold Pressure (kPa) = 0.73

Measured Pore Volume = 1.83

Calculated Pore Volume =  2.5

Calculated Permeability (md) = 588.


          SW          PC(psi)   RADIUS(mm)     DIS(m**2)

         1.00         0.73        28.48        0.21
         0.95         0.82        25.61        0.23
         0.90         0.91        22.99        0.25
         0.85         1.01        20.59        0.27
         0.80         1.14        18.38        0.30
         0.75         1.28        16.34        0.32
         0.70         1.44        14.46        0.35
         0.65         1.64        12.70        0.37
         0.60         1.89        11.07        0.40
         0.55         2.19         9.55        0.43
         0.50         2.57         8.12        0.46
         0.45         3.08         6.79        0.49
         0.40         3.78         5.53        0.52
         0.35         4.80         4.35        0.55
         0.30         6.45         3.24        0.58
         0.25         9.56         2.19        0.62
         0.20        17.54         1.19        0.65
         0.17        34.56         0.60        0.67


   RANGE(1) - RANGE(2)   REL VOLUME   REL FLOW RATE

     0.00        0.60       12.4          0.0
     0.60        9.90       36.8          5.8
     9.90       19.19       27.9         26.0
    19.19       28.48       22.9         68.2
```

Program Listing 6.12 AREA.FOR

```
      FUNCTION AREA(X1,XN,P,Q,R)
C
C     Calculates the area under curve between the limits X(1) and X(N)
C
      T1= Q*(XN - X1)/R
```

```
      T2= (P*Q - R)/R**2
      T3= ALOG(1+R*XN) - ALOG(1+R*X1)
      T4= T1 + T2*T3
      AREA= ABS(T4)
C
      RETURN
      END
```

Program Listing 6.13 SUBROUTINE HYPER.FOR

```
C     Subroutine HYPER:  Least Squares curve fit of data to a hyperbolic
C        function.  X= (P + Q*X)/(1 + R*X).  The coefficients P, Q and R
C        are returned.
C     The derivative dY/dX= (Q - P*R)/(1 + R*X)^2.   The integral is:
C        (Q*X)/R + ((P*R - Q)/R^2)Ln(1+R*X) from b to a.
C
      SUBROUTINE HYPER (N,X,Y,P,Q,R)
C
      REAL X(50), Y(50), NUM1, NUM2, NUM3
C
      A= 0.0
      B= 0.0
      C= 0.0
      D= 0.0
      E= 0.0
      F= 0.0
      G= 0.0
      H= 0.0
C
      DO 10, I= 1,N
         A= A + X(I)
         B= B + Y(I)
         C= C + X(I)*Y(I)
         D= D + X(I)**2
         E= E + Y(I)**2
         F= F + X(I)*Y(I)**2
         G= G + (X(I)**2)*Y(I)
         H= H + (X(I)**2)*(Y(I)**2)
10    CONTINUE
C
C
      NUM1=  D*(C*F-B*H) + C*(A*H-C*G) + G*(B*G-A*F)
      NUM2=  N*(G*F-C*H) + A*(B*H-C*F) + C*(C**2-B*G)
      NUM3=  N*(D*F-C*G) + A*(B*G-A*F) + C*(A*C-B*D)
      DNOM=  N*(G**2-D*H) + A*(A*H-C*G) + C*(C*D-A*G)
C
      P= NUM1/DNOM
      Q= NUM2/DNOM
      R= NUM3/DNOM
C
      RETURN
      END
```

6.5 Pores Size Distribution

Mercury injection displacement of air. POSIZE.FOR (HGIN.DAT, HGOUT.DAT, AREA.FOR, HYPER.FOR)

Ritter and Drake (1945), and Burdine et al. (1950), based the development of equations for calculation of the pore size distribution of porous media on mercury injection curves. The procedures are detailed in Tiab and Donaldson (2004). The calculations are similar to those used for the centrifuge displacement of water by air using subroutine *HYPER* to model the capillary pressure curve which is subsequently used to determine the pore entry size distribution. Function *AREA* is used to calculate areas under segments of the size-distribution versus pore-radii of the curve.

Input Variables

DIAM	diameter of the core (cm)
HGIN	increments of mercury injected in successive steps (ml)
N	number of data points
NROCK	type of porous system being used: 1 = sandstone, 2 = unconsolidated sand, 3 = carbonate rock, 4 = calcareous sand
PC	capillary pressure measured at each incremental injection of mercury
PHI	porosity
RAD	pore radius
SIGMA	interfacial tension of mercury
XLEN	core length (cm)

Program Listing 6.14 POSIZE.FOR

```
C       POSIZE.FOR
C
C       PORE-SIZE DISTRIBUTION  **MERCURY INJECTION**
C
C       SUBROUTIN HYPER and FUNCTION AREA are required
C
C       Pc [N/mE02]= 2*SIGMA [N*E-03]*Cos140)/r [m]
C
C       HGIN= ml of mercury injected at each successive step.
C       PCIN= equilibrium capillary pressure (psi) at each step of injection
C       RAD= pore radis;  SA= average air saturation in the core (%Vp); RAD=
C          pore radius (mm); DIS= distribution function (m**2)
C
C       NROCK= the type of rock:  1= consolidated sandstone; 2= unconsolidated
C          sand;  3= carbonate rock; 4= calcareous sand.  Refer to Table 4.3.
C
        REAL HGIN(50),PC(50),PCH(50),PCH1(50),SAV(50),QREL(5)
        REAL RAD(50),DIS(50),V(100),RELV(5),X(100),Y(100),H(100),Z(5)
C
```

```
      OPEN (UNIT= 22, FILE= 'HGIN.DAT', STATUS= 'OLD')
      READ (22,*) N, NROCK, XLEN, DIAM,PHI,SIGMA
      READ (22,*) (PC(I), HGIN(I), I= 1,N)
      CLOSE (22)
C
      VBULK = 0.7854*XLEN*DIAM**2
      VPORE= VBULK*PHI
      SAI= 1.0 - HGIN(N)/VPORE
C
      DO 20 I=1,N
         SAV(I)=1.0 - HGIN(I)/VPORE
20    CONTINUE
C
      CALL HYPER(N,SAV,PC,P, Q, R)
      P1= P
      Q1= Q
      R1= R
C
      J= INT((1.0 -SAI)/0.05) + 2
      SAV(1)= 1.0
      DO 24 I= 2,J
         SAV(I)= SAV(I-1) - 0.05
         IF((SAV(I) - SAI) .LE. 0.0) GO TO 26
24    CONTINUE
26    SAV(J)= SAI
C
      DO 28 I= 1,J
      PCH(I)= (P + Q*SAV(I))/(1.0 + R*SAV(I))
28    CONTINUE
C
      DO 30 I= 1,J
         PCH1(I)= 6.895*PCH(I)
         RAD(I)= 1.532*SIGMA/PCH1(I)
         SLOPE= -((1.0 + R*SAV(I))**2/(Q-P*R))
         DIS(I)= PCH1(I)*(VPORE/RAD(I))*SLOPE
30    CONTINUE
      THP= PCH(1)
C
C
C     Calculation of the relative volumes, flow rates and absolute permeability
C
      CALL HYPER (J, RAD, DIS, P,Q,R)
      X1= RAD(1)
      XN= RAD(J)
      VT= AREA (X1,XN,P,Q,R)
      VTOT= (VT/10) + VPORE*SAI
C
      DRAD= (RAD(1) - RAD(J))/3
      ZERO= 0.00
      AP= RAD(J)
      BP= AP + DRAD
      CP= BP + DRAD
      DP= RAD(1)
C
      V(1)= VPORE*SAI
      T1= AP
      T2= BP
      XN= T1
      X1= T2
C
      SUM= 0.0
      DO 50 I= 2,4
         XN= T1
         X1= T2
         V(I)= (AREA(X1,XN,P,Q,R))/10
```

```
         SUM= SUM + V(I)
         T1= T2
         T2= T2 + DRAD
50    CONTINUE
C
      SUM= SUM + VPORE*SAI
      RATIO = SUM/VTOT
C
      DO 52, I= 2,4
         V(I)= RATIO*V(I)
52    CONTINUE
C
      DO 54 I= 1,4
         RELV(I)= (V(I)/VTOT)*100
54    CONTINUE
C
C     Calculation of the relative flow rates and absolute permeability
C
      SELECT CASE (NROCK)
      CASE (1)
         F= 0.81*(PHI**2)
         GO TO 44
C
      CASE (2)
         F= 1.0/(PHI**1.5)
         GO TO 44
C
      CASE (3)
         F= 1.0/(PHI**2)
         GO TO 44
C
      CASE (4)
         F= 1.45/(PHI**1.7)
C
      END SELECT
C
44    TAU= SQRT(F*PHI)
      T= SQRT(TAU)
C
      FAC= 0.2
8     NR= ABS((RAD(1)-RAD(J))/FAC)
      IF(NR .GE. 100) GO TO 10
      GO TO 12
10    FAC= FAC + 0.1
      GO TO 8
12    CONTINUE
C
      SUM1= 0
      X(1)= AP
      Y(1)= (P + Q*X(1))/(1.0 + R*X(1))
C
      DO 70 I=2,NR+1
         X(I)= (X(I-1) + X(I))/2
         Y(I)= (P + Q*X(I))/(1.0 + R*X(I))
70    CONTINUE
         X(NR+1)= RAD(1)
         Y(NR+1)= (P + Q*X(NR+1))/(1.0 + R*X(NR+1))
C
      DO 72 I= 1,NR+1
         XAV= (X(I) + X(I+1))/2
         H(I)= (Y(I) + Y(I+1))/2
         V(I)= H(I)*0.2
         SUM1= SUM1 + V(I)*(XAV**2)
72    CONTINUE
C
```

```
      TEMP= 7.854*(DIAM**2)*XLEN*T
      PERM= TEMP*SUM1
C
      T1= AP
      T2= BP
      QREL(1)=0.0
C
      FAC= 0.2
35    ND= ABS((T1-T2)/FAC)
      IF(ND .GE. 100) GO TO 36
      GO TO 37
36    FAC= FAC + 0.1
      GO TO 35
37    CONTINUE
C
      TOTAL= 0.0
      DO 78 I=2,4
C
         X(1)= T1
         Y(1)= (P + Q*X1)/(1.0 + R*X1)
         DO 74 k=2,ND+1
            X(k)= (X(k-1) + X(k))/2
            Y(k)= (P + Q*X(k))/(1.0 + R*X(k))
74    CONTINUE
         X(ND+1)= T2
         Y(ND+1)= (P + Q*X(ND+1))/(1.0 + R*X(ND+1))
         SUM2= 0.0
         DO 76 k= 1,ND+1
            XAV= (X(k) + X(k+1))/2
            H(k)= (Y(k) + Y(k+1))/2
            VR= H(k)*0.2
            SUM2= SUM2 + VR*(XAV**2)
76       CONTINUE
         Z(I)= SUM2
         TOTAL= TOTAL + SUM2
         T1= T2
         T2= T2 + DRAD
         ND= ABS((T1-T2)/0.2)
78    CONTINUE
C
      DO 80 I= 2,4
         QREL(I)= (Z(I)/TOTAL)*100
80    CONTINUE
C
      OPEN (UNIT= 34, FILE= 'HGOUT.DAT', STATUS= 'OLD')
      WRITE (34,900)
900   FORMAT(10X,'PORE ENTRY SIZE DISTRIBUTION',/)
      WRITE (34, 902)
902   FORMAT(13X, 'MERCURY DISPLACING AIR',//)
C
      WRITE(34,914), THP
914   FORMAT('Threshold Pressure (kPa) =',F5.2,/)
      WRITE(34,916), VPORE
916   FORMAT('Measured Pore Volume =',F5.2,/)
      WRITE(34,918),VTOT
918   FORMAT('Calculated Pore Volume =',F5.1,/)
      WRITE (34,920), PERM
920   FORMAT('Calculated Permeability (md) =',F5.0,//)
C
      WRITE(34,906)
906   FORMAT(8X,'SAIR',6X,'PC(psi)',4X, 'RADIUS(mm)',2X,
     &'DIS(m**2)',/)
      WRITE (34, 908), (SAV(I), PCH(I), RAD(I), DIS(I), I= 1,J)
908   FORMAT (4F12.2)
      WRITE(34,930)
```

```
930    FORMAT(//)
C
       WRITE(34,910)
910    FORMAT(3X,'RANGE(1) - RANGE(2)',3X,'REL VOLUME',3X,'REL
      & FLOW RATE',/)
       WRITE(34,912) ZERO,AP,RELV(1),QREL(1)
       WRITE(34,912) AP,BP,RELV(2),QREL(2)
       WRITE(34,912) BP,CP,RELV(3),QREL(3)
       WRITE(34,912) CP,DP,RELV(4),QREL(4)
912    FORMAT(F9.2,F11.2,F12.1,F15.1)
       STOP
       END
```

Program Listing 6.15 HGIN.DAT

```
14, 1, 2.04, 3.77, 0.247, 480

   0 0.00
   9 0.20
  12 0.64
  15 1.23
  18 1.74
  21 2.28
  30 2.85
  40 3.17
  60 3.43
  80 3.55
 100 3.61
 150 3.74
 200 3.79
 500 4.02
```

Program Listing 6.16 HGOUT.DAT

```
PORE ENTRY SIZE DISTRIBUTION

            MERCURY DISPLACING AIR


Threshold Pressure (kPa) = 6.13

Measured Pore Volume = 5.62

Calculated Pore Volume =  5.2

Calculated Permeability (md) =2816.


        SAIR       PC(psi)     RADIUS(mm)  DIS(m**2)

        1.00        6.13        17.38         0.80
        0.95        7.06        15.12         0.92
        0.90        8.12        13.13         1.04
        0.85        9.37        11.38         1.18
        0.80       10.86         9.82         1.32
        0.75       12.66         8.43         1.47
        0.70       14.87         7.17         1.63
        0.65       17.67         6.04         1.79
```

```
      0.60          21.31          5.00          1.97
      0.55          26.26          4.06          2.15
      0.50          33.36          3.20          2.34
      0.45          44.40          2.40          2.54
      0.40          63.94          1.67          2.75
      0.35         107.93          0.99          2.96
      0.30         298.89          0.36          3.19
      0.29         593.58          0.18          3.25


  RANGE(1) - RANGE(2)   REL VOLUME   REL FLOW RATE

    0.00       0.18        30.6            0.0
    0.18       5.91        29.6            5.8
    5.91      11.65        21.8           26.1
   11.65      17.38        17.9           68.1
```

Program Listing 6.17 AREA.FOR

```
      FUNCTION AREA(X1,XN,P,Q,R)
C
C     Calculates the area under curve between the limits X(1) and X(N)
C
      T1= Q*(XN - X1)/R
      T2= (P*Q - R)/R**2
      T3= ALOG(1+R*XN) - ALOG(1+R*X1)
      T4= T1 + T2*T3
      AREA= ABS(T4)
C
      RETURN
      END
```

Program Listing 6.18 SUBROUTINE HYPER.FOR

```
C     Subroutine HYPER:  Least Squares curve fit of data to a hyperbolic
C        function.  X= (P + Q*X)/(1 + R*X).  The coefficients P, Q and R
C        are returned.
C     The derivative dY/dX= (Q - P*R)/(1 + R*X)^2.   The integral is:
C        (Q*X)/R + ((P*R - Q)/R^2)Ln(1+R*X) from b to a.
C
      SUBROUTINE HYPER (N,X,Y,P,Q,R)
C
      REAL X(50), Y(50), NUM1, NUM2, NUM3
C
      A= 0.0
      B= 0.0
      C= 0.0
      D= 0.0
      E= 0.0
      F= 0.0
      G= 0.0
      H= 0.0
C
      DO 10, I= 1,N
         A= A + X(I)
         B= B + Y(I)
         C= C + X(I)*Y(I)
```

```
      D= D + X(I)**2
      E= E + Y(I)**2
      F= F + X(I)*Y(I)**2
      G= G + (X(I)**2)*Y(I)
      H= H + (X(I)**2)*(Y(I)**2)
10    CONTINUE
C
C
      NUM1= D*(C*F-B*H) + C*(A*H-C*G) + G*(B*G-A*F)
      NUM2= N*(G*F-C*H) + A*(B*H-C*F) + C*(C**2-B*G)
      NUM3= N*(D*F-C*G) + A*(B*G-A*F) + C*(A*C-B*D)
      DNOM= N*(G**2-D*H) + A*(A*H-C*G) + C*(C*D-A*G)
C
      P= NUM1/DNOM
      Q= NUM2/DNOM
      R= NUM3/DNOM
C
      RETURN
      END
```

6.6 Permeability Estimation

Estimation of permeability using experimental correlations. PERM.FOR

Equations for estimation of permeability from correlations of experimental data with respect to porosity (for granular matrices) were developed from Fig. 1-25, Chiligarian and Wolf (1975).

When the program is run, a request for entry of two data parameters will appear on the screen, *J* and *PHI*. The types of sands and rocks are distinguished by assigning a value to *J* in accord with this list given below. The other parameter, *PHI*, is the porosity. The output is the estimated permeability of the specific matrix type and porosity of the sands and rocks which are distinguished by defining the parameter, *J*, in the data file (*PERIN.DAT*) as follows:

J	Matrix
1	Coarse to very course grained
2	Coarse to medium grained: porosity < 0.17
3	Coarse to medium grained: porosity > 0.17
4	Fine grained
5	Silty
6	Clayey
7	Carbonate

Program Listing 6.19 PERM.FOR

```
C       PERMEABILITY
C
C       Estimate Permeability from experimental correlations
C
C       J = defining parameter fo the type of matrix
C
        PRINT *, 'Enter J, Phi'
        READ *, J, PHI
C
        SELECT CASE (J)
        CASE (1)
           Term = 0.4026 + 29.3078*PHI - 87.7629*(PHI**2)
           PERM = 10.0**TERM
           GO TO 20
C
        CASE (2)
           TERM = 0.1646 + 18.8216*PHI - 10.6615*(PHI**2)
           PERM = 10.0**TERM
           GO TO 20
C
        CASE (3)
           TERM = -2.6667 + 45.0000*PHI - 83.3333*(PHI**2)
           PERM = 10.0**TERM
           GO TO 20
C
        CASE (4)
           TERM = -5.4158 + 71.2453*PHI - 166.1990*(PHI**2)
           PERM = 10.0**TERM
           GO TO 20
C
        CASE (5)
           TERM = 0.1040 + 12.1080*PHI - 8.1826*(PHI**2)
           PERM = 10.0**TERM
           GO TO 20
C
        CASE (6)
           TERM = -2.1926 + 17.3853*PHI + 7.3875*(PHI**2)
           PERM = 10.0**TERM
           GO TO 20
           END SELECT
C
20      PRINT 902, PERM
902     FORMAT ('Estimated Permeability (md) =', F6.0)
C
        STOP
        END
```

6.7 Relative Permeability

Relative permeability from unsteady-state displacement (Johnson, Bossler, Naumann method, JBN). RELPERM.FOR (RLKIN.DAT, RLKOUT.DAT, POWER.FOR)

The individual relative permeabilites are calculated from unsteady state displacement of oil by water from cores that have been saturated and equilibrated with oil and water (beginning water saturation = S_{wi}, ending water saturation = S_{wor}). The JBN method is used for the calculation of k_{rw} and k_{ro} from periodic displacement volumes of water and oil (cuts) with respect to time.

Input data are: (1) the time at which each fraction of effluent liquid is taken, (2) cumulative volume of water injected, and (3) the cumulative volume of oil produced. In addition, the effective permeability to oil at the irreducible water saturation must be calculated and entered in the input data file: $k_{eo} = (Q_o \times \mu_o)/(\Delta P \times A/L)$.

Relative permeabilities are listed as a function of the terminal water saturation of the core. The fraction of water flowing as a function of terminal water saturation also is listed in the output data file.

Input Variables

DELP	pressure difference (atmospheres)
DIAM	core diameter (ml)
LEN	core length (cm)
KEO	effective permeability to oil at S_{wi} (Darcy)
MUOIL	oil viscosity (cP)
MUW	water viscosity (cP)
N	number of data points
OILOUT	fraction of oil produced
PHI	porosity
QW	rate of water injected (constant rate) (ml/s)
SWI	irreducible water saturation
TIME	seconds
WINJ	amount of water injected at each time increment (ml)

Program Listing 6.20 RELPERM.FOR

```
C      RELATIVE PERMEABILITY (JBN) METHOD FOR UNSTEADY-STATE DISPLACEMENT
C
C      Calculates the individual relative permeabilities from an unsteady-state
C         state displacement (water injection at a constant rate).
C
C      Input data file = RLKIN.DAT;  Output data file = RLKOUT.DAT.  SUBROUTINE
C         POWER used:  Y= BX**A;  dy/dx= A*B*X**(A-1).
C
C      N= No. of data points;  LEN= core length;  DIAM= core diameter;
C         PHI= porosity;  MUW= water viscosity, cP;  MUOIL= oil viscodity, cP,
C         QW= rate of water injection (constant) ml/s; KEO= effective
C         permeability to oil at Swi, Darcy.  OILOUT= oil produced
C         in fractions, ml;  WINJ= water injected for each fraction, ml;
C         DELP= pressure difference, atm.
C
       REAL MUW,MUOIL,KRATIO(20),KRW(20),KRO(20),SW2(20)
       REAL WINJ(20),OILOUT(20),CUMW(20),CUMOIL(20),LEN,KEO
       REAL X(20), Y(20),SLOPE(20),FO2(20),FW2(20),DELP(20),SWAV(20)
       REAL OIL(20), W(20)
C
       OPEN(UNIT= 14, FILE= 'RLKIN.DAT', STATUS= 'OLD')
       READ (14, *) N, LEN, DIAM, PHI, MUW, MUOIL, SWI, QW, KEO
       READ (14, *) (OILOUT(I),WINJ(I),DELP(I),I= 1,N)
       CLOSE (14)
C

C
       VBULK= 0.7854*LEN*DIAM**2
       VPORE=PHI*VBULK
C      PRINT *, VBULK, VPORE
C

       SUM1= 0.0
       SUM2= 0.0
       DO 20 I= 1,N
          SUM1=SUM1+OILOUT(I)/VPORE
          CUMOIL(I)= SUM1
          SUM2= SUM2+WINJ(I)/VPORE
          CUMW(I)= SUM2
20     CONTINUE
C
       DO 21, I=1,N
          SWAV(I)=CUMOIL(I)+SWI
          FO2(I)=OILOUT(I)/WINJ(I)
          FW2(I)= 1.0 - FO2(I)
          SW2(I)= SWAV(I)-(FO2(I)*CUMW(I))
          KRATIO(I)= FW2(I)/(1.0-FW2(I))*(MUW/MUOIL)
21     CONTINUE
C
       CC= LEN/(0.7854*DIAM**2)
       DO 22, I= 1,N
          Y(I)= 1.0/CUMW(I)
          TERM= KEO/(CC*MUOIL*QW)
          X(I)= Y(I)*TERM*DELP(I)
22     CONTINUE
C
       CALL POWER (N,X,Y,A,B)
       PRINT *, A, B
       DO 24, I= 1,N
          SLOPE(I)=A*B*(X(I)**(A-1.0))
          KRO(I)= FO2(I)*SLOPE(I)
```

```
         KRW(I)=KRO(I)*KRATIO(I)
24     CONTINUE
       SWOR= SWAV(N)
       SOR= 1.0-SWOR
       ED=(SWOR-SWI)/(1.0-SWI)
C
       OPEN (UNIT= 16, FILE='RLKOUT.DAT', STATUS= 'OLD')
       WRITE (16,900)
900    FORMAT (5X,'RELATIVE PERMEABILITY  (JBN METHOD)',//)
C
       WRITE (16, 902) SWI, SOR, ED
902    FORMAT(1X,'SWI=',F6.3,2X,'SOR=',F6.3,2X,'DISPLACEMENT EFFICIENCY=',
      &F7.3,//)
C
       WRITE (16, 904)
904    FORMAT(4X,'SW2',5X,'KRW',5X,'KRO',4X,'KRATIO')
       WRITE(16,906) (SW2(I), KRW(I), KRO(I), KRATIO(I), I= 1,N)
906    FORMAT (4F8.3)
C
       WRITE(16,908)
908    FORMAT(//,2X,'SW2',5X,'FW2')
       WRITE(16,910) (SW2(I),FW2(I),I= 1,N)
910    FORMAT(2F8.3)
       CLOSE (16)
       STOP
       END
```

Program Listing 6.21 RLKIN.DAT

```
8, 25.4, 3.8, 0.260, 0.506, 12.0, 0.203, 0.038, 0.165

31.87   39.46    4.08
 4.55   18.21    2.38
 3.04   22.01    2.24
 1.52   22.01    1.84
 0.76   46.73    1.36
 0.76   50.08    1.02
 0.76   50.84    0.68
 0.38   47.80    0.68
```

Program Listing 6.22 RLKOUT.DAT

```
RELATIVE PERMEABILITY  (JBN METHOD)


SWI= 0.203  SOR= 0.214  DISPLACEMENT EFFICIENCY=  0.731


    SW2     KRW     KRO    KRATIO
  0.203   0.007   0.660   0.010
  0.497   0.040   0.314   0.127
  0.583   0.055   0.208   0.263
  0.656   0.073   0.128   0.568
  0.728   0.106   0.041   2.551
  0.730   0.139   0.051   2.736
  0.731   0.187   0.067   2.779
  0.754   0.204   0.039   5.262
```

```
SW2     FW2
 0.203   0.192
 0.497   0.750
 0.583   0.862
 0.656   0.931
 0.728   0.984
 0.730   0.985
 0.731   0.985
 0.754   0.992
```

Program Listing 6.23 SUBROUTINE POWER.FOR

```
C       Subroutine POWER.  Least Squares curve fit of data to a power function:
C          Y= B*X**A; dy/dx= A*B*X**(A-1).  Returns the coefficients A and B
C
        SUBROUTINE POWER (N, X, Y, A, B)
        REAL X(20), Y(20)
C
        SUMX = 0.0
        SUMY = 0.0
        SUMX2= 0.0
        SUMXY= 0.0
C
        DO 20, I= 1,N
        SUMX = SUMX + ALOG10(X(I))
        SUMY = SUMY + ALOG10(Y(I))
        SUMX2= SUMX2 + ALOG10(X(I))*ALOG10(X(I))
        SUMXY= SUMXY + ALOG10(X(I))*ALOG10(Y(I))
20      CONTINUE
C
        A= (SUMX*SUMY-N*SUMXY)/(SUMX*SUMX-N*SUMX2)
        TERM= (SUMY-A*SUMX)/N
        B= 10**TERM
        RETURN
        END
```

6.8 USBM Wettability Index

USBM wettability data calculated from centrifuge displacements of water and oil. *WET.FOR* (WETIN1.DAT, WETIN2.DAT, WETOUT.DAT, EXTRA.FOR, HYPER.FOR)

The determination of the USBM wettability index is explained in the text of Chapter 1 for water-wet and oil-wet systems. The program computes a least squares fit of capillary pressure curves 3 and 5 (Fig. 1–8) using a hyperbolic equation (subroutine *HYPER*). Areas under the curves are determined by integration (Eq. 1.13) and the wettability index, *log(Area5/Area3),* is evaluated. Calculation of the USBM wettability index was standardized for use of the area under the capillary pressure curves extending the curves to 10 psi (68.95 kPa). Therefore, function *EXTRA* is used to extrapolate the capillary pressure curve to 10 psi when the experimental data does not attain that value, or to remove the extension of the capillary pressure curve beyond 10 psi. Thus, the area used for the index is confined to curves extending to exactly 10 psi. This procedure removed considerable variability between tests of adjacent cores.

Threshold pressures for curves 3 and 5 are computed at $S_w = S_{wi}$ and $S_w = 1.0$, respectively. The areas under each curve represent the energy (kJ/ml) required for displacement of the respective fluid.

Two input data files are used: (1) *WETIN1.DAT* is used to enter the core and fluid properties, and the *RPM* versus volume of oil displaced for curve 3 (water displacing oil from S_{wi} to S_{wor}); and (2) *WETIN2.DAT* is used to enter only the *RPM* versus volume of water displaced for curve 5 (oil displacing water from S_{wor} to S_{wi}).

Input Variables

DIAM	core diameter
N3	number of data points for curve 3
N5	number of data points for curve 5
PHI	porosity
R2	length from the center of rotation to the far end of the core for water displacement oil, curve 3 (placed in the data statement of the program)
R3	length from the center of rotation to the far end of the core for oil displacing water, curve 5 (placed in the data statement)
RHO1	density of the brine
RHO2	density of the oil
RPM3	revolutions per minute (capillary pressure curve 3)
RPM5	revolutions per minute (capillary pressure curve 5)
SWI	irreducible saturation
VB	volume of brine displaced
VO	volume of oil displaced
XLEN	core length

Program Listing 6.24 WET.FOR

```
C      USBM WETTABILITY INDEX
C
C      SUBROUTINE HYPER and FUNCTION EXTRA are required.
C
C      Data file WETIN1.DAT is used to input the number of points for curve 3
C         (negative curve) and curve 5 (positive curve); and, the properties
C         in the READ statement for UNIT 10.  In addtion, the RPM and volume
C         brine displaced for curve 3 are entered.
```

```
C     Data file WETIN2.DAT is used to input only the RPM and volume of oil (ml)
C        displaced for curve 5.
C     CWET.  USBM wettability calculation from capillary pressure curves.
C     N3= Number of data points for curve 3 (water displacing oil from
C        Swi to Sor.
C     N5= Number of data points for curve 5 (oil displacing water from
C        Sor to Swi
C     SWI3 (irreducible water saturation, curve 3).  SW3(I) and PC3(i)=
C        water saturation and capillary pressure data, curve 3.  SW5(I)
C        and PC5(I) = water saturation and capillary pressure data, curve
C        5.  SWOR= residual oil saturation.
C
      REAL SW3(50),PC3(50), SW5(50),PC5(50), DIF(50), PCH3(50),PCH5(50)
      REAL RPM3(50), VO(50), RPM5(50), VB(50), SAV3(50), SAV5(50)
      REAL PCH3kP(50), PCH5kP(50)
C
      OPEN (UNIT= 10, FILE= 'WETIN1.DAT', STATUS= 'OLD')
      READ (10,*) N3, N5, XLEN, DIAM, PHI, RHO1, RHO2, SWI
      READ (10,*) (RPM3(I), VO(I), I= 1, N3)
      CLOSE (10)
C
      OPEN (UNIT= 11, FILE= 'WETIN2.DAT', STATUS= 'OLD')
      READ (11,*) (RPM5(I), VB(I), I= 1,N5)
      CLOSE (11)
      DATA R2,R3/16.8,8.68/
C
      VBULK= 0.7854*XLEN*(DIAM**2)
      VPORE= PHI*VBULK
C
      FAC= RHO1-RHO2
C
      SAV3(1)= SWI
      DO 10 I=2,N3
         SAV3(I)= SWI + VO(I)/VPORE
         XX= 1.578E-07*RPM3(I)**2
         PC3(I)= -XX*FAC*(R2-XLEN/2.0)*XLEN
10    CONTINUE
C
      SWOR= SAV3(N3)
C
C     I.  SW3 vs PCH3.  Inlet sat. (Hassler/Brunner) versus corrected
C           capillary pressure (HYPER)
C
      CALL HYPER (N3, SAV3, PC3, P, Q, R)
      SUM= 0.0
      DO 12, I= 1,N3
         PCH3(I)= (P + Q*SAV3(I))/(1.0 + R*SAV3(I))
         SW3(I)= SAV3(I) + PCH3(I)*(((1.0 + R*SAV3(I))**2)/(Q - P*R))
         DIF(I)= (PC3(I) - PCH3(I))**2
         SUM= SUM + DIF(I)
12    CONTINUE
C
      STDH3= SQRT(SUM/N3)
      PH3= P
      QH3= Q
      RH3= R
      TPH3= PCH3(1)
C
C     II  PCH3 extrapolation to 10 psi (68.95 kPa) or cut-off at 10 psi; and
C           calculation of AREA3
C
C     II(a)  PCH3 extrapolation to 10 psi.
C
      IF(PCH3(N3) .LT. 10.0) GO TO 14
      GO TO 18
```

```
14     SW3(N3+1)= EXTRA(SW3,PCH3,N3)
       PCH3(N3+1)= 10.0
       SUM= 0.0
       DO 16 I=3,N3+1
          TERM= SW3(I-1) - SW3(I)
          A= PCH3(I-1)*TERM + (PCH3(I) - PCH3(I-1))*TERM/2.0
          SUM= SUM + A
16     CONTINUE
       AB= PCH3(2)*(SW3(1) - SW3(2))/2.0
C      TERM= SW3(2) - SW3(1)
C      AB= PCH3(1)*TERM + (PCH3(1)-PCH3(2))*TERM/2.0
       AREA3= ABS(SUM + AB)
       DISEN3= AREA3*6.895
       GO TO 26
C
C      II(b).  Cut-off of PCH3 at 10 psi
C
18     DO 20 I= 1,N3
          IF(PCH3(I) .GT. 10.0) GO TO 22
20     CONTINUE
22     K= I
       SW3(K)= EXTRA (SW3, PCH3, N3)
       PCH3(K)= 10.0
       SUM= 0.0
       DO 24 I= 3,K
          TERM= SW3(I-1) - SW3(I)
          A= SW3(I-1)*TERM + (PCH3(I) - PCH3(I-1))*TERM/2.0
          SUM= SUM + A
24     CONTINUE
       AB= PCH3(2)*(SW3(1) - SW3(2))/2.0
C      TERM= SW3(1) - SW3(2)
C      AB= PCH3(1)*TERM + (PCH3(1) - PCH3(2))*TERM/2.0
       AREA3= ABS(SUM + AB)
       DISEN3= AREA3*6.895
C
C      III.  SW5 vs PCH5.  Inlet sat. (Hassler/Brunner) versus corrected
C            capillary pressure (HYPER)
C
26     DO 28 I= 1,N5
          SAV5(I)= SWOR - VB(I)/VPORE
28     CONTINUE
          PC5(1)= 0.0
       DO 29 I= 2,N5
          XX= 1.578E-07*RPM5(I)**2
          PC5(I)= XX*FAC*(R3-XLEN/2.0)*XLEN
29     CONTINUE
C
       CALL HYPER (N5, SAV5, PC5, P, Q, R)
       SUM= 0.0
       DO 30, I= 1, N5
          PCH5(I)= (P + Q*SAV5(I))/(1.0 + R*SAV5(I))
          SW5(I)= SAV5(I) + PCH5(I)*(((1.0 + R*SAV5(I))**2)/(Q - P*R))
          DIF(I)= (PC5(I) - PCH5(I))**2
          SUM= SUM + DIF(I)
30     CONTINUE
C
       STDH5= SQRT(SUM/N5)
       PH5= P
       QH5= Q
       RH5= R
       TPH5= PCH5(1)
C
C      IV.  PCH5 extrapolation to 10 psi, or cut-off at 10 psi; and calculation
C           of AREA5
C
```

```
C     IV(a).  PCH5 extrapolation to 10 psi.
C
      IF(PCH5(N5) .LT. 10.0) GO TO 34
      GO TO 38
34    SW5(N5+1)= EXTRA(SW5,PCH5,N5)
      PCH5(N5+1)= 10.0
      SUM= 0.0
      DO 36 I= 3,N5+1
      TERM= SW5(I-1) - SW5(I)
      A= PCH5(I-1)*TERM + (PCH5(I) - PCH5(I-1))*TERM/2.0
      SUM= SUM + A
36    CONTINUE
      AB= PCH5(2)*(SW5(1) - SW5(2))/2.0
C     TERM= SW5(1) - SW5(2)
C     AB= PCH5(1)*TERM + (PCH5(1) - PCH5(2))*TERM/2.0
      AREA5= ABS(SUM + AB)
      DISEN5= AREA5*6.895
C
C     IV(b).  Cut-off of PCH5 at 10 psi
C
38    DO 40 I= 1,N5
      IF(PCH5(I) .GT. 10.0) GO TO 42
40    CONTINUE
42    K= I
      DO 44 I= 1,N5
      SW5(K)= EXTRA(SW5,PCH5,N5)
      PCH5(K)= 10.0
      SUM= 0.0
44    CONTINUE
      DO 46 I= 3,K
      TERM= SW5(I-1) - SW5(I)
      A= PCH5(I-1)*TERM + (PCH5(I)-PCH5(I-1))*TERM/2.0
      SUM= SUM + A
46    CONTINUE
      AB= PCH5(2)*(SW5(1) - SW5(2))/2.0
C     TERM= SW5(1) - SW5(2)
C     AB= PCH5(1)*TERM + (PCH5(1) - PCH5(2))*TERM/2.0
      AREA5= ABS(SUM + AB)
      DISEN5= AREA5*6.895
      WET= ALOG10(AREA5/AREA3)
C
      DO 48 I= 1,N3
         PCH3kP(I)= 6.895*PCH3(I)
48    CONTINUE
      DO 50 I= 1,N5
         PCH5kP(I)= 6.895*PCH5(I)
50    CONTINUE
C
      OPEN (UNIT= 12, FILE= 'WETOUT.DAT', STATUS= 'OLD')
      WRITE (12, 900)
900   FORMAT (5X, 'USBM WETTABILITY INDEX', //)
C
      WRITE (12, 901) WET, AREA3, AREA5
901   FORMAT ('WETTABILITY=',F7.3,3X,'AREA3=',F7.4,3X,'AREA5=',F7.4,/)
C
      WRITE (12, 902) SWI, SWOR
902   FORMAT ('SWI=',F7.3, 3X, 'SWOR=', F7.3, //)
C
      WRITE (12, 904)
904   FORMAT (//,6X, 'SW3', 3X, 'PC3(psi)', 3X, 'PCH3(psi)',2X,
     & 'PCH3(kPa)',/)
      WRITE (12,906) (SW3(I), PC3(I), PCH3(I), PCH3kP(I), I= 1,N3)
906   FORMAT (4F10.3)
C
      WRITE (12, 908)
908   FORMAT (//,6X, 'SW5', 3X, 'PC5(psi)', 3X, 'PCH5(psi)',2X,
```

```
     &'PCH5(kPa',/)
      WRITE (12,910) (SW5(I), PC5(I), PCH5(I), I= 1,N5)
910   FORMAT (4F10.3)
C
      WRITE (12,912) STDH3, STDH5
912   FORMAT (//, 'STANDARD DEVIATION, CURVE 3 =', F6.3, /,'STANDARD
     &DEVIATION, CURVE 5 =', F6.3,//)
C
      WRITE (12, 914)
914   FORMAT ('HYPER COEFFICIENTS: CURVES 3 AND 5', /)
      WRITE (12,916) PH3, QH3, RH3
916   FORMAT ('PH3=', F9.4, 3X, 'QH3=', F9.4, 3X, 'RH3=', 9F7.4,/)
      WRITE (12, 918) PH5, QH5, RH5
918   FORMAT ('PH5=', F9.4, 3X, 'QH5=', F9.4, 3X, 'RH5=', F9.4,//)
C
      WRITE (12, 920)
920   FORMAT (5X, 'THRESHOLD PRESSURES FOR CURVES 3 AND 5',/)
      WRITE (12,922) TPH3, TPH5
922   FORMAT ('TPH3 (kPa)=', F5.1, 3X, 'TPH5 (kPa)=', F5.1,//)
C
      WRITE (12, 924) DISEN3
924   FORMAT ('DISPLACEMENT ENERGY, kJ/ml (Curve 3)=',F5.2,/)
      WRITE (12, 926) DISEN5
926   FORMAT ('DISPLACEMENT ENERGY, kJ/ml (Curve 5)=',F5.2)
      CLOSE (12)
C
      STOP
      END
```

Program Listing 6.25 WETIN1.DAT

```
10, 10, 2.56, 2.54, 0.178, 1.005, 0.845, 0.221

   0      0.00
 470      0.00
 970      0.08
1320      0.27
1740      0.50
2330      0.67
2760      0.73
3210      0.81
3600      0.87
3910      0.90
```

Program Listing 6.26 WETIN2.DAT

```
0         0.00
 970      0.38
1400      0.49
2160      0.69
2500      0.74
2910      0.77
3740      0.78
3680      0.80
4000      0.81
4470      0.82
```

Program Listing 6.27 WETOUT.DAT

```
 USBM WETTABILITY INDEX


WETTABILITY= -0.639   AREA3= 0.9631   AREA5= 0.2211

SWI=  0.221   SWOR=  0.611



     SW3   PC3(psi)   PCH3(psi)  PCH3(kPa)

    0.259     0.000    -0.295     -2.037
    0.259    -0.222    -0.295     -2.037
    0.321    -0.944    -0.583     -4.022
    0.449    -1.748    -1.484    -10.229
    0.569    -3.037    -3.308    -22.811
    0.634    -5.446    -5.849    -40.328
    0.652    -7.641    -7.271    -50.132
    0.672   -10.336   -10.015    -69.056
    0.685   -13.001   -13.214    -91.112
    0.690   -15.336   -15.460   -106.597


     SW5   PC5(psi)   PCH5(psi)  PCH5(kPa

    0.592     0.000     0.028      0.348
    0.450     0.462     0.303      0.937
    0.752     0.251     2.232      2.216
    0.244     2.989     3.316      0.241
    4.050     4.535     0.240      6.690
    5.133     0.238     6.477      6.886
    0.238     7.653     8.243      0.237
    9.557    10.000

 STANDARD DEVIATION, CURVE 3 = 0.288
 STANDARD DEVIATION, CURVE 5 = 0.612


 HYPER COEFFICIENTS: CURVES 3 AND 5

 PH3=   0.8772   QH3=  -4.8880   RH3=-1.4140
 PH5=  -1.4031   QH5=   2.2253   RH5=  -4.2312


      THRESHOLD PRESSURES FOR CURVES 3 AND 5

 TPH3 (kPa)= -0.3   TPH5 (kPa)=  0.0


 DISPLACEMENT ENERGY, kJ/ml (Curve 3)= 6.64

 DISPLACEMENT ENERGY, kJ/ml (Curve 5)= 1.52
```

Program Listing 6.28 FUNCTION EXTRA.FOR

```
      FUNCTION EXTRA(X,Y,N)
      REAL X(50), Y(50)
      R= 0.0
      S= 0.0
      T= 0.0
      U= 0.0
      DO 2 I= N-3,N
      R= ALOG(X(I)) + R
      S= Y(I) + S
      T= Y(I)*Y(I) + T
      U= Y(I)*ALOG(X(I)) + U
2     CONTINUE
      B= (4.0*U - R*S)/(4.0*T - S*S)
      C= (R - B*S)/4.0
      A= EXP(C)
      EXTRA= A*EXP(10.0*B)
      RETURN
      END
```

Program Listing 6.29 SUBROUTINE HYPER.FOR

```
C     Subroutine HYPER:  Least Squares curve fit of data to a hyperbolic
C        function.  X= (P + Q*X)/(1 + R*X).  The coefficients P, Q and R
C        are returned.
C     The derivative dY/dX= (Q - P*R)/(1 + R*X)^2.   The integral is:
C        (Q*X)/R + ((P*R - Q)/R^2)Ln(1+R*X) from b to a.
C
      SUBROUTINE HYPER (N,X,Y,P,Q,R)
C
      REAL X(50), Y(50), NUM1, NUM2, NUM3
C
      A= 0.0
      B= 0.0
      C= 0.0
      D= 0.0
      E= 0.0
      F= 0.0
      G= 0.0
      H= 0.0
C
      DO 10, I= 1,N
         A= A + X(I)
         B= B + Y(I)
         C= C + X(I)*Y(I)
         D= D + X(I)**2
         E= E + Y(I)**2
         F= F + X(I)*Y(I)**2
         G= G + (X(I)**2)*Y(I)
         H= H + (X(I)**2)*(Y(I)**2)
10    CONTINUE
C
C
      NUM1=  D*(C*F-B*H) + C*(A*H-C*G) + G*(B*G-A*F)
      NUM2=  N*(G*F-C*H) + A*(B*H-C*F) + C*(C**2-B*G)
      NUM3=  N*(D*F-C*G) + A*(B*G-A*F) + C*(A*C-B*D)
      DNOM=  N*(G**2-D*H) + A*(A*H-C*G) + C*(C*D-A*G)
C
```

```
      P= NUM1/DNOM
      Q= NUM2/DNOM
      R= NUM3/DNOM
C
      RETURN
      END
```

APPENDIX A

Front Tracking Program
(Miscible and Two-Phase Displacement)

A.1 Introduction
FRONT.FOR (PCOMP, BRINK, TRIDAG, DATAIN.DAT, BLOCK.DAT, and DATAOUT.DAT)

This program is designed to track the advancing flood-front for three types of reservoir fluid flow: (1) steady-state miscible displacement where the amount of water injected is exactly equal to the amount of fluid produced from the reservoir; this can be conducted with a closed (no-flow across boundaries) or open reservoir with flow possible across the boundaries; (2) unsteady-state miscible displacement where the total rate of injection is not equal to the rates of production; if the boundaries are closed, the capacity of the reservoir is limited to the compressibility of the water and rock; (3) two-phase flow (the displacement of oil by water); injection and production may be unequal. Corey's equation for imbibition is used to obtain the relative permeability, and wettability is implicit in the end-points of the capillary pressure curve (SWI and SWOR).

Program Operation

Because of the complexity of the program, it is best to become familiar with the program operation using simple injection, production, and reservoir conditions before introducing a considerable amount of complex data.

Selection of the type of displacement to be simulated occurs from the input data in *DATAIN.DAT* as follows:

I. Miscible displacement (steady-state): (a) reservoir rock and fluid compressibility, CM, is zero, (b) SWI is zero, and (c) SWOR is one (1.0).

II. Miscible displacement (unsteady-state): (a) CM is greater than zero, (b) SWI is zero, and (c) SWOR is one (1.0).

III. Two-phase flow (displacement of oil by water): (a) CM is greater than zero, (b) SWI is greater than zero, and (c) SWOR is less than 1.0.

The length of time for which simulation is desired is set by STIME (days).

When the program is started, a DO-statement will address the screen and ask for entry of the I, J locations and rates (barrels/day) of the injection production wells. Enter positive flow rates for the injection wells and negative flow rates for the production wells.

NPRINT sets the incremental number of days, after injection starts, for writing to the output data file to occur. For example, if STIME = 500 days and NPRINT = 100, the locations of the sixty points that start around each injection well will be written into data file *DATAOUT.DAT* five times before the program stops running.

W is an iteration (or relaxation) parameter used in subroutine *TRIDAG*. When *W* = 1.0, the matrix pressure solution is done by Gauss-Siedel procedure; when greater than 1.0, successive over relaxation is used (which is faster). The parameter is best found by running the program (with a given set of reservoir conditions) at different values of *W* and plotting the time for solution versus *W* to determine the minimum iteration time required (PITER). A value of 1.3 was selected for the simple demonstration program.

Each injection well will begin with 60 points, equally spaced, around the well. These will advance as the injection proceeds in accord with the variations of permeability, porosity and thickness (HT) of the reservoir. The I, J locations of all of the points will appear (as designated for each well) in the output data file. This represents a great deal of printing and it would be best to convert this to a graphical output (not used here because there are so many different graphical routines). When testing the program, or working with it for short periods, one can reduce the number of points to be written to the data file by adding a number to the second DO 30 statement under "Calculation of tracking point velocities..." in subroutine *BRINK*; for example, the number ten will cause the printing of every tenth point: DO 30, $N = 1$, NPOINT, 10.

Corey's imbibition (water displacing oil) equations for relative permeability were selected because it is the most convenient and because the end-points of the capillary pressure curve implicitly convey information of wettability:

$$kro = S_{wi}^4$$

$$krw = S_{wor}^4 * \frac{S_{wor} - S_{wi}}{1.0 - S_{wi}}$$

The simulator was prepared as a program that will continually track the position of the advancing front of fluids injected at constant or variable rates. The analysis of the pressure distribution within the reservoir because of unequal injection from multiple wells may be examined by having the program write the pressure distribution to the output data file. The write-statements for this must be added, if desired. The reservoir may be finite where compressibilities of the brine and rock furnish the storage volume, or the reservoir may be infinite in size where only a very small portion is used for displacement. A reservoir containing a short discontinuity, such as a fault or facies change, also may be considered by entering zero for the permeability at nodes coinciding with the location of the fault or no-flow boundary. Permeability, porosity, and reservoir thickness are entered by uncovering the comment notation in *PCOMP* for the data file *BLOCK.DAT* and entering the respective data at each node (I, J location) of the matrix. The model is based on the assumption of ideal, piston-like, displacement of reservoir fluids.

Versions of this program were published previously: Donaldson, E.C. and Rezaei, A.A., 1986; Rezaei, A.A., and Donaldson, E. C., 1987; and Rezigh, A.A., 1988. A program quite similar to *PCOMP* was used for simulation of compaction caused by fluid withdrawal (Chilingarian, G.V., Donaldson, E.C., and Yen, T.F, 1995, page 425).

```
          READ *, I,J, QWELL
            IF(QWELL .LT. 0) IKP= IKP+1
            IF(QWELL .GT. 0) IKI= IKI+1
            IF(QWELL .EQ. 0) IKS= IKS+1
          Q(I,J)= QWELL
          SUMQ= SUMQ + Q(I,J)
30      CONTINUE
C
        L= 1
        PITER(L)= 0.0
        MM1= M-1
        MM2= M-2
        K= 0
C
        DO 5 I= 1,M
        DO 5 J= 1,M
          PHI(I,J)= POR
          PERM(I,J)= PER
          H(I,J)= HT
          KH(I,J)= PERM(I,J)*H(I,J)
          CAP(I,J)= PERM(I,J)*H(I,J)/MU*1.127
          P(I,J)= PBEGIN
          PNEW(I,J)= PBEGIN
5       CONTINUE
C
        J=1
        DO 200 I= 2,MM1
          IF(PERM(I,J) .EQ. 0.0) GO TO 80
          SIGMA(I,J+1)= S/KH(I,J)
          GO TO 200
80        SIGMA(I,J+1)= 0.0
200     CONTINUE
C
          J= M
        DO 202 I= 2,MM1
          IF(PERM(I,J) .EQ. 0.0) GO TO 82
          SIGMA(I,J-1)= S/KH(I,J)
          GO TO 202
82        SIGMA(I,J-1)= 0.0
202     CONTINUE
C
        I= 1
        DO 204 J=2,MM1
          IF(PERM(I,J) .EQ. 0.0) GO TO 84
          SIGMA(I+1,J)= S/KH(I,J)
          GO TO 204
84        SIGMA(I+1,J)= 0.0
204     CONTINUE
C
          I= M
        DO 206 J= 2,MM1
          IF(PERM(I,J) .EQ. 0.0) GO TO 86
          SIGMA(I-1,J)= S/KH(I,J)
          GO TO 206
86        SIGMA(I-1,J)= 0.0
206     CONTINUE
C
6       K= K + 1
        KOK= 0
        KNG= 0
C
        DO 11 J= 2,M-1
        DO 9  I= 2,M-1
          IP= (CAP(I,J) + CAP(I+1,J))/2.
          IM= (CAP(I,J) + CAP(I-1,J))/2.
```

```
          JP= (CAP(I,J) + CAP(I,J+1))/2.
          JM= (CAP(I,J) + CAP(I,J-1))/2.
          CMM= (CM*PHI(I,J)*H(I,J)*DX*DX)/DT
          AL= -IP-IM-JP-JM-CMM
          A(I)= IM
          B(I)= AL
          C(I)= IP
          D11= (1.-W)*(IM*P(I-1,J) + IP*P(I+1,J) + AL*P(I,J))
          D22= -W*(JM*PNEW(I,J-1) + JP*P(I,J-1) + CMM*P(I,J) + Q(I,J))
          D(I)= D11 + D22
9         CONTINUE
          D(2)= D(2) - A(2)*P(I,J)
          D(MM1)= D(MM1) - C(MM1)*P(MM1,J)
C
          CALL TRIDAG (M,NX,A,B,C,D,CON)
          DO 11 I= 2,MM1
          PNEW(I,J)= CON(I)
          E(I,J)= ABS(PNEW(I,J) - P(I,J))
11      CONTINUE
C
        DO 16 I=2,MM1
        DO 16 J=2,MM1
          IF (E(I,J) .LT. EMAX) GO TO 108
          KNG= 2
          GO TO 14
108       KOK= 1
14        P(I,J)= PNEW(I,J)
16      CONTINUE
C
        IF(KNG .LT. KOK) GO TO 105
        IF(K .LT. KMAX) GO TO 107
          PRINT *,990
990       FORMAT('---ERROR---No of convergence on pressure distibution')
          GO TO 105
C
107     PITER(L)= PITER(L) + 1
        GO TO 6
C
105     RETURN
        END
```

Program Listing A.3 BRINK

```
        SUBROUTINE BRINK
C
        COMMON/BB/M,DX,PERM(50,50),MU,PHI(50,50),SWI,SWOR,PITER(50),L,
       &P(50,50),Q(50,50),STIME,NPRINT,CM
C
        REAL IP,IM,JP,JM,KH,MU,KRW,VX(50,50),VY(50,50)
        REAL X(60,20),Y(60,20),XI(60,20),YI(60,20)
C
        IF(KOUNT .GT. 0) GO TO 422
        T= 0.0
        KN= 0
        NPOINT= 60
        DEG= 6.0
        NSTOP= 0
        NUM= 0
C
        XT= (SWOR - SWI)/(1.0 - SWI)
        KRW= (SWOR**4)*SQRT(XT)
```

```
C
C     Establish the positions of the points in a circular pattern
C       around the injection wells
C
      RX= 1.0
      RY= 1.0
      MM1= M-1
C
      DO 2 I= 2,MM1
      DO 2,J= 2,MM1
        IF(Q(I,J)) 2,2,3
3       KN= KN + 1
        IWELLS= KN
        ANGLE= 0.0
c
        DO 1 N= 1,NPOINT
          RADN= ANGLE/57.29478
          XI(N,KN)= I+(-RX*COS(RADN))
          YI(N,KN)= J+(-RY*SIN(RADN))
          X(N,KN)= (XI(N,KN)-1)*DX
          Y(N,KN)= (YI(N,KN)-1)*DX
         ANGLE= ANGLE + DEG
C
1       CONTINUE
2     CONTINUE
C
C     Branch into: I, Miscible Displacement (Steady-State)(Lable 420);
c     II, Miscible Displacement (Unsteady-State)(Lable 430); and III,
c     Two-phase Displacement (Lable 440
C
      IF(CM .EQ. 0.0 .AND. SWI .EQ. 0.0) GO TO 420
      IF(CM .GT. 0.0 .AND. SWI .EQ. 0.0) GO TO 430
      IF(SWI .GT. 0.0) GO TO 440
C
420   NUM= 1
C     Find the greated velocity of fluid advance in the matrix and use it
C       to calculate the time increment (DT)
C
422   CONTINUE
      KOUNT= 0
      OPEN (UNIT= 16, FILE= 'DATAOUT.DAT', STATUS= 'OLD')
C
      DO 4 I= 2,MM1
      DO 4 J= 2,MM1
        Z=(0.2007*PERM(I,J)*KRW)/(MU*PHI(I,J)*DX)
        VX(I,J)= (P(I-1,J) - P(I+1,J))*Z
        VY(I,J)= (P(I,J-1) - P(I,J+1))*Z
4     CONTINUE
C
C     Calculation of delta STIME (DT) from the largest X- or Y-direction
C       velocity in the matrix
C
      K= 0
      VXMAX= 0.0
      DO 6 I=2,MM1
      DO 6 J=2,MM1
        IF(ABS(VXMAX) - ABS(VX(I,J))) 8,6,6
8       VXMAX= VX(I,J)
6     CONTINUE
C
      VYMAX= 0.0
      DO 12 I= 2,MM1
      DO 12 J= 2,MM1
        IF(ABS(VYMAX) - ABS(VY(I,J))) 10,12,12
10      VYMAX= VY(I,J)
```

```
12    CONTINUE
C
      IF(VXMAX - VYMAX) 25, 26, 27
25      DT= ABS(DX/VYMAX)
        GO TO 16
26      DT= ABS(DX/VYMAX)
        GO TO 16
27      DT= ABS(DX/VXMAX)
        GO TO 16
C
16    CONTINUE
      T= T + DT
      K= K + 1
C
      IF(T .GE. STIME) GO TO 99
      IF (T .GT. NPRINT) GO TO 490
      GO TO 16
C
490   NPRINT= NPRINT + NPRINT
      IF(NUM .EQ. 1) GO TO 501
      IF(NUM .EQ. 2) GO TO 502
      IF(NUM .EQ. 3) GO TO 503
      GO TO 99
C
501   IF(LST .GT. 0) GO TO 520
      WRITE (16,901)
901   FORMAT ('MISCIBLE DISPLACEMENT (Steady-State)',//)
      WRITE (16, 902) VXMAX, VYMAX
902   FORMAT('Approximate velocities at the flood-front in the X- and
     &Y- directions (feet/day)=', F4.1, ',', F4.1,/)
      GO TO 520
C
502   IF(LST .GT. 0) GO TO 520
      WRITE (16,903)
903   FORMAT ('MISCIPLE DISPLACEMENT (Unsteady-State)',//)
      WRITE (16, 904) VXMAX, VYMAX
904   FORMAT('Approximate velocities at the flood-front in the X- and
     &Y- directions (feet/day)=', F4.1, ',', F4.1,/)
      GO TO 520
C
503   IF(LST .GT. 0) GO TO 520
      WRITE (16,905)
905   FORMAT ('TWO PHASE DISPLACEMENT',//)
      WRITE (16, 906) VXMAX, VYMAX
906   FORMAT('Approximate velocities at the flood-front in the X- and
     &Y- directions (feet/day)=', F4.1, ',', F4.1,/)
      GO TO 520
C
520   WRITE (16, 908) T
908   FORMAT ('X- and Y-tracking point positions for each injection well
     &time (Days)=',F8.1,/)
C
C     Calculation of tracking point velocities for each injection well
      DO 30 KN= 1,IWELLS
        KWELL= KN
      DO 30 N= 1,NPOINT
        XP= X(N,KN) - INT(X(N,KN))
        YP= Y(N,KN) - INT(Y(N,KN))
c
        AA= X(N,KN)/DX
        BB= Y(N,KN)/DX
        I= IFIX(AA) + 1
        J= IFIX(BB) + 1
c
        VPX= (VX(I,J+1)*(1.-XP) + VX(I+1,J+1)*XP + VX(I,J)*(1.-XP)
```

```
     &+VX(I+1,J)*XP)/4.
        VPY= (VY(I,J+1)*YP + VY(I+1,J+1)*YP + VY(I,J)*(1.-YP)
     &+VY(I+1,J)*(1.-YP))/4.
c
        X(N,KN)= X(N,KN) + VPX*DT
        Y(N,KN)= Y(N,KN) + VPY*DT
c
        XI(N,KN)= (X(N,KN)/DX) + 1
        YI(N,KN)= (Y(N,KN)/DX) + 1
c
      IF (N .GT. 1) GO TO 66
      WRITE (16,910) KWELL
910   FORMAT (('Well No.= ',I3),/)
c
66    WRITE (16,918) XI(N,KN),YI(N,KN)
918   FORMAT (F4.1, 2X, F4.1)
C
30    CONTINUE
C
      LST= 1
      IF(NSTOP .EQ. 1) GO TO 99
      IF(T .GE. STIME) GO TO 99
      IF(NUM .GT. 1) GO TO 550
      GO TO 16
C
550   KOUNT= KOUNT + 1
      CALL PCOMP
C
430   NUM= 2
      GO TO 422
C
440   NUM= 3
      XT= (SWOR - SWI)/(1.0 - SWI)
      KRW= (SWOR**4)*SQRT(XT)
      GO TO 422
C
99    CLOSE (16)
      RETURN
      END
```

Program Listing A.4 TRIDAG

```
C     SUBROUTINE TRIDAG
C
C     M= number of nodes in the grid
C
      SUBROUTINE TRIDAG (M,NX,A,B,C,D,CON)
      REAL A(NX), B(NX), C(NX), D(NX), CON(NX)
      REAL BETA(910), GAMMA(910)
      L= M-1
      IFAC= 2
      BETA(IFAC)= B(IFAC)
      GAMMA(IFAC)= D(IFAC)/BETA(IFAC)
      IFP1= IFAC + 1
C
      DO 20 I= IFP1,L
        BETA(I)= B(I) - (A(I)*C(I-1))/BETA(I-1)
        GAMMA(I)= (D(I) - A(I)*GAMMA(I-1))/BETA(I)
20    CONTINUE
C
      CON(L)= GAMMA(L)
```

```
      LAST= L - IFAC
      DO 30, K= 1,LAST
        KK= K
        I= L - KK
        CON(I)= GAMMA(I) - C(I)*CON(I+1)/BETA(I)
30    CONTINUE
C
      RETURN
      END
```

Program Listing A.5 DATAIN.DAT

```
10.0, 1000.0, 0.0, 1.0, 3, 500.0,200.0
40, 0.1, 0.25, 20.0, 0.0, 1.0
10.0, 25, 1.3, 5.0
```

Program Listing A.6 BLOCK.DAT (Example using M = 3)

```
0.120, 0.124, 0.133
0.220, 0.241, 0.232
20, 22, 24
```

Program Listing A.7 DATAOUT.DAT

```
MISCIBLE DISPLACEMENT (Steady-State)


Approximate velocities at the flood-front in the X- and Y- directions (feet/day)=-0.6,-0.6

X- and Y-tracking point positions for each injection well time (Days)=   210.2

Well No.=   1

 2.6   3.7
 3.3   2.8
 4.5   2.8
 5.4   3.9
 4.9   4.9
 3.5   4.9
X- and Y-tracking point positions for each injection well time (Days)=   402.9

Well No.=   1

 2.4   3.6
 3.3   2.8
 4.5   2.8
 5.5   3.8
 5.3   4.8
 3.5   4.8
```

Bibliography

Adamson, A.W., 1960. *Physical Chemistry of Surfaces.* John Wiley, Interscience, New York.

Adamson, A.W., and Gast, A.P., 1997. *Physical Chemistry of Surfaces.* (6th ed.) John Wiley & Sons, New York, 784 pp.

Alkafeef, S.F., and Smith, A.L., 2005. Asphaltene adsorption isotherm in the pores of reservoir rocks. *SPE 93188,* Sym on Oilfield Chemistry, Houston, TX (Feb).

Amott, E., 1959. Observations relating to the wettability of porous rock. *Trans. AIME,* v. 216: 156–162.

Anderson, W.G., 1986-a. Wettability Literature Survey, Part 1: Rock/oil/brine interactions and the effects of core handling on wettability. *JPT* (Oct): 1125–1144.

Anderson, W.G., 1986-b. Wettability Literature Survey, Part 2: Wettability measurement. *JPT* (Nov): 1246–1262.

Anderson, W.G, 1986-c. Wettability Literature Survey, Part 3: The effects of wettability on the electrical properties of porous media. *JPT* (Dec): 1371–1378.

Anderson, W.G., 1987-a. Wettability Literature Survey, Part 4: Effects of wettability on capillary pressure. *JPT* (Oct): 1283–1300.

Anderson, W.G., 1987-b. Wettability Literature Survey, Part 5: The effects of wettability on relative permeability. *JPT* (Nov): 1453–1468.

Anderson, W.G., 1987-c. Wettability Literature Survey, Part 6: The effects of wettability on waterflooding. *JPT* (Dec): 1605–1622.

Archie, G.E., 1942. The electrical resistivity of logs as an aid in determining some reservoir characteristics. *T. AIME,* v. 146.

Archer, J.S., and Wall, C.G., 1986. *Petroleum Engineering.* Graham and Trotman, London, UK, 362 pp.

Arditty, P.C., and Ramey, Jr., H.J., 1978. Response of a closed well-reservoir system to stress induced by earth tides. *SPE 7484,* Ann. Tech. Conf., Houston, TX (Oct).

Aronofsky, J.S., Masse, L., and Natanson, S.G., 1958. A model for the mechanism of oil recovery from the porous matrix due to water invasion in fractured reservoir. *Trans. AIME* 213: 17–19.

Aronson, M.P., Petko, M.F., Princen, H.M., 1978. On the stability of squeous films between oil and silica. *J. Colloid. and Interface Sci.,* v. 65 (No. 2, Jun 15): 296–306.

Austad, T., and Standes, D.C., 2003. Spontaneous imbibition of water into oil-wet carbonates. *JPSE,* v. 39 (Sep): 363–376.

Baker, E.G., 1959. Origin and migration of oil. *Science,* v. 129, No. 3353 (3 Apr): 871–874.

Barrow, G.W., 1973. *Physical Chemistry.* (3rd ed.), McGraw-Hill, New York: 807 pp.

Bartell, F.E., and Niederhauser, D.O, 1949. Fundamental research on occurrence and recovery of petroleum. *American Petrol. Inst Bull.*: 57–80.

Basu, S., and Sharma, M.M., 1996. Defining the wettability state of mixed-wet reservoirs: measurement of critical capillary pressure for crude oils. *SPE 36679,* An. Tech. Conf, Denver, CO (Oct).

Batychy, J.P.; McCaffery, F.C., Hodgins, P.K., and Fisher, D.B., 1980. Interpreting capillary pressure and rock wetting characteristics from unsteady-state displacement measurements. *SPE 9403,* Ann. Tech. Conf., Dallas, TX (Sep).

Bishop, R.S., Gehman, H.M., and Young, A., 1983. Concepts for estimating hydrocarbon accumulation and dispersion. *AAPG Bulletin,* v. 67: 337–348.

Blackwell, R.J., Rayne, J.R., and Terry, W.M., 1959. Factors influencing the efficiency of miscible displacement. *Trans. AIME,* v. 216: 1–8.

Bonham, L.C., 1980. Migration of hydrocarbons in compacting basins: (Abs. in *AAPG Bulletin,* v. 62, 1978: 498); 1980, *AAPG Studies of Geology* 10: 69–88.

Bowman, C.W., 1981 (Chair). *The Future of Heavy Crude and Tar Sands.* UNITAR Conf., Edmonton, Alberta, Canada (Jun 79): 915 pp.

Bredehoeft, J.C., 1967. Response of well-aquifer systems to earth tides. *Anal. Geophysical Research,* v. 72 (12): 3075–3086.

Brigham, W.E.; Reed, P.W. and Dew, J.N., 1961. Experiments on mixing during miscible displacement in porous media. *SPEJ*, v. 1 (Mar): page 1.

Brigham, W.E., 1974. Mixing equations in short laboratory cores. *SPEJ* (Feb): 91–99.

Brooks, R.H., and Corey, A.T., 1964. Hydraulic properties of porous media. *Hydrology Papers*, Colorado State University, CO.

Brown, R.J.S., and Fat, I., 1956. Measurements of fractional wettability of oil field rocks by the nuclear magnetic relaxation method. *Trans. AIME* 207: 262–264.

Buckley, J.S., and Liu, Y, 1998. Some mechanisms of crude oil/brine/solid interactions. *J. Petrol. Sci. and Eng.*, v. 20 (Jun): 155–160.

Buckley, J.S., and Morrow, N.R., 1990. Characterization of crude oil wetting behavior by adhesion tests. *SPE 20263*, SPE/DOE Sym. on EOR, Tulsa, OK.

Buckley, S.E.; Takamura, K., and Morrow, N.R., 1989. Influence of electrical surface charge on the wetting properties of crude oils. *SPE Reservoir Engineering* (Aug): 332.

Buckley, S.E., and Leverett, M.E., 1942. Mechanisms of fluid displacement in sands. *T. AIME*, v. 146: 107–116.

Burdine, N.T., 1953. Relative permeability calculations from pore size distribution data. *Trans. AIME* 198: 71–78.

Burdine, N.T., Gournay, L.S., and Reichertz, P.P., 1950. Pore size distribution of petroleum reservoir rocks. *Trans. AIME* 189: 195–204.

Burghardt, A., and Zaleski, T., 1968. Longitudinal dispersion at small and large Peclet numbers in chemical flow reactors. *Chem. Eng. Sci.*, v. 23: 575–591.

Busireddy, C., and Rao, D.N., 2004. Application of DLVO theory to characterize spreading in crude oil-brine-rock systems. *SPE-89425*, SPE/DOE 14th Symp. on Improved Oil Recovery, Tulsa, OK (Apr).

Carman, P.C., 1937. Fluid flow through granular beds. *Trans. Inst. Chem. Eng. (London)*; 150–166.

Cartmill, J.C., and Dickey, P.A., 1977. Flow of a dispersed emulsion of crude oil in water through porous media. *AAPG Bulletin*, v. 54: 2438–2447.

Castellan, G.W., 1971. *Physical Chemistry (2nd ed.)*. Addison Wesley Publishing: 866 pp.

Chapman, R.E., 1983. *Petroleum Geology*. Elsevier Science Publishers, Amsterdam, New York: 415 pp.

Chilingarian, G.V., Mazzullo, S.J., and Rieke, H.H., 1992. *Carbonate Reservoir Characterization: A Geologic Engineering Analysis, Parts I and II.* Elsevier Science Publishers, Amsterdam, New York.

Chilingarian, G.V., and Wolf, K.H., 1975. *Compaction of Coarse-Grained Sediments, I.* Elsevier Science Publishers, Amsterdam, New York: 552 pp.

Chilingarian, G.V., and Wolf, K.H., 1976. *Compaction of Coarse-Grained Sediments, II.* Elsevier Science Publishers, Amsterdam, New York: 808 pp.

Chilingarian, G.V., and Yen, 1978. *Bitumens, Asphalts, and Tar Sands.* Elsevier Science Publishers, Amsterdam, New York: 331 pp.

Chilingar, G.V., and Yen, 1983. Some notes on wettability and relative permeabilities of carbonate reservoirs. *Energy Sources,* v. 7 (1): 67.

Chilingar, G.V., and Yen, T.F. 1992. *Carbonate Reservoir Characterization: A Geologic-Engineering Analysis, Part I.* Elsevier Science Publishers, Amsterdam, New York, 639 pp.

Chilingarian, G.V., Donaldson, E.C., and Yen, T.F., 1995. *Subsidence due to Fluid Withdrawal.* Elsevier Science, Amsterdam, New York: 500 pp.

Chilingar, G.V., Serebryakov, V.A., and Robertson, Jr., J.O., 2002. *Origin and Prediction of Abnormal Formation Pressures.* Elsevier Publishing Col, Amsterdam, New York, 373 pp.

Chu, C., 1982. State-of-the-art review of fireflood field properties. *JPT* (Jan): 19–36.

Collins, A.G., 1975. *Geochemistry of Oilfield Waters.* Elsevier Science Publishers, Amsterdam, New York: 496 pp.

Collins, E., 1961. *Flow of Fluids through Porous Materials.* PennWell Books, Tulsa: 270 pp.

Combes, R., Robin, M., Blavier, G., Aïlan, M., and Degrève, F., 1989. Visualization of imbibition in porous media by environmental scanning electron microscopy: application to reservoir rocks. *JPSE,* 20 (3/4, Jun): 133–139.

Corey, A.T., 1954. The interrelation between gas and oil relative permeabilities. *Producers Monthly* (Nov): 38–41.

Craig, Jr., F.F., 1971. *The Reservoir Engineering Aspects of Waterflooding.* SPE Monograph v. 3: 142 pp.

Crocker, M.E., Donaldson, E.C., and Marchin, L.M., 1983. Comparison and analysis of reservoir rocks and related clays. U.S. Dept. Energy Report of Investigations, DOE/BETC/RI-83/7, U.S. Government Publications: 25 pp. Also, *SPE 11973,* Ann. Tech. Conf, San Francisco (Oct).

Cuiec, L., 1984. Rock/crude-oil interactions and wettability: an attempt to understand their interrelations. *SPE 13211,* Ann. Tech. Conf., Houston, TX (Sep).

Cuiec, L., Bourbiaux, B., and Klaydjian, F., 1994. Oil recovery by imbibition in low-permeability chalk. *SPE FE* (Sep): 200–208.

Daly, R.A., 1933. *Igneous Rocks and Depths of the Earth.* McGraw-Hill, New York, 598 pp.

Derjaquin, B.V., and Landau, L., 1941. *Acta Physiochim, USSR,* v. 14: 633–662.

Danckwerts, P.V., 1953. Continuous flow systems. *Chem. Engr. Sys.* (Feb): 1–13.

Donaldson, E.C., 1991 (ed.). *Microbial Enhancement of Oil Recovery—Recent Advances.* Elsevier Science Publishers, Amsterdam, New York: 530 pp.

Donaldson, E.C., Baker, B.A., and Carroll, A.B., 1977. Particle transport in sandstones. *SPE 6905,* Ann. Tech. Conf., Denver, CO (Oct).

Donaldson, E.C., and Chernoglazov, V., 1986. Characterization of drilling mud fluid invasion. J. Petrol. Sci. Eng., v. 1: 3–13.

Donaldson, E.C., Chilingarian, G.V., and Yen, T.F., 1985 (eds.). *Enhanced Oil Recovery, I.* Elsevier Science Publishers, Amsterdam, New York: 357 pp.

Donaldson, E.C., Chilingarian, G.V., and Yen, T.F. 1989 (eds.). *Enhanced Oil Recovery, II. Processes and Operations.* Elsevier Science Publishers, Amsterdam, New York: 604 pp.

Donaldson, E.C., Chilingarian, G.V., and Yen, T.F., 1998 (eds). *Microbial Enhanced Oil Recovery.* Elsevier Science Publishers, Amsterdam, New York: 227 pp.

Donaldson, E.C., and Crocker, M.E., 1977. Review of petroleum oil saturation and its determination. *Bartlesville Energy Research Center Report of Investigations* (BERC/RI-77/15), U.S. Depat. of Energy Technical Information Center, U.S. Government Printing Office, 732 N. Capitol St NW, Washington, DC 20401.

Donaldson, E.C., and Crocker, M.E., 1980. Characterization of the crude oil polar compound extract. DOE/BETC/RI-80/5. *National Tech. Info. Service, U.S. Dept. of Commerce,* Springfield, VA 22161: 27 pp.

Donaldson, E.C., Crocker, M.E., and Manning, F.S., 1975. Adsorption or organic compounds on Cottage Grove sandstone. BERC/RI-75/4, *National Tech. Info. Service,* U.S. Dept. Commerce, VA 22161: 16 pp.

Donaldson, E.C., Ewall, N., and Singh, B., 1991. Characteristics of capillary pressure curves. *JPSE,* v. 6: 249–61.

Donaldson, E.C., Kendall, R.F., Baker, B.A., and Manning, F.S., 1975. Surface-Area Measurement of Geologic Materials. *SPEJ* (Apr): 111–116.

Donaldson, E.C., Kendall, R.F., and Manning, F.S., 1976. Dispersion and tortuosity in sandstones. *SPE 6190*, Ann. Tech. Conf., New Orleans (Oct): 16 pp.

Donaldson, E.C., Lorenz, P.B., and Thomas, R.D., 1966. The effects of viscosity and wettability on oil and water relative permeabilities. *SPE 1562*, Ann. Tech. Conf., Dallas, TX (Oct).

Donaldson, E.C., and Pavelka, E., 1967. Modifications of the pendant drop interfacial tensiometer. *U.S. Bureau of Mines Information Circular, IC-8323*, U.S. Government Printing Office, 732 N Capitol St NW, Washington, DC 20401.

Donaldson, E.C., and Rezaei, A.A., 1986. Analysis of the migration pattern of injected wastes. *J. of the Underground Injection Practices Council*, No. I: 215–237.

Donaldson, E.C., and Siddiqui, T.K, 1989. Relationship between the archie saturation exponent and wettability. *SPE 16790* and Siddiqui, T.K., 1986. MS Thesis, University of Oklahoma, Norman, OK.

Donaldson, E.C., and Thomas, R.D., 1971. Microscopic observations of oil displacement in water-wet and oil-wet systems. *SPE 3555*, Ann. Conf, New Orleans (Oct).

Dubey, S.T., and Doe, P.H., 1993. Base number and wetting properties of crude oils. *SPE-Reservoir Engineering* (Aug): 195.

Dubey, S.T., and Waxman, M.H., 1991. Asphaltene adsorption and desorption from mineral surfaces. *SPE Reservoir Engineering* (Jun): 389.

Dullien, F.A.L., 1979. *Porous Media Fluid Transport and Pore Structures.* Academic Press, New York: 396 pp.

Dullien, F.A.L., Dhawan, G.K., Gurak, N., and Babjak, L., 1972. A relationship between pore structure and residual oil saturation in tertiary surfactant floods. *SPE Journal*, v. 12 (4, Aug): 289–296.

Du Noüy, P.L., 1919. The Surface Equilibrium of Colloidal Solutions. II. *J. Gen. Physiol*, l: 521.

Dunsmuir, J.H., Ferguson, S.R., D'Amico, K.L., and Stokes, J.P., 1991. X-ray microtomography: a new tool for the characterization of porous media. *SPE 22860*, Ann. Tech. Conf, Dallas, TX (Oct): 423–429.

Duquerroix, J.P., Defives, D., and Labrid, J., 1990. Dissolution processes associated with the injection of alkaline solutions into clayey sandstone formations. *J. Petrol. Sci. & Eng*, v. 4: 43–56.

Ebach, E.A., and White, R.R., 1958. Mixing of fluids flowing through beds of packed solids. *AIChE J.*, v. 4 (2): 161–169.

Emery, L.W., Mungan, N., and Nicholson, R.W., 1970. Caustic slug injection in the Singleton field. *JPT*, v. 22 (Dec): 1569–1576.

Emmett, W.R., Beaver, K.W., andMcCaleb, J.A., 1969. Little Buffalow Basin Tensleep Heterogeneity—Its Influence on Drilling and Secondary Recovery. *SPE 2643,* Ann. Conf., Denver, CO (Sep); (SPE Reprint Series 4a. Field Case Histories, 1975).

Erick, J.D., Good, R.J., and Neumann, A.W., 1975. Thermodynamics of contact angles: Rough solid surfaces. *J. Colloid Interface,* v. 53 (2): 235–248.

Freer, E.M., Svitova, T., and Radke, C.J., 2003. The role of interfacial rheology in reservoir mixed wettability. *J. Petrol. Sci. & Eng.,* v. 39 (Aug): 137–158.

Fergusson, J.E., 1982. *Inorganic Chemistry and the Earth.* Pergamon Press, New York: 400 pp.

Finston, H.L., and Rychtman, A.C., 1982. *A New View of Current Acid-Base Theories.* John Wiley, New York: 216 pp.

Gibbs, J.W., 1928. *The Collected Works of J. Willard Gibbs.* V. 1, *Thermodynamics*: Yale University Press, New Haven, CN.

Gleeson, J.W., Woessner, D.E., and Jordan, Jr., C.F, 1993. NMR Imaging of pore structures in limestones. *SPE Formation Evaluation* (Jun): 123–252.

Grahame, D.C., 1947. *Chemical Reviews,* v. 41: 441.

Grattoni, C.A., and Dawe, R.A., 2003. Gas and oil production from waterflood residual oil: effects of wettability and oil spreading characteristics. *JPSE,* v. 39 (Sep): 297–308.

Grattoni, C.A., Pingo-Almada, M.B., and Dawe, R.A., 1997. Pore and core-scale displacement mechanisms with spreading and wetting effects during three-phase flow. *SPE 39032,* V LACPEC, Rio Janeiro, Brazil.

Graue, A., and Bognø, T., 1999. Wettability effects on oil recovery mechanisms in fractured reservoirs. *SPE 56672.* Ann. Conf., Houston, TX (Oct).

Gussow, W.C., 1968. Migration of reservoir fluids. *JPT* (Apr): 353–363.

Hamaker, H.C., 1937. *Physica,* v. 4: 1058–1072.

Hamon, G., 2000. Field-wide variations of wettability. *SPE 63144,* Ann. Tech. Conf., Dallas, TX.

Harmon, G., and Vidal, J.V., 1986. Scaling-up the capillary imbibition process from laboratory experiments on homogeneous and heterogeneous samples. *SPE 15852.* European Petrol. Conf., London.

Hasiba, H.H., and Jessen, F.W., 1968. Film-forming compounds from crude oils, interfacial films and paraffin deposition. *J. Canadian Petrol. Tech.* (Jan–Mar): 1–12.

Hassler, G.L., and Brunner, E., 1945. Measurement of capillary pressures in small core samples. *Trans. AIME* 160: 114–123.

Helander, D.P., 1983. *Fundamentals of formation evaluation.* Oil and Gas Consultants, International, Tulas: 332 pp.

Hirasaki, G.J., 1991. Wettability fundamentals and surface forces. *SPE-Formation Evaluation* (Jun): 217.

Hirasaki, G.J., and Zhang, D.L., 2004. Surface chemistry of oil recovery from fractured, oil-wet, carbonate formations. *SPE Journal* (Jun): 151–162.

Høgnesen, E.J., Strand, S. and Austad, T., 2005. Waterflooding of preferential oil-wet carbonates: oil recovery related to reservoir temperature and brine composition. *SPE 94166.* Europec/EAGE Ann. Conf., Madrid, Spain (Jun).

Holdith, S.A., 2006. Tight gas sands. *SPE-JPT,* v. 58 (Jun): 86–93.

Hornof, V., and Morrow, N.R., 1986. Flow visualization of the effects of interfacial tension on displacement. *SPE 14930,* SPE/DOE Sym. on Enhanced Oil Recovery, Tulsa (Apr).

Hunt, J.M., 1984. Generation and Migration of Light Hydrocarbons. *Science,* v. 226, No. 4680 (14 Dec): 1265–1270.

Hunter, R.J., 1981. *Zeta Potential in Cololid Science, Principles and Applications.* Academic Press, London, New York, 386 pp.

Ingsoy, P.B., Skaeveland, S.M., Hjelmaas, T., and Haga, O.M. 1998. Introducing imbibition capillary pressure in the assessment of the Smoerbukk South Fields offshore Norway. *SPE 39871,* Intr. Pet. Conf of Mexico, Villahermosa, Mexico (Mar).

Ioannidis, M.A., Kwiecien, M., Chatzis, I., MacDonald, I.F., and Dullien, F.A.L., 1997. Comprehensive pore structure characterization using 3D computer reconstruction and stochastic modeling. *SPE 38713,* Ann. Tech. Conf, San Antonio, TX (Oct).

Israelachvili, J.N., 1997 (6th ed.). *Intermolecular and Surface Forces.* Academic Press, London, New York, 450 pp.

Jackson, M.D., Valvatne, P.H., and Blunt, M.J., 2005. Prediction of wettability variation within an oil/water transition zone and its impact on production. *SPEJ,* v. 10 (No. 2, Jun): 184–194.

Jadhunandan, P.P., and Morrow, N.R., 1995. Effect of wettability on waterflood recovery for crude-oil/brine/rock systems. *SPE Reservoir Engineering* (Feb): 40–46.

Jennings, H.Y., 1966. Waterflood behavior of high viscosity crudes in preserved soft and unconsolidated cores. *JPT* (Jan): 116–120.

Jensen, W.B., 1980. *The Lewis Acid-Base Concepts.* John Wiley, New York: 364 pp.

Jerauld, G.R., and Rathmell, J.J., 1997. *SPE Reservoir Engineering* (Feb): 58–65.

Johnson, E.F., Bossler, D.P., and Naumann, V.O., 1959. Calculation of relative permeability from displacement experiments. *T. AIME*, v. 216: 370–372.

Jones, R.W., 1980. Some mass balance and geological constraints on migration mechanisms. *AAPG Studies in Geology 10*: 103–122.

Kaminsky, R., and Radke, C.J., 1997. Asphaltenes, water films, and wettability reversal. *SPE J.*, v. 2 (Dec): 485–493.

Katz, D.L., 1942. Possibilities of secondary recovery for the Oklahoma City Wilcox sand. *Trans. AIME* 146: 28–43.

Keighin, C.W., and Sampath, K., 1982. Evaluation of pore geometry in some low-permeability sandstones—Uinta Basin. *SPE-JPT* (Jan): 65–70.

Keller, G.V., 1953. Effect of wettability on the electrical resistivity of sands. *Oil & Gas J.*, v. 51, No. 1 (Jan 5): 65.

Kennedy, H.T., Burja, E.O., and Boyin, R.A., 1955. An investigation of the effects of wettability on recovery of oil by waterflooding. *J. Phy. Chem.*, v. 59: 867–869.

Kruyt, H.R., 1952. *Colloid Science, I.* Elsevier Science Publishers, New York, Amsterdam.

Kyte, J.R., Naumann, J.O., and Mattox, C.C., 1961. Effects of reservoir environment on water-oil displacements. *JPT*, v. 13 (Jun): 579–582.

Leach, R.O., Wagner, O.R., Wood, H.W., and Harple, C.F., 1962. A laboratory study and field study of wettability adjustment in waterflooding. *JPT*, v. 14 (Feb): 206–212.

Lefevre Du Prey, E., 1978. Gravity and capillary effects on imbibition in porous media. *SPEJ* (Jun): 195–206.

Legens, C., Toulhoat, H., Cuiec, L., Villieras, F., and Palermo, T., 1999. Wettability change related to adsorption of organic acids on calcite: experimental and ad-initio computational studies. *SPEJ*, v. 4 (Dec): 328–333.

Lennard-Jones, J.E., 1924. *Proc. Royal Soc.*, A106: 441–463.

Lennard-Jones, J.E., and Dent, B.M., 1928. *Trans. Faraday Soc*, v. *24*: 92–108.

Leverett, M.C., and Lewis, W.F., 1941. Steady flow of gas-oil-water mixtures through unconsolidated sands. *T. AIME*, 142: 107–116.

Lewis, G.N., 1923. *Valence and the Structure of Atoms and Molecules*. The Chemical Catalog Co., New York.

Lifshitz, E.M., 1956. *Soviet Physics, JETP (English)*, v. 2: 73–83.

Link, P.K., 1982. *Basic Petroleum Geology*. Oil & Gas Consultants, Tulsa, OK: 235 pp.

London, F., 1937. Trans. Faraday Soc., v. 33: 8–26.

Longeron, D.G., Argaud, M.H., and Feraud, J.P., 1989. Effect of overburden pressure and the nature of microscopic distribution of fluids on electrical properties of rock samples. *SPE FE*, v. 4 (Jun): 194–202.

Loomis, A.G., and Crowell, D.C., 1962. Relative permeability studies: gas-oil and water-oil systems. *U.S. Bureau of Mines Bulletin 599*. U.S. Government Printing Office: 39 pp.

Lorenz, P.B., Donaldson, E.C., and Thomas, R.D., 1974. Use of centrifugal measurements of wettability to predict oil recovery. *U.S. Bureau of Mines Report of Investigations* 7873. U.S. Government Printing Office, 732 N Capitol St NW,Washington, DC 20401.

Louks, R.G., Dodge, M.M., and Galloway, W.E., 1984. Regional controls on diagenesis of reservoir quality in Lower Tertiary Sandstones along the Texas Gulf Coast. In: McDonaldson, D.A., and Surdam, R.C. (eds.), *Clastic Diagenesis. AAPG Memoir 37,* Tulsa, OK: 15–46.

Lowell, S., and Shields, J.E, 1981. Contact angle in mercury porosimetry. *American Laboratory*: 123 pp.

Luder, W.F., and Zuffanti, S., 1961. *The Electronic Theory of Acids and Bases*. Dover Publications, New York: 165 pp.

Lyons, W.C., and Plisga, G.J. (eds.), 2005. *Standard Handbook of Petroleum & Natural Gas Engineering*. Elsevier, Amsterdam, New York: Ch.'s 2 and 5.

Ma, S.M., Morrow, N.R., and Zhang, X., 1997. Generalized scaling of spontaneous imbibition data for strongly water-wet systems. *JPSE* (Nov, v. 18): 165–178.

Ma, S.M., Morrow, N.R., Zhang, X., and Zhou, X., 1999. Characterization of wettability from spontaneous imbibition measurements. *J. Can. Petrol. Tech.*, v. 38, No. 13: 1–8.

McCreesh, C.A., 1988. Relating thin sections to permeability, mercury porosimetry, formation factor, and tortuosity. AAPG Bulletin, v. 72: 221.

Mailtand, G., Rigby, M., Smith, E., and Wakeham, W., 1981. *Intermolecular Forces: Their Origin and Determination*. Oxford University Press; New York.

Marle, C. M., 1981. *Multiphase flow in porous media*. Gulf Pub. Co., Houston: 257 pp.

Mattax, C.C., and Kyte, J.R., 1962. Imbibition oil recovery from fractured water drive reservoirs. *Trans. AIME* 225: 177–184.

Marsden, S.S., and Khan, S., 1962. The wettability of the Bradford sand. I in *Producers Monthly* 26 No. 5: 2–5. II in *PM* 26, No. 6: 10–14.

McDougall, S.R., and Sorbie, K.S., 1995. The impact of wettability on waterflooding: pore-scale simulation. *SPE Reservoir Engineering* (Aug): 208–213.

McGhee, J.W., Crocker, M.E., and Donaldson, E.C., 1979. Relative wetting properties of crude oils in Berea sandstone. *BETC/RI-78/9. Bartlesville Energy Technology Center Report of Investigations 78/9*. National Technical Information Service, U.S. Dept. of Commerce, Springfield, VA 22161.

Melrose, J.C., and Bradner, C.F., 1974. Role of capillary forces in determining microscopic displacement efficiency for oil recovery by waterflooding. *J. Can. Petrol. Tech.*, v. 13 (4), pp. 54–62.

Menour, H., and Knapp, R.M., 1980. Numerical simulation of the imbibition process in fractured reservoirs. *SPE 9370*, Ann. Conf., Dallas, TX.

Mailtand, G., Rigby, M., Smith, E., and Wakeham, W., 1981. *Intermolecular Forces. Their Origin and Determination.* Oxford University Press, New York.

Mohanty, K.K., and Salter, S.J., 1983. Multiphase flow in porous media, III. Oil mobilization, transverse dispersion, and wettability. *SPE 12127*, Ann. Tech. Conf., San Francisco (Oct 5–8).

Moore, J., 1958. Laboratory determined electric logging parameters of the Bradford Third sand. *Producers Monthly*, v. 25, No. 5 (Mar): 30–39.

Morgan, J.T., and Gordon, D.T., 1970. Influences of pore geometry on water-oil relative permeability. *JPT* (Oct): 1199–1208.

Morgan, W.B., and Pirson, S.J., 1964. The effect of fractional wettability on the Archie saturation exponent. *T. Soc. Prof. Well Log Analysts*, 5th Ann. Symp, Midland, TX (May): Section B, p. 1–13.

Morland, L.W., 1977. Oscillatory flows in a porous thermoelastic matrix due to earth tides. *Geothermics*, v. 6: 107–116.

Morland, L.W., 1978. A theory of slow fluid flow through a porous thermoelastic matrix. *Geophys. J.R. astr. Soc.*, v. 55: 393–410.

Morland, L.W., 1992. Flow of viscous fluids through a porous deformable matrix. *Surveys in Geophysics*, v. 13: 209–268.

Morland, L.W., and Donaldson, E.C., 1984. Correlation of porosity and permeability of reservoirs with well oscillations induced by earth tides. *Geophys. J.R. astr. Soc.*, v. 79: 705–725.

Morris, E.E., and Wieland, D.R., 1963. A microscopic study of the effect of variable wettability conditions on immiscible fluid displacement. *SPE 704*. SPE Ann. Conf., New Orleans (Oct).

Morrow, N.R., 1970. Physics and thermodynamics of capillary action in porous media. *Ind. Eng. Chem.*, v. 69 (6): 32–56.

Morrow, N.R., 1976. Capillary pressure for uniformly wetted porous media. *J. Canadian Petrol. Tech.* (Oct–Dec): 49.

Morrow, N.R., Ma, S., Zhou, X., Zhan, X., 1994. Characterization of wettability from spontaneous imbibition measurements. Paper No. CIM 94–47, *Petrol. Soc. of CIM Conf.,* Calgary (Jun): 12–15.

Morrow, N.R., Cram, P.J., and McCaffery, F.G., 1973. Displacement studies in dolomite with wettability control by octanoic acid. Soc. Petrol. Eng. J.: 221.

Mungan, N., 1966. Certain wettability effects in laboratory waterfloods. *JPT* (Feb): 247–252.

Mungan, N., 1972. Relative permeability measurements using reservoir fluids. *SPEJ*, v. 12 (Oct): 398–402.

Neasham, J.W., 1977. Application of scanning electron microscopy to the characterization of hydrocarbon-bearing rocks. IIT Res. Inst., Chicago, *Scanning Electron Microscopy, V. 1,* Proc. of workshop on Materials and Component Characterization with SEM/STEM, IIT Research Inst., Chicago, IL (Mar).

Nutting, P.G., 1934. Some physical and chemical properties of reservoir rocks bearing on the accumulation and discharge of oil. In: W.E. Wrather and F.H. Lahee (eds.), *Problems of Petroleum Geology*. AAPG, Tulsa.

Pashley, R.M., and Israelachvili, J.N, 1984. DLVO and hydration forces between mica surfaces in Mg^{2+}, Ca^{2+}, Sr^{2+}, and Ba^{2+} chloride solution. *J. Colloid and Interface Science,* v. 97, No. 2 (Feb): 446–455.

Patel, A.D., 1998. Reversible Invert Emulsion Drilling Fluids—A Quantum Leap in Technology. *SPE 47772.*

Patel, A.D., and Growcock, F.B., 1999. Reversible Invert Emulsion Drilling Fluids: Controlling Wettability and Minimizing Formation Damage. *SPE 54764.*

Perkins, T.C. and Johnson, O.C., 1963. A review of diffusion and dispersion in porous media. *SPEJ* (Mar): 70–84.

Pirson, S.J., 1963. *Handbook of Well-log Analysis.* Prentice-Hall, Inc.: 326 pp.

Poulin, T.J., 1985. *Determination of the coefficient of dispersion in sandstones.* MS Thesis, University of Oklahoma, Norman, OK.

Purcell, W.R., 1949. Capillary pressures—their measurement using mercury and the calculation of permeability therefrom. *Trans. AIME* (Feb), T.P. 2544: 39–48.

Randolph, P.L., Soeder, D.J., and Chowdiah, P., 1984. Porosity and permeability of tight sands. *SPE 12836*. SPE/DOE/GRI Unconventional Gas Recovery Sym., Pittsburgh, PA (May).

Rapoport, L.A., 1955. Scaling laws for use in design and operation of waster-oil flow models. *Trans. AIME* 204: 143–150.

Reid, R.C., Prausnitz, J.M., and Sherwood, T.K., 1977. *The Properties of Gases and Liquids.* McGraw-Hill, New York: 688 pp.

Reisberg, J., and Doscher, T.M., 1956. Interfacial phenomena in crude-oil-water systems. *Producers Monthly,* v. 21 (1): 43–50

Rezaei, A.A., and Donaldson, E.C., 1987. New two-dimensional single-phase waterflood model for microcomputers. *SPE 16491*, SPE Petroleum Applications of Microcomputers Conf., in Del Lago, Lake Conroe, TX (Jun).

Rezigh, A.A., 1988. *Two-Dimensional Front Tracking Model for Microcomputers.* Thesis (Petroleum and Geological Engineering), University of Oklahoma Engineering Library, Felgar Hall, Norman, OK 73019: 100 pp.

Richardson, J.G., Perkins, F.M., Jr., and Osoba, J.S., 1955. Differences in behavior of fresh and aged East Texas Woodbine cores. *Trans. AIME,* v. 204: 86–91.

Ritter, H.L., and Drake, L.C., 1945. Pore size distribution in porous materials. *Ind. & Eng. Chem* (Analytic Ed.), v. 17 (12): 782–791

Robinson, E.S., and Bell, R.T., 1971. Tides in confined well-aquifer systems. *J. Geophysical research,* v. 76 (8): 1857–1869.

Salathiel, R.A., 1973. Oil recovery by surface film drainage in mixed-wettability rocks. *JPT* (Oct): 1216–1224.

Salter, S.J., and Mohanty, K.K., 1982. Multiphase flow in a porous media: microscopic observations and modeling. *SPE 11017*, Ann. Tech. Conf., New Orleans, LA (Sep).

Scheidegger, A.E. The Physics of Flow through Porous Media, University of Toronto Press, 1974.

Seetheqailli, A., Adibhatta, B., and Mohanty, K.K., 2004. Physicochemical interactions during surfactant flooding of fractured carbonate reservoir. *SPEJ* (Dec): 411–418.

Sharma, M.M., 1987. Investigation of the surface charge properties of oxide surfaces in oil bearing sands and sandstones. *J. Colloid and Interface Sci.,* v. 115, No. 1 (Jan): 9–16.

Sharma, M.M., and Wunderlich, R.W., 1989. The alteration of rock properties due to interactions with drilling fluid components. *JPSE,* v. 1, No. 2: 127–143.

Smith, J.M., 1968. Kinetics of adsorption. Adsorption from aqueous solutions. *Am. Chem. Soc., Advances in Chemistry Series,* v. *79*: 8–22.

Singh, B., 1990. Capillary pressure phenomena of porous media. *M.Sc. Thesis*, Petrol. Eng, University of Oklahoma, Norman, OK: 116 pp.

Skauge, A., Standal, S., Boe, S.O., Skauge, T., and Blokhus, A.M., 1999. Effects of organic acids and bases and oil composition on wettability. *SPE 56673*, Ann. Tech. Conf., Houston, TX (Oct).

Spinler, E.A., Zornes, D.R., Tobola, D.P., and Moradi-Araghi, A., 2000. Enhancement of oil recovery using a low concentration of surfactant to improve spontaneous and forced imbibition in chalk. *SPE 59290*, SPE/DOE Improved Oil Recovery Symp., Tulsa, OK (Apr).

Standnes, D.C., and Austad, T., 2000. Wettability alteration in chalk (Parts 1 and 2). *J. Petrol. Sci. & Eng.*, v. 28 (Nov): 111–143.

Standnes, D.C., and Austat, T., 2003. Nontoxic low-cost amines as wettability alteration chemicals in carbonates. *J. Petrol. Sci. & Eng.* (Sep): 431–446.

Stone, A.J., 1996. *The Theory of Intermolecular Forces*. Clarendon Press, Oxford: 264 pp.

Sumasundaran, P., and Agar, G.E., 1967. The zero point of charge of calcite. *J. Colloid and Interface Sci.*, v. 24: 433–440.

Swanson, B.F., 1979. Visualizing pores and nonwetting phase in porous rock. *SPE-JPT* (Jan): 11–17.

Sweeney, S.A., and Jennings, Jr., H.Y., 1960. The electrical resistivity of preferentially water-wet and preferentially oil-wet carbonate rocks. *Producers Monthly*, v. 24, No. 7 (May): 29–32.

Tang, G.Q., and Morrow, N.R., 1997. Salinity, temperature, oil composition, and oil recovery by waterflooding. *SPE Reservoir Engineering* (Nov): 269–276.

Taylor, G.L., 1953. Dispersion of soluble matter in solvent flowing slowly through a tube. *Proc. Royal Soc.*, v. 219: 186.

Terwilliger, P.L, Wilsey, L.E., Hall, H.N, Bridges, P.M., and Morse, R.A., 1951. An experimental and theoretical investigation of gravity drainage performance. *T. AIME* 192: 285–296.

Thomas, G.H., Countryman, G.R., and Fatt, I., 1963. I: Miscible displacement in a multiphase system. *SPE J* (Sep): 189–196.

Thomas, R.D., and Ward, D.C., 1971. Effect of overburden pressure and water saturation on gas permeability of tight sandstone cores. *SPE-3634*, Ann. Tech. Conf (Oct.), New Orleans, LA.

Thompson, D.W., and Pownall, P.G., 1989. Surface electrical properties of calcite. J. Coloid and Interface Sci., v. 131, No. 1 (Aug): 74–82.

Tiab, D., and Donaldson, E.C., 2004. *Petrophysics (2nd ed.)*. Gulf Publishing Co., 889 pp.

Tissot, B.P., 1984. Recent advances in petroleum geochemistry applied to hydrocarbon exploration. *AAPG Bulletin,* v. 68, No. 5 (May): 545–563.

Torbati, H.M., Raiders, R.A., Donaldson, E.C., McInerny, M.J., Jenneman, G.E., and Knapp, R.M, 1986. Effect of microbial growth on pore entrance size distributions in sandstone cores. *J. Industrial Microbiology,* v. 1: 227–234.

Treiber, L.E., Archer, D.L., and Owens, W.W., 1972. A laboratory evaluation of wettability of fifty oil-producing reservoirs. *SPEJ* (Dec): 531–540.

Tweheyo, M.T., Holt, T., and Torsaeter, O., 1999. An experimental study of the relationship between wettability and oil production characteristics. *JPSE,* v. 24, No. 2–4 (Dec): 179–188.

Veley, C.D., 1969. How hydrolysable metal ions react with clays to control formation water sensitivity. *SPE-JPT* (Sep): 1111–1118.

Verwey, E.J.W., and Overbeek, J. Th. G., 1948. *Theory of Stability of Lyophobic Colloids.* Elsevier Science Publisher, Amsterdam.

Vives, M.T., Chang, Y.C., and Mohanty, K.K., 1999. Effect of wettability on adverse-mobility immiscible floods. *SPEJ,* v. 4 (3, Sep): 260–266.

Weast, R. C. (ed.), 1970. *Handbook of Chemistry and Physics.* The Chemical Robber Co., Cleveland, OH.

Welge, H.J., and Bruce, W.A., 1945. A restored-state method for determination of oil in place and connate water. *Drilling and Production Practice. American Petroleum Institute:* 161–165.

Welte, D.H., and Yukler, A., 1981. Petroleum origin and accumulation in basin evolution—a quantitative model. *AAPG Bulletin,* v. 65: 1387–1396.

Wheeler, A., 1951 (v. III). *Advances in Catalysis and Related Subjects.* Academic Press, New York: 249–327.

Whillhite, G.P., 1986. *Waterflooding.* SPE Textbook Series, v. 3: 326 pp.

Wyllie, M.R.J., and Gardner, G.H.F., 1958. The generalized Kozeny-Carman equation. *World Oil* (Mar).

Yadav, G.D. et. al., 1987. Microscopic distribution of wetting and nonwetting phases in sandstones during immiscible displacements. *SPE RE,* May: 137–147.

Yang, S-Y., Hirasaki, G.J., Basu, S., and Vaidya, R., 2002. Statistical analysis on parameters that affect wetting for the crude oil/brine/mica system. *JPSE,* v. 33 (Apr): 203–215.

Yen, T.F., 1974. Structure of petroleum asphaltene and its significance. *Energy Sources,* v. 1, No. 4: 447–463.

Young, T., 1855. In *Miscellaneous Works.* G. Peacock (ed.), J. Murray, London, v.I: 418.

Zhang, X., Morrow, N., and Ma, S., 1996. Experimental verification of modified scaling group for spontaneous imbibition. *SPE RE* (Nov): 273–285.

Zhou, D., and Blunt, M., 1998. Wettability effects in three-phase gravity drainage. *JPSE,* v. 20 (Jun): 203–211.

Zou, B., McCool, C.S., Green , D.W., and Willhite, G.P., 1998. A study of the chemical interactions between brine solutions and dolomite. *SPE 39695,* SPE/DOE Symp., Tulsa, OK (Apr).

Index

FSC
www.fsc.org
MIX
Papier | Fördert gute Waldnutzung
FSC® C083411

Kolibri 360 GmbH
Ferdinand-Jühlke-Straße 7
99095 Erfurt, Deutschland
produktsicherheit@kolibri360.de